Green Building Management and Smart Automation

Arun Solanki
Gautam Buddha University, India

Anand Nayyar
Duy Tan University, Vietnam

A volume in the Advances in Civil and Industrial Engineering (ACIE) Book Series

IGI Global
DISSEMINATOR OF KNOWLEDGE

Published in the United States of America by
IGI Global
Engineering Science Reference (an imprint of IGI Global)
701 E. Chocolate Avenue
Hershey PA, USA 17033
Tel: 717-533-8845
Fax: 717-533-8661
E-mail: cust@igi-global.com
Web site: http://www.igi-global.com

Library of Congress Cataloging-in-Publication Data

Names: Solanki, Arun, 1985- editor. | Nayyar, Anand, editor.
Title: Green building management and smart automation / Arun Solanki and Anand Nayyar, editors.
Description: Hershey, PA : Engineering Science Reference, an imprint of IGI Global, [2020] | Includes bibliographical references and index.
Identifiers: LCCN 2019007882| ISBN 9781522597544 (hardcover) | ISBN 9781522597551 (softcover) | ISBN 9781522597568 (ebook)
Subjects: LCSH: Sustainable buildings. | Intelligent buildings. | Buildings--Environmental aspects. | Building management.
Classification: LCC TH880 .G7436 2020 | DDC 696--dc23 LC record available at https://lccn.loc.gov/2019007882

This book is published in the IGI Global book series Advances in Civil and Industrial Engineering (ACIE) (ISSN: 2326-6139; eISSN: 2326-6155)

British Cataloguing in Publication Data
A Cataloguing in Publication record for this book is available from the British Library.

All work contributed to this book is new, previously-unpublished material. The views expressed in this book are those of the authors, but not necessarily of the publisher.

For electronic access to this publication, please contact: eresources@igi-global.com.

Advances in Civil and Industrial Engineering (ACIE) Book Series

Ioan Constantin Dima
University Valahia of Târgovişte, Romania

ISSN:2326-6139
EISSN:2326-6155

Mission

Private and public sector infrastructures begin to age, or require change in the face of developing technologies, the fields of civil and industrial engineering have become increasingly important as a method to mitigate and manage these changes. As governments and the public at large begin to grapple with climate change and growing populations, civil engineering has become more interdisciplinary and the need for publications that discuss the rapid changes and advancements in the field have become more in-demand. Additionally, private corporations and companies are facing similar changes and challenges, with the pressure for new and innovative methods being placed on those involved in industrial engineering.

The **Advances in Civil and Industrial Engineering (ACIE) Book Series** aims to present research and methodology that will provide solutions and discussions to meet such needs. The latest methodologies, applications, tools, and analysis will be published through the books included in **ACIE** in order to keep the available research in civil and industrial engineering as current and timely as possible.

Coverage

- Structural Engineering
- Quality Engineering
- Optimization Techniques
- Operations research
- Earthquake engineering
- Materials Management
- Urban Engineering
- Engineering Economics
- Transportation Engineering
- Construction Engineering

IGI Global is currently accepting manuscripts for publication within this series. To submit a proposal for a volume in this series, please contact our Acquisition Editors at Acquisitions@igi-global.com or visit: http://www.igi-global.com/publish/.

The Advances in Civil and Industrial Engineering (ACIE) Book Series (ISSN 2326-6139) is published by IGI Global, 701 E. Chocolate Avenue, Hershey, PA 17033-1240, USA, www.igi-global.com. This series is composed of titles available for purchase individually; each title is edited to be contextually exclusive from any other title within the series. For pricing and ordering information please visit http://www.igi-global.com/book-series/advances-civil-industrial-engineering/73673. Postmaster: Send all address changes to above address.

Titles in this Series

For a list of additional titles in this series, please visit: www.igi-global.com/book-series

Recycled Waste Materials in Concrete Construction Emerging Research and Opportunities
Jahangir Mirza (Research Institute of Hydro-Québec, Canada) Mohd Warid Hussin (Universiti Teknologi Malaysia, Malaysia) and Mohamed A. Ismail (Miami College of Henan Universiy, China)
Engineering Science Reference • ©2019 • 169pp • H/C (ISBN: 9781522583257) • US $165.00

Ambient Urbanities as the Intersection Between the IoT and the IoP in Smart Cities
H. Patricia McKenna (AmbientEase, Canada)
Engineering Science Reference • ©2019 • 333pp • H/C (ISBN: 9781522578826) • US $195.00

Optimizing Current Strategies and Applications in Industrial Engineering
Prasanta Sahoo (Jadavpur University, India)
Engineering Science Reference • ©2019 • 382pp • H/C (ISBN: 9781522582236) • US $245.00

Big Data Analytics in Traffic and Transportation Engineering Emerging Research and Opportunities
Sara Moridpour (RMIT University, Australia)
Engineering Science Reference • ©2019 • 188pp • H/C (ISBN: 9781522579434) • US $165.00

Optimization of Design for Better Structural Capacity
Mourad Belgasmia (Setif 1 University, Algeria)
Engineering Science Reference • ©2019 • 283pp • H/C (ISBN: 9781522570592) • US $215.00

Reusable and Sustainable Building Materials in Modern Architecture
Gülşah Koç (Yildiz Technical University, Turkey) and Bryan Christiansen (Global Research Society, LLC, USA)
Engineering Science Reference • ©2019 • 302pp • H/C (ISBN: 9781522569954) • US $195.00

Measuring Maturity in Complex Engineering Projects
João Carlos Araújo da Silva Neto (Magnesita SA, Brazil) Ítalo Coutinho (Saletto Engenharia de Serviços, Brazil) Gustavo Teixeira (PM BASIS, Brazil) and Alexandro Avila de Moura (Paranapanema SA, Brazil)
Engineering Science Reference • ©2019 • 277pp • H/C (ISBN: 9781522558644) • US $245.00

Big Data Analytics for Smart and Connected Cities
Nilanjan Dey (Techno India College of Technology, India) and Sharvari Tamane (Jawaharlal Nehru Engineering College, India)
Engineering Science Reference • ©2019 • 348pp • H/C (ISBN: 9781522562078) • US $225.00

IGI Global
DISSEMINATOR OF KNOWLEDGE

701 East Chocolate Avenue, Hershey, PA 17033, USA
Tel: 717-533-8845 x100 • Fax: 717-533-8661
E-Mail: cust@igi-global.com • www.igi-global.com

Editorial Advisory Board

Table of Contents

Detailed Table of Contents

Non-sustainable buildings have threatened the ecosystem globally. In this chapter, a comprehensive discussion on the green and smart building is presented, considering how the buildings are made green and smart and how they support in developing sustainable cities. Though smart buildings are the positive catalyst towards sustainability, the excessive use of electronic devices puts a check in attaining the overall green goal. This chapter suggests merging green and smart technologies to have green smart building (GSB) with the aim of offering the populations a smart and eco-friendly living. Promises and challenges in attaining this goal are meticulously explored. The GSB concept is discussed in detail, suitably supported with the architectural models of overall and the various components of a GSB. The communication architecture is also presented emphasizing on various entities and activities in different levels of communication between various digital components of a GSB. A few cases have been presented showing practical applications of green and smart technologies in buildings.

This chapter presents the role of renewable energy techniques to design and develop a sustainable framework for green building. The first viewpoint is identified with the earlier structure and the low encapsulated energy building materials for the design and development of a framework for green building. The primary perspective is to manage energy protection using renewable energy techniques in the green building. Green building interchangeably can be used with the term's sustainable construction or green construction. So, durable construction means using environmentally responsible and resource-efficient procedures in development to be ensured of sustainability throughout the lifetime of the building. This chapter also presents the combination of renewable, energy-based technology for green building construction and sustainability with the economics of renewable energy.

Internet of things (IOT) and Vertical Cavity Surface Emitting Laser (VCSEL) future can be seen together, since VCSEL technology-based 3D sensors are introduced for IoT applications. The improved VCSEL structure design with fixed wavelength using a thermally actuated cantilever structure is presented. This improved structure of VCSEL will help us in realizing athermal VCSEL. In athermal VCSEL the dependency of VCSEL on temperature will be much less because it will not require temperature controllers. Realizing fully temperature-independent VCSEL (i.e., athermal VCSEL) is still a challenge but we can reduce it to some extent. In this chapter, recent diversification of application of VCSEL technology from data communication to sensing has been discussed. This proposed VCSEL structure may give us an opportunity to improve the VCSEL technology. Therefore, smart 3D sensors based on VCSEL will help in making internet of things applications more reliable and will directly or indirectly serve the concept of smart homes and smart cities.

There is a growing universal awareness of protecting the living and non-living environment and making enlightened decisions to achieve a sustainable development without destruction of the natural resources. In this point of view, selecting building materials according to their energy and health performances gains importance in sustainable design. 3Rs (reducing, reusing, recycling), and supplying a healthy, non-hazardous indoor air for building occupants are two important parameters of environmental life-cycle assessment for materials. Information on exposure to gases and vapors from synthetic materials made from petrochemicals, to heavy metals and pesticides, and to some combustion pollutants that cause acid rain should be determined by analyzing environmental product declarations or material specifications. After studying on building materials individually, they are analyzed in the form of tables for four different stages; manufacturing, application, usage, demolition phase. Consequently, this chapter can guide the designer and engineer to think on the elements of design and construction activity.

Green building defines the way buildings are designed, constructed, operated, and maintained in such a way that the negative impacts are reduced and eliminated. The green building ensures to impart the sustainability of a building environment for quality living of citizens. It is to be noted that, not only the structure of the building, but its entire life cycle—including designing, constructing, operating, maintaining, renovating, and demolishing—must ensure responsibility towards environmental and natural resources. However, there are numerous factors that influence the effect of green building. Further factors such as

a large consumption of natural resources. Due to the exploitation of natural resources on a large scale through these buildings, it has become necessary to have a better-designed building for the efficient use of resources. The concept of "green building" solves the aforementioned issues apart from promoting eco-friendly activities. IoT makes the idea of having buildings that are energy sufficient possible through networked sensors that not only help in managing the assets better but also reducing harmful impacts on human health and the environment. This chapter talks about the concept of the green building and the smart automation achieved through IoT as well as cloud architecture for the green building also referred to as green cloud. While it explains the basic cloud architecture in green building, it also proposes future challenges for the aforementioned subject.

Rachna Jain, Bharati Vidyapeeth's College of Engineering, India
Vaibhav Goel, Bharati Vidyapeeth's College of Engineering, India
Jusleen Kaur Rekhi, Bharati Vidyapeeth's College of Engineering, India
Jafar A. Alzubi, Al-Balqa Applied University, Jordan

Among the various domains of IoT, one domain that is highly emerging in recent years is the application of IoT in green buildings. With the advent of IoT, the concept of green buildings has taken an even broader perspective. Incorporating intelligence into the current building management systems could revolutionize the buildings in terms of energy efficiency. The chapter explores some sound benefits of integrating IoT into a green building. It offers insight into the various technologies used in green construction, followed by some IoT-based architectures. Some machine learning algorithms that can be used to boost the efficiency of IoT devices are also discussed. Finally, the chapter dives into the future of IoT-enabled green buildings, and explores the challenges in achieving zero-energy buildings, while addressing the questions it raises. It focuses on how a green building, together with the internet of things, may lead to zero-energy buildings, thus carving our path towards a secure and energy-efficient future.

Suman Kumar, Delhi Technological University, India
Gurjit Kaur, Delhi Technological University, India

There are various environmental problems (i.e., "global warming," air and water pollution), which need to be prevented. Construction of buildings plays a significant role in pollution. To reduce the harmful effects in constructing buildings, it is necessary to move on to sustainable architecture. In this chapter, different advantages and standards for green buildings will be discussed. Different organizations are contributing towards a green environment. There are even different sensors that are able to detect wastage of energy and can predict the requirement of energy. Machine learning, a hot topic these days, can also play its role in demand prediction. In this chapter, role of network communication and sensing to optimize the energy of green buildings and machine learning-based demand prediction to optimize the energy of green buildings are discussed. Further predicting energy harvesting from weather forecasts, return on investment of green buildings, and potential benefits of energy-efficient green buildings are also discussed.

The world faces many environmental crises such as increased threat of climate change, the depletion of key natural resources, increasing air and water pollution, and growing levels of solid wastes. These issues are becoming the major aspects of value in real estate and a key driver in the decision-making processes. The strategic sustainability process called "the halo effect" was more worldwide which is affected by the popularity in environmental actions criteria. It showed consequence that green concept has to not only focus on technical or moral issues, but also need to base more on the economic and financial imperative.

A green building is a sustainable building that has minimal impacts on the environment throughout its life. For the purposes of this report, "green building" is understood to mean construction that makes efficient use of energy and resources in every aspect. This includes the production of building materials, and the design, use, and eventual demolition of a building in any sector (commercial, residential, industrial, public buildings) and at all stages, from new buildings to "retrofitting" or adapting existing ones. The construction sector, which accounts for 10% of global GDP, has direct and indirect impacts on the environment. It produces 23% of global greenhouse gas (GHG) emissions, and buildings are responsible for between 30% and 40% of all material flows. A green building is a sustainable building that has minimal impacts on the environment throughout its life.

Foreword

There is a revolution going all over the planet earth, and it's about time! It is transforming the marketplace for buildings, homes and communities, and it is part of a larger sustainability revolution that will transform everything the mankind knows, and experience over the next few decades. This revolution is called "Green Building". There are essential challenges which are required to be addressed for the future, like responsible approach towards nature. Also, there is search for environmentally-friendly energy supply that is easy on resources and climate. In addition of this, serious challenge which has to be combatted is clean sources of drinking water. Aside from novel and more efficient technologies that are currently in place, additional efforts are required for energy consumption reduction, clean water requirements without decreasing comfort level or standard living.

The worldwide building and construction sector uses up to 40% of energy requirements and even considerable amount of water. Hence, these buildings influence envisioned energy and water needs for the next 50 to 80 years. This is utmost requirement today, to plan, the construction of these buildings to be done on principles of energy efficiency, climatic conditions and water conservation. Buildings that show strong signs of these attributes with regard to sustainability are called "*Green Buildings*". Green Buildings or Green Construction or Sustainable Building is referred to both a structure and the application of processes that are environmentally responsible and resource-efficient throughout a building's life-cycle: from planning to design, construction, operation, maintenance, renovation and demolition. The green building expands and complements the classical building design concerns of economy, utility, durability and comfort.

In 1993, the U.S. Green Building Council (USGBC) was founded and is committed to a sustainable, prosperous future through LEED, the leading program for green buildings and communities worldwide. Leadership in Energy and Environmental Design (LEED) is a set of rating system for the design, construction, operation and maintenance of green buildings and incorporate emerging green building technologies. LEED v4 was introduced in November, 2013. Until October 31, 2016, new projects could choose between LEED 2009 and LEED v4. LEED 2009 encompasses ten rating systems for the design, construction and operation of buildings, homes and neighborhoods. That suite consists of: LEED for New Construction, LEED for Core & Shell, LEED for Schools; LEED for Retail: New Construction and Major Renovations; LEED for Healthcare; LEED for Commercial Interiors; LEED for Retail: Commercial Interiors; LEED for existing buildings: Operations & Maintenance; LEED for Neighborhood Development and LEED for homes.

Through LEED, business professionals, policy makers, developers, designers, scientists and citizens are joining together to find solutions to most complex problems of present to have safe future. As today's building construction account for 18% of global emissions today, or the equivalent of 9 billion tons of

CO_2 annually. If new technologies in construction are not adopted during this time of rapid growth, emissions could double 2050, according to United Nations Environment Program. The International Energy Agency released a publication that estimated that existing buildings are responsible for more than 40% of the world's total primary energy consumption and for 24% of global carbon dioxide emissions. Green building practices aim to reduce the environmental impact of building. Since construction almost always degrades a building site, not building at all is preferable to green building, in terms of reducing environmental impact. The second rule is that every building should be as small as possible. The third rule is not to contribute to sprawl, even if the most energy-efficient, environmentally sound methods are used in design and construction.

The green building movement in the U.S. originated from the need and desire for more energy efficient and environmentally friendly construction practices. There are a number of motives for building green, including environmental, economic, and social benefits. However, modern sustainability initiatives call for an integrated and synergistic design to both new construction and in the retrofitting of existing structures. Also known as sustainable design, this approach integrates the building life-cycle with each green practice employed with a design-purpose to create a synergy among the practices used.

Green building brings together a vast array of practices, techniques, and skills to reduce and ultimately eliminate the impacts of buildings on the environment and human health. It often emphasizes taking advantage of renewable resources, e.g., using sunlight through passive solar, active solar, and photovoltaic equipment, and using plants and trees through green roofs, rain gardens, and reduction of rainwater run-off. Many other techniques are used, such as using low-impact building materials or using packed gravel or permeable concrete instead of conventional concrete or asphalt to enhance replenishment of ground water.

Green buildings often include measures to reduce energy consumption – both the embodied energy required to extract, process, transport and install building materials and operating energy to provide services such as heating and power for equipment.

As high-performance buildings use less operating energy, embodied energy has assumed much greater importance – and may make up as much as 30% of the overall life cycle energy consumption.

To reduce operating energy use, designers use details that reduce air leakage through the building envelope (the barrier between conditioned and unconditioned space). They also specify high-performance windows and extra insulation in walls, ceilings, and floors. Another strategy, passive solar building design, is often implemented in low-energy homes. Designers orient windows and walls and place awnings, porches, and trees to shade windows and roofs during the summer while maximizing solar gain in the winter. In addition, effective window placement (daylighting) can provide more natural light and lessen the need for electric lighting during the day. Solar water heating further reduces energy costs.

As a result of the increased interest in green building concepts and practices, a number of organizations have developed standards, codes and rating systems that let government regulators, building professionals and consumers embrace green building with confidence. In some cases, codes are written so local governments can adopt them as bylaws to reduce the local environmental impact of buildings.

Green building rating systems such as BREEAM (United Kingdom), LEED (United States and Canada), DGNB (Germany), CASBEE (Japan), and VERDEGBCe (Spain), GRIHA (India) help consumers determine a structure's level of environmental performance. They award credits for optional building features that support green design in categories such as location and maintenance of building site, conservation of water, energy, and building materials, and occupant comfort and health. The number of credits generally determines the level of achievement.

Green building codes and standards, such as the International Code Council's draft International Green Construction Code, are sets of rules created by standards development organizations that establish minimum requirements for elements of green building such as materials or heating and cooling.

The critical success factors to achieve green building can generally fall into three categories, i.e. technical, managerial and behavioural.

- **Under Technical Category**: Utilization of renewable energy technological innovations has been pivotal for achieving green building objectives and accreditation. The common renewable energy resources used in buildings include: solar heat water, solar PV, small scale wind turbine, geothermal heat pump, etc. Construction and demolition (C&D) waste control also plays a critical role to achieve green building. One of key elements of sustainable building design is to reduce the consumption of resources and to improve the resource utilization efficiency.

The life cycle assessment (LCA) approach is one of most popular method to analyse the technical aspects of green buildings. In essence, LCA considers a building as a system, while quantifying the material flow and energy consumption flow across various stages of the life cycle. The advantage of LCA is to go beyond the traditional study of focusing on a single stage by extending the investigation to other stages as well, e.g. manufacturing and transportation of materials; energy consumption, water consumption and GHG emissions during the operation stage. Since 1990, LCA has achieved wide implementation in building assessments. According to ISO 14040 and ISO 14044, LCA consists of four phases, i.e. goal and scope definition, inventory analysis, impact assessment, and results interpretation. LCA has been utilized to analyse the water consumption, energy consumption, carbon emission and cost of buildings

- **Under Managerial Category**: There are multiple managerial aspects of green buildings, i.e. project level, company level and market level. At project level, specific project management skill sets are required for managing green buildings. At company level, the implementation of environmental management system (EMS) help to save 90% of energy consumption, reduce 63% of C&D waste, reduce 70% of water consumption, lower 20% of accident rate and 80% of quality complaints. In addition, the cost predictability is enhanced which in turn eases the cost management pressure.

Managerial issues also exist at the market level, mainly focusing on the health of the entire green building market. Similar to other sustainability initiatives, green building is to a large extent dependent on public policies. The function of public policy can be either positive or negative incentives (i.e. penalties and compensations). As an initiative of the Council of Australian Governments (COAG), the Commercial Building Disclosure (CBD) Program mandates the disclosure of building performance information (e.g. energy efficiency and greenhouse gas emission) to buyers or tenants.

Similar to other sustainability initiatives, green building is to a large extent dependent on public policies. The function of public policy can be either positive or negative incentives (i.e. penalties and compensations). As an initiative of the Council of Australian Governments (COAG), the Commercial Building Disclosure (CBD) Program mandates the disclosure of building performance information (e.g. energy efficiency and greenhouse gas emission) to buyers or tenants.

- **Under Behavioural Category**: The behavioural and cultural factors are also crucial for green-building developments. Therefore, it is critical to raise the level of awareness of all stakeholders (e.g. clients, designers, contractors, and end users) on concepts of sustainable development and green buildings.

The future of green building technology is full of opportunities for us to undo some of the damage we have done to the planet and the environment. The green building materials market is expected to be more than $254 billion in 2020, according to Navigant Research. Timothy C. Mack of AAI Foresight predicts the future of green building materials will hold air cleaning materials, micro-grids, net zero buildings, and smart glass.

The book "Green Building Management and Smart Automation" is a valuable primer that covers the basis of green building from single-family homes to large buildings in a clear and concise manner. The examples in this book show that a building can indeed be run according to the principles of energy and resource conservation when – from the base of an integrated energy concept – usage within a given establishment is being consistently tracked and optimized. The contents of this book are based on the extensive experience of the authors, researchers and scientists in planning, construction and operation of Green Buildings. It documents, through examples, innovative architectural and technical solutions for both planning and operation. This book is directed primarily for researchers, architects, construction planners and building operators, looking for an energy approach that is easy on resources. It is meant as a guideline for planning, building and operation of sustainable and energy-efficient buildings.

We would be pleased if, by means of this book, we succeeded in raising the level of motivation for erecting Green Buildings anywhere in the world, whether from scratch or as renovation projects. Engineering solutions to make this happen are both available and economically viable. Our sustainability approach goes even further, incidentally.

We would now like to invite you to join us on a journey into the world of Green Buildings, to have fun while reading about it, and above all, to also discover new aspects that you can then use for your own buildings in future.

Hanaâ Hachimi
Ibn Tofail University, Morocco

Preface

Automation is termed as "Futuristic Technology". All the plans of development laid under automation are ahead of time. It legitimizes the motivation towards green living. Houses and business structures, nowadays, are raised on green building ideas. Eco friendly disposed structure material and green building ideas are being utilized in this procedure. Automation focuses on bringing in smart alternatives for natural fuel resources with man-made creative options. This is a clear indication of sustainable development for a considerable length of time. The dangers floating over planet earth on account of the disturbed environmental balance can be well dealt via automated green living. In addition to the developing enthusiasm for manageable green buildings in the course of recent years, another energy ecosystem system is rising i.e. Smart Green Buildings. A 'Green building' is a structure that, in its plan, development or task, lessens or disposes of negative effects, and can make positive effects, on the earth's atmosphere and indigenous habitat. Green buildings safeguard valuable normal assets and improve our personal satisfaction. Be that as it may, it is important that not every single green building is and should be the equivalent. Diverse nations and locales have an assortment of attributes, for example, distinctive climatic conditions, one of a kind societies and conventions, various building types and ages, or wide extending environmental, monetary and social needs – all of which shape their way to deal with green building.

Green Buildings carefully bolsters installation of vitality effective appliances in every nook of automated homes. Subsequently, a green home surely has something much beyond the greenery in its environment. With the more noteworthy effect of automated solar energy panels and predominant setups, SMEs and greater businesses are becoming environmentally friendly. Reliance of businesses on synthetic power is diminishing or to be exact, it is moving towards common vitality assets for example solar energy. Earlier than presentation of such more prominent parts of automation, it was almost difficult to perceive and use this exponential intensity of the sun. The exponential development of the lodging division over the most recent couple of years has profited the Indian economy colossally. Be that as it may, simultaneously, it has represented some genuine difficulties to nature. To guarantee comparative continued development in years to come, selection of green home building ideas and strategies in lodging part is the need of great importance.

Green Smart Buildings are associated with a smart grid to optimize energy flows. As the advancements of brilliant building structures and smart grids meet, they'll give immense advantages in terms of more efficient energy use, incorporation of on location energy demand and generation with the grid, and better-working structures that are better and provide a more secure work environment and to live. Automated green structures open a door for energy efficiency and productivity and mass-scale sustainable generation, as well as automated Demand-Response (DR) systems: While some interest is moved to bring lower-cost, off-peak times, the peak power generation that's avoided often comes from the most polluting power plants. The frameworks that empower DR are a foundation of, by and large energy-efficiency programs. They give itemized energy use data that settles for smart energy decisions overall. Up to this point, a building manager gets a call from the utility and actually strolls around to turn off hardware and appliances. Smart green buildings will have advanced control frameworks that automate the procedure and the outcome will be more environmentally friendly, energy efficient building.

Smart automated green building energy efficiency control innovation is backed by a software protocol called 'ZigBeeTM'. ZigBeeTM depends on IEEE Standard 802.15.4, which was sanctioned in 2003. ZigBeeTM characterizes a new open-architecture, Wireless Personal Area Network (WPAN) technology, which is planned explicitly for connecting gadgets that perform observation and control. It gives a standard to low-controlled computerized Radio Frequency (RF) transceivers that work basically in the unregulated modern, logical, and medical (ISM) band at 2.4 GHz. For green buildings, these gadgets can incorporate anything from a simple light switch or thermostat to more complex security and life-safety gadgets. Since ZigBee items depend on an industry standard and no other organization's restrictive framework and any control gadget that is ZigBee agreeable can be utilized. This will open the entryway for the ZigBee wireless building monitoring and controlling gadgets. ZigBee innovation characterizes how these gadgets cooperate and gives an affirmation program to help guarantee consistence. ZigBee can be considered as the next big thing in building automation and building owners cum managers need to be aware of this advanced wireless technology.

The automation of smart green building is not possible without the Internet of Things. IoT is enhancing green buildings in following ways:

- **Enabling Edge Analytics**: Greater instrumentation and sensing, embedded computing and storage in various building devices and equipment, and embedded fault detection and diagnostics in connected systems.
- **Creating Adaptive Building System**: Self-optimization of building equipment and systems, responsiveness to changing environmental and other factors, and the ability to implement multiple outcome-based strategies, like model predictive controls or "extreme seeking controls," machine learning and artificial intelligence.
  •Life and reliability modeling of various building equipment and systems.
- **Predictive Maintenance and Operations**: Earlier predictive interventions were schedule and issue-based; now they are event cum model-based due to real-time data monitoring and analytics.
- Easily solve difficult integration issues between varied building devices and management systems.
- Overcomes security and privacy issues regarding building technologies.

All of these changes will impact significant growth in building technologies. The commercial building systems market is expected to grow beyond $75 billion in the next five years. Out of this, 39 percent will come from automation, metering, and lighting control systems. Building Energy Management Systems (BEMS) which comprise of software, hardware, and services is an intersection point in this space. Traditionally, the buildings industry has transformed slowly. Rapid changes in technology, stakeholder expectations, and environmental concerns will drive much faster changes. Social media has created more engagement between people and their environments, which is also impacting the industry. IoT will provide the foundation for the next big-step change in making green buildings smart. To help realize the full potential of Green Building Management and Smart Automation, this book addresses the numerous aspects and challenges for integrating them and discusses the conceptual and technological solutions for tackling them. The challenges include the deployment of Cloud-based IoT model, security cum privacy issues, novel architectures integrating Cloud and IoT, renewable energy distribution and techniques for sustainable design and development of Green Buildings. The book contains 12 chapters authored by several leading academicians, researchers and experts in the field of Green Buildings, IoT and Automation. The book is presented in a coordinated and integrated manner starting with the fundamentals, and followed by the technologies that implement them. The content of the book is organized as follows:

CHAPTER 1: GREEN SMART BUILDING – REQUISITES, ARCHITECTURE, CHALLENGES, AND USE CASES

In this chapter, a comprehensive discussion on the green and smart building is presented, considering how the buildings are made green and smart and how they support in development of sustainable cities. Though smart buildings are the positive catalyst towards sustainability, the excessive use of electronic devices requires a check in attaining the overall green goal. This chapter suggests merging green and smart technologies to have Green Smart Building (GSB) with the aim of offering the population a smart and eco-friendly living.

CHAPTER 2: ROLE OF RENEWABLE ENERGY TECHNIQUES TO DESIGN AND DEVELOP SUSTAINABLE GREEN BUILDING

This chapter presents the role of renewable energy techniques to design and develop a sustainable framework for green building. The first viewpoint is identified with the earlier structure and the low encapsulated energy building materials for the design and development of a framework for green building. The primary perspective is to manage energy protection using renewable energy techniques in the green building. Green building interchangeably can be used with the term's sustainable construction or green construction. So, durable construction means using environmentally responsible and resource-efficient procedures in development to be ensure sustainability throughout the lifetime of the building. The chapter also presents the combination of renewable energy-based technology for green building construction and sustainability with the economics of renewable energy.

CHAPTER 3: IMPROVED DESIGN OF VERTICAL CAVITY SURFACE EMITTING LASER USING 3D SENSING TECHNOLOGY FOR INTERNET OF THINGS APPLICATIONS

Internet of Things (IoT) and Vertical Cavity Surface Emitting Laser (VCSEL) future can be seen together, since VCSEL technology-based 3D sensors are introduced for IoT applications. This chapter presents the improved VCSEL structure design with fixed wavelength using 'a thermally' actuated cantilever structure. This improvised structure of VCSEL will support realizing 'athermal' VCSEL. In 'athermal' VCSEL the dependency of VCSEL on temperature will be very less because it will not require temperature controllers. Realizing fully temperature independent VCSEL i.e. 'athermal' VCSEL is still a challenge but we can reduce it to some extent. In this chapter, recent diversification of application of VCSEL technology from data communication to sensing has been discussed. The proposed VCSEL structure provides an opportunity to improvise the existing VCSEL technology. Therefore, smart 3D sensors based on VCSEL will help in making Internet of Things (IoT) applications more reliable and will directly or indirectly serve the concept of smart homes, smart cities and green smart buildings in the near future.

CHAPTER 4: ENVIRONMENTAL ANALYSIS OF CONSTRUCTION MATERIALS – MATERIAL SPECIFICATIONS FOR GREEN BUILT ENVIRONMENT

The chapter discusses the role of sustainable development without any sort of destruction to the natural resources. In this point of view, selecting building materials according to their energy and health performances gains importance in sustainable design of Green buildings. 3Rs reducing, reusing, recycling and supplying a healthy, non-hazardous indoor air for building occupants are two important parameters of environmental life-cycle assessment for materials.

CHAPTER 5: A COMPREHENSIVE OVERVIEW OF IOT-BASED GREEN BUILDINGS – ISSUES, CHALLENGES, AND OPPORTUNITIES

The chapter summarizes that green building assures sustainability for better living of citizens. It is to be noted that, not only the structure of the building, but throughout the life cycle of the building like designing, constructing, operating, maintaining, renovating and demolishing must ensure responsibility towards environmental and use of natural resources. However, there are numerous factors that influence the effect of Green Building. Further factors such pollution control, air quality monitoring, adaptability to evolving environment, materials used also need to be handled by Green Building. Hence, this chapter focuses on exploring the various issues, challenges and opportunities of Green Building concepts. Further, this chapter addresses how IoT based Green building will assist in achieving it using cloud computing techniques.

CHAPTER 6: GREEN BUILDING EFFICIENCY AND SUSTAINABILITY INDICATORS

This chapter presents indicators to define, manage, measure, and enhance efficiency and sustainability phenomena for proposing a green building. A primary objective of this study is to identify influencing factors and set forth viable indicators and framework in term of energy-efficient green building from different standpoints hiring innovative tangible and non-tangible tools and techniques.

CHAPTER 7: GREEN AND SMART BUILDINGS – A KEY TO SUSTAINABLE GLOBAL SOLUTIONS

The aim of this chapter is to present a comprehensive and significant research conducted till date with regard to Green Buildings. The chapter provides comprehensive analysis of design technologies i.e.

Passive and Active technologies which lay a strong foundation for Green Building. The chapter also highlights the smart automation technologies that supports energy conservation along with various performance metrics.

CHAPTER 8: CLOUD-BASED IOT ARCHITECTURE IN GREEN BUILDINGS

The chapter highlights that the concept of "Green building" as a strong solution to varied critical environmental issues apart from promoting eco-friendly activities. IoT lays the foundation of having buildings that are energy efficient enough via networked sensors that not only helps in managing the assets better but also reduces harmful impacts on Human health and the environment. The chapter talks regarding the concept of the Green Building and the smart automation achieved through IoT as well as Cloud architecture termed as "Green Cloud". In addition to this, significant future challenges are also rounded off in this regard.

CHAPTER 9: IOT-BASED GREEN BUILDING – TOWARDS AN ENERGY-EFFICIENT FUTURE

The chapter shows the application of IoT in Green Buildings. With the advent of IoT, the concept of green buildings has taken a broader perspective. Incorporating intelligence into the current Building Management Systems can revolutionize the legacy buildings into energy efficient buildings. The chapter explores some sound benefits of integrating IoT into a green building. It offers insight into the various technologies used in Green construction, followed by some IoT based architectures. The chapter points out various machine learning algorithms that can lay a strong base for IoT devices efficiency. Finally, the chapter dives into the future of IoT enabled green buildings, and explore the challenges in achieving zero-energy buildings, while addressing the questions it raises.

CHAPTER 10: RENEWABLE ENERGY DISTRIBUTION AND MANAGEMENT IN GREEN BUILDINGS

Construction of building plays also a significant role in pollution. To reduce the harmful effects in construction buildings it is necessary to move on to sustainable architecture. In this chapter different advantages of green buildings are enlisted. The green building structure is compared with traditional conventional buildings. In addition to this, different organization contributing towards green environment are mentioned along with various sensors assisting towards detection of energy wastage and prediction.

CHAPTER 11: A STUDY OF GREEN BUILDING PROSPECTS ON SUSTAINABLE MANAGEMENT DECISION MAKING

This chapter lays a strong foundation regarding discussion of sustainable management decision making with regard to prospects of green building. Nowadays, the world faces many environmental crises such as increased threat of climate change, the depletion of key natural resources, increasing air and water pollution and growing levels of solid wastes. These issues are becoming the major aspects of value in real estate and a key driver in the decision-making processes. The strategic sustainability process as called "the halo effect" was more worldwide which is affected by the popularity in environmental actions criteria. It showed consequence that green concept has not only focused on technical or moral issue but also need to base more on the economic and financial imperative.

CHAPTER 12: SUSTAINABILITY OF GREEN BUILDING PRACTICES IN RESIDENTIAL PROJECTS

A green building is a sustainable building that has minimal impacts on the environment throughout its life. This chapter elaborates that green building makes efficient use of energy and resources in every aspect which includes the production of building materials, and the design, use and eventual demolition of a building in any sector (commercial, residential, industrial, public buildings) and at all stages, from new buildings to 'retrofitting' or adapting existing ones.

Chapter 1
Green Smart Building:
Requisites, Architecture, Challenges, and Use Cases

Pijush Kanti Dutta Pramanik
https://orcid.org/0000-0001-9438-9309
National Institute of Technology Durgapur, India

Bulbul Mukherjee
Bengal Institute of Technology, India

Saurabh Pal
Bengal Institute of Technology, India

Tanmoy Pal
Bengal Institute of Technology, India

Simar Preet Singh
https://orcid.org/0000-0002-2443-7835
Thapar Institute of Engineering and Technology, India

ABSTRACT

Non-sustainable buildings have threatened the ecosystem globally. In this chapter, a comprehensive discussion on the green and smart building is presented, considering how the buildings are made green and smart and how they support in developing sustainable cities. Though smart buildings are the positive catalyst towards sustainability, the excessive use of electronic devices puts a check in attaining the overall green goal. This chapter suggests merging green and smart technologies to have green smart building (GSB) with the aim of offering the populations a smart and eco-friendly living. Promises and challenges in attaining this goal are meticulously explored. The GSB concept is discussed in detail, suitably supported with the architectural models of overall and the various components of a GSB. The communication architecture is also presented emphasizing on various entities and activities in different levels of communication between various digital components of a GSB. A few cases have been presented showing practical applications of green and smart technologies in buildings.

DOI: 10.4018/978-1-5225-9754-4.ch001

INTRODUCTION

As a result of continuous urbanization globally, the number of buildings (residential and commercial) is on the steep rise, be it in small cities or the metropolis. In some of the global cities, nearly 50 residents turn up to settle down every hour (State of Green, 2019). This shows the high-rate increase in urban population and the demand and need for buildings, especially the residential buildings. Because of the high density of population in the cities, they have become one of the primary sources of CO_2 emissions. Particularly, the buildings are accountable for more than 40 percent of global energy consumption and an almost equal amount of CO_2 emissions (State of Green, 2019). This is way beyond that of the other two major culprit's - the industrial sector and transportation, which account for 30% and 29% of total energy consumption respectively. In the U.S., about 73% of the country's electricity consumption is attributed to the buildings (BossControls, 2019).

Not only the high energy consumption and CO_2 emissions, but also the inefficiency of today's building plans and constructions, in terms of optimized use of resources and recycling, has posted a great environmental and societal challenge. For instance, in the European Union (EU), nearly 30 percent of total waste generated is accounted to the cast off and waste materials from construction and demolition (State of Green, 2019). According to the Environmental Protection Agency (EPA), in the U.S., around 170,000 commercial buildings are built every year, while 44,000 are demolished (BossControls, 2019). This vast amount of waste materials put pressure on the overall ecology and the environment of the urban structure. The absence of sustainable urban planning sets a big threat to the city's natural resources like water, energy, and clean air as these resources are wanted in large volumes by the buildings (construction and maintenance) and their occupants (operation and use).

The traditional building construction concepts have a significant negative impact on the global environment due to the huge emission of greenhouse gas from this kind of building, huge amount of water uses (13.6% of all potable water is consumed by buildings (BossControls, 2019)) as well as wastewater, using non-sustainable construction materials which produces CO_2 and the building wastes such as plastic, concrete, glass, wood, metals, etc.

To have green and sustainable cities, we need to make buildings green and sustainable. Not only in the new buildings, but it is important to adopt and implement green and sustainable technologies in the existing buildings as well. Green building means to apply green technologies to a building throughout its life cycle, i.e., from construction to operation to demolition so that the negative environmental impact of buildings can be minimized as far as possible. Using sustainable building materials and recycling and reusing these materials, reduce waste significantly. A planned and strategic approach in designing buildings, selecting building materials, and the use of modern, efficient construction technologies is the key to have sustainable buildings.

Besides using sustainable materials, curbing the energy consumption of buildings and reducing CO_2 emissions are the crucial factors for green buildings. Focusing on energy efficiency solutions and operations should be widely implemented in all categories of buildings across the globe. Several existing solutions such as energy-efficient windows, proper insulation, heat/cold regulators, ventilation systems, efficient pumps, smart meters, intelligent management systems, etc., if properly implemented, can reduce energy consumption by 50 percent (State of Green, Energy efficiency in buildings, 2019). Also, a sensible choice of building materials in constructing green buildings can reduce the emission of greenhouse and other harmful gases considerably.

The idea of constructing smart buildings has fueled the green building goals towards minimizing the environmental hazards of buildings. Smart building technologies with advanced digital services and analytics make a building management system more sustainable and cost-effective. The state-of-the-art sensors for automated controlling of lighting, air quality, and climate, along with the smart meters and energy management systems have enabled smart buildings in achieving intelligent control of energy consumption that results in considerable energy savings, flexibility, and comfort, benefitting the people as well as the climate. The smart buildings can also play a vital role in power management by integrating with the energy networks from beyond the levels (e.g., state or country level) through the smart grid (Tuballa & Abundo, 2016). The efficient distribution of electricity among each building and also throughout the city leads to a sustainable energy efficient city.

The emergence of the Internet of Things (IoT) and associated technologies not only have eased our lifestyle greatly (Pramanik, Pal, & Choudhury, 2018) but also enabled structuring smart buildings while unlocking the viable potential to build sustainable buildings. IoT (Pramanik & Choudhury, 2018) is the principal architect of automated, safe, secure, comfortable, energy efficient building. The continuous use of IoT that includes smartphones, smart gadgets, and various sensors enhances our building experience more comfortable, more people friendly, but with a cost of increased power consumption and other environmental hazards such as radiations and rising carbon footprints from the electronic gadgets and the mobile towers (Pramanik, Pal, & Choudhury, 2019). Hence, the smart buildings need to be made further green, giving rise to the green smart building (GSB). Green design of a smart building is now becoming a challenge to maintain sustainability with nature, to conserve non-renewable sources of energy and to make our earth cool and green. The principle of 3Rs (Reduce Reuse Recycle) is applied in design, construction, and operation of a smart building to make it green which helps to sustain the world pollution-free. The goal of a GSB is not only to minimize the adverse effects of buildings to the environment by adopting energy efficiency measures and intelligent design, but also to provide the citizens with a good indoor climate that can be adapted to the users' varying activities and needs over time. The vision of smart buildings is to provide a smart, livable, and sustainable city.

In view of the statements as mentioned above, this chapter aims to provide a comprehensive overview of the GSB considering the different aspects of it. The rest of the chapter is organized as follows. Section 2 discusses green building highlighting its benefits and challenges. The key factors of the green building along with different means of making buildings, green are also explained. The lifecycle of a green building is presented along with the different dimensions that make buildings green and sustainable are also discussed. Likewise, Section 3 talks about smart buildings covering the components of a smart building and the benefits and challenges of smart buildings. The different approaches to make a smart building are also appropriately elaborated. Section 4 asserts the need for merging of these two concepts (green and smart), i.e., the importance of GSB. Its emphases on the vision and benefits of GSBs. Section 5 presents the architectural model of a standard GSB while discussing in detail the different components of a GSB with suitable model diagrams. The communication architecture of GSBs has been elaborated appositely in Section 6. It describes the layered framework of a GSB and the requirements, components, and services offered by each layer. Several assessment tools and standards for GSB are also cited. Section 6 points out the benefits of GSBs. The technical challenges in implementing GSB are discussed in Section 7 while Section 8 draws attention to other associated issues. Several real-life examples of existing buildings where green smart technology are adopted are mentioned in Section 9. Section 10 lists some building assessment tools and standards followed by various countries to assess the green and smart buildings. Finally, Section 11 concludes the chapter.

BASICS OF GREEN BUILDING

What Is a Green Building?

Though the idea of the green building is transfiguring day by day, a green building in generic could be said as one whose life-cycle focuses on reducing the negative impact on nature through better design, siting, construction, operation, maintenance, and better waste management. Due to self-sufficiency and less energy consumption, the green buildings are often said as a sustainable or high-performance building (Howe, 2010).

Key Factors of a Green Building

Making green building needs to consider certain key factors like the material used for construction, energy sources, and management. Each key factor has effective roles which are being discussed subsequently.

Materials Used in Green Building

Materials used for green building are generally obtained from natural and renewable sources, keeping in mind some important factors like durability, energy cost in transportation, waste minimization, etc. Table 1 shows some recycled products which are being used in green building construction along with their sources of production.

Table 1. Recycled materials used in green building construction

Product Name	Sources & Uses
Newspaper wood	Every year in Norway, 1 million tons of paper and card wood are recycled.
Nappy roofing	Polymers are extracted from the thrown away nappies and sanitary products and used for manufacturing different construction materials.
Recy blocks	Colorful bricks produced from plastic bags which are difficult to recycle. These plastic bags are forced together to form blocks by applying heat.
Blood brick	Animal blood, which is considered to be a waste product is one of the powerful bio-adhesive. According to British architecture student Jack Munro, a mix of frozen and dried blood powder with sand can be used for making bricks. This is very helpful to communities in remote place, where animal slaughter blood is easily available.
Bottle bricks	Cuboid or other tessellation shaped bottles are used for forming the building structure as these are easily transportable. The bottleneck slots of the bottle are placed into the base of another one to form an interlocking line which makes it a more rigid structure.
Smog insulators	One of the biggest waste repositories is air. In Bangkok, peoples have designed a system which consists of an electrically charged metal mesh which is capable of attracting smog particles and sticks them to form a silvery fur over the building.
Mushroom walls	Mycelium bacteria, found in a rotten organism like a tree trunk and many agricultural by-products, if placed in some mold these grow to the desired shape which is used as an insulating and packing materials.
Plasphalt	One of the major building components produced from plastic waste. The idea is recycling the plastic by mixing with asphalt which makes it more rigid, with water-resistant quality.
Wine cork panels	Granulated cork combined with wine bottle cork form this kind of panels, which can be used as floor tiles.

(CityMetric, 2015)

Energy Systems

The efficient design of the energy system is an important clause for making a building green. If natural light is utilized methodically, it can reduce the electricity cost and ameliorate the people's health and productivity. Low energy appliances, energy efficient lighting and also some renewable energy technologies can be incorporated in the green building. Figure 1 shows the natural lighting and the heating and cooling effect of the building by the natural conduction, convection, and heat radiation process.

Figure 1. Heating and cooling effect by conduction, convection, and radiation of heat wave in a building

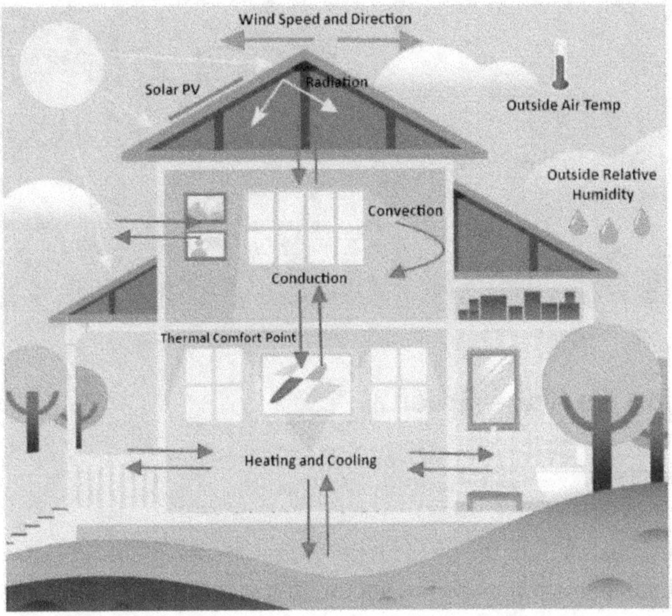

Passive Solar Design: The concept of passive solar design is used to heat, cool and light the homes without the intervention of any kind of electrical or mechanical devices. Placement of windows and skylights are part of the design for making the building green. Figure 2 represents a passive solar design of a typical house model where the elements like a sunshade, panels, heat absorbents, etc. are used in capturing and distribution of natural heat and light.

Passive Solar Heating: Capturing the sun's heat in building materials and use that heat inside the building when the sun is not shining. The absorbed heat also used for later usage like making the space comfortable without overheating.

Health Components of Green Building

Nontoxic, emission-free, less volatile organic compounds (VOC) content and moisture resistant materials and products are used as building materials which can reduce the asthma rate, allergy by improving the indoor air quality. The proper ventilation system of the building and using humid control materials can also enhance the building's health.

Figure 2. Elements of passive solar design
(Williams, 2012)

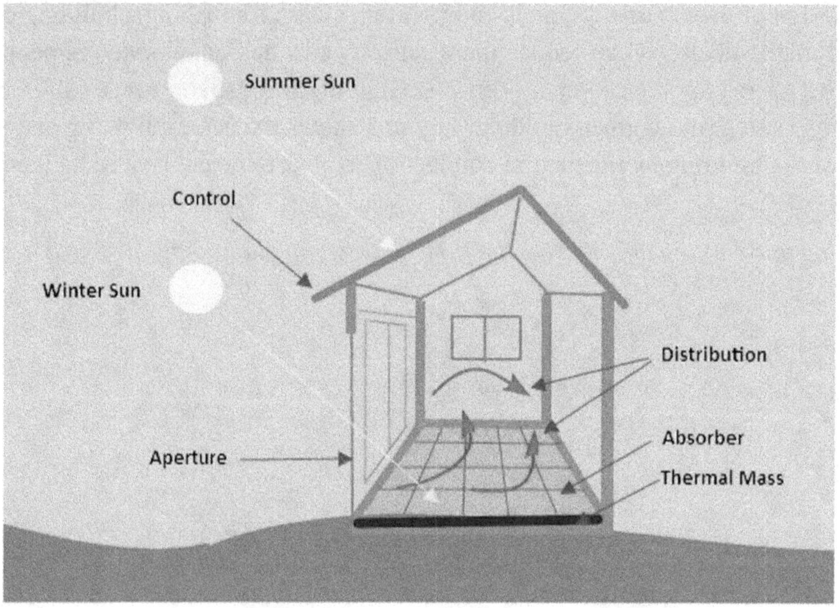

How Can a Building Be Made Green?

Technological Aspects

As discussed in the previous section, the use of renewable sources of energy, proper waste management, and natural resource management are the key factors for making a building green. But besides these, there are other factors – technological aspects which plays a major role in making a building green.

Energy Savings

The electrical appliance used in daily jobs consumes a lot of energy, proper selection of standard electrical appliance and their careful operations can lead to minimize energy consumption. Use of electrical appliances with energy star rating though, looks small but is a major step in decreasing energy consumption. Besides using energy optimizing, techniques can lead to energy conservation and preservation. Using modern technologies like occupancy sensors, dimmers, and daylight controllers can reduce the energy consumption up to 22% (Grainger, 2019). Besides these, using open column systems for air circulation, cable management, and newer power distribution systems help in reducing overheating and thus depends on the air conditioner and thereby reducing the power consumption (Grainger, 2019).

Space Efficiency

Space utilization is one of the important key factors which need to be considered when designing the building. Open floor plan raised floor system and reduced building height are some of the ways for efficient use of space that could lead to making a building green. The open space help in good air circulation and spread of light. This reduces the heating and lighting problem and thus lessens the dependency on air conditioners and lights.

Other Associated Factors

Rain Water Harvesting

Water is one of the essential commodities in our daily life. Water plays a significant role in the existence of animal kind. But for the last few years, the shortage of this natural resource raised a question mark on our survival. Multiple reasons like rapid climate change, wastage of water, unplanned usage, and deforestation have caused the water shortage problem. Different techniques have been found useful in water conservation. Among the different types of water conservation techniques, one of the easiest, eco-friendly, and cost-effective method is rainwater harvesting (Go Smart Bricks, 2017). In this method, rainwater is stored efficiently for future use. Generally, water is collected from different surface layers and stored for future use.

Objectives of Rainwater Harvesting

We have already discussed that a green smart concept not only makes a building green, but it is also responsible for making a pollution free environment. Following are some major objectives of rainwater harvesting:

- Recharging and raising the groundwater level.
- To prevent soil erosion or flooding of the surrounding area.
- Prevent the rainwater from becoming polluted.
- Meeting the demand for water in the dry season.

Different Methods of Rainwater Harvesting

There are two types of water harvesting techniques (Padmanabhan, 2018):

1. **Surface Runoff Harvesting:** In this method, rainwater flowing away as surface runoff, are stored and kept for later use.
2. **Rooftop Rainwater Harvesting:** In the rooftop rainwater harvesting technique, rainwater is collected over a rooftop of a house and then it can be channelized through small PVC pipes to underground pits or sumps or wells. The stored water can then be later used on the requirement.

Different Components

The different components of rainwater harvesting are:

1. **Catchments:** Catchment of the rainwater harvesting system is the surface which receives the rainfall directly. It can be our terrace or courtyard. Paved or unpaved open ground could be used as a catchment. Flat RCC or stone roof or sloping roof is ideal for catchments. In short, we can consider the catchment as the supplier of the rainwater harvesting system.

2. **Transportation:** Collected rainwater can be transported through UV resistant pipes to the harvesting system. Sloping roof water can also be caught through gutters.
3. **First Flush:** The First flush is a kind of device which is used to flush off first shower water to avoid harmful components of the first rainwater. It is also useful to clean up the roof by cleaning the deposited materials during dry seasons. For the first flushing, first rain separator should be present at the outlet of each drainpipe.
4. **Filter:** There is always some chance that rainwater can contaminate the groundwater. To avoid this kind of situation, the proper filtration mechanism should be installed. Another important factor that should be taken care of is to regular checking of underground sewage drain which has chances of leakage. In spite of that proper water treatment process should be carried out to remove the turbidity, color, and microorganisms.

Wall Gardening

Wall gardening is another approach to green smart building. In this approach, plants are grown vertically from the wall. This is an alternative to potted plants. But potted plants have some benefits like it can be placed anywhere inside the building. The disadvantage is that it acquires some space. Following are some benefits of wall gardening (Plant Connection, 2019):

- It can reduce the urban heat effect and smog.
- By absorbing outside air of pollutants and CO_2 gas, dust, and fuel emissions, it purifies the atmosphere.
- It also removes harmful toxins like benzene and formaldehyde.
- It also acts as a soundproofing wall.
- Water flowing through the wall can also be filtered by this natural plant filter.
- Wall gardening is a good insulator which keeps the building cool and also protects from harmful elements.
- It has a major role in increasing biodiversity by creating habitats for birds and insects.
- Wall gardening cans also be used for growing food in urban areas which support and control the local food source.
- It speeds up the recovery process for patients through Biophilia.
- Reduces the people stress and boost them.

Rooftop Agriculture

Rooftop Agriculture is also called as 'vegetated roofs,' or 'eco-roofs' is agriculture for the production of fresh vegetables over building rooftop. This consists of a waterproofing membrane, along with the growing medium (soil) and plants overlying a traditional roof. This rooftop agriculture not only supports the local vegetable demands, but also help to reduce the usage of artificial cooling components due to absorption of heat by the bare roof and also absorbs the fuel emission produced by industries and vehicles and act as an air filter.

Properly designed and maintained green roofs have multiple social and environmental benefits. It is an important tool to enhance sustainability and biodiversity by reducing energy consumption, greenhouse gas effect and urban heat island impacts.

Benefits of Rooftop Agriculture

The benefits of rooftop agriculture can be categorized in two ways (United States General Services Administration, 2011):

1. Environmental Benefits:
    - Habitat promoting biodiversity has been increased due to the green roof concept.
    - It absorbs harmful radiation.
    - Cooling of the building roof and surrounding air.
    - It acts as a good sound absorber.
    - Improves the air quality by absorbing pollutants from the air.
2. Economic Benefits:
    - The cooling effect of plant respiration and the insulation property of plants, shading and thermal mass of the plant and soil layers lowers energy costs for cooling.
    - Less frequent replacement of roofs due to greater resilience than conventional roofs.
    - Reduced stormwater management costs.

Energy Efficiency and Using Renewable Energy

Energy efficiency is one of the important factors in the environment. Not only it saves our natural resources, but also saves the environment by reducing pollution. For greater energy efficiency the dependency of the conventional energy source like coal, petroleum should be reduced. Alternate to conventional energy resources, natural, renewable energy is environment-friendly and produces no greenhouse gas (GHG). For a building to be more energy efficient and hence reducing carbon emission, the maximum use of renewable energy is the smart and pollution-free solution. Renewable energy is produced from natural sources like wind, solar heat, sunlight, etc. The sources of renewable energy are abundant and never exhaust.

Conditions for Energy Efficient Building: The following criteria should be taken into consideration to achieve energy efficiency while constructing a new building or renovating it:

- Comprehensive, integrated design process.
- Fix the goals and strategies.
- Smart implementation of an energy-efficient system.

Types of Renewable Energy: There are mainly three types of renewable energy resources (Venkata, Bhargav, Choudhary, & Sharma, 2015):

- **Solar Energy:** Sunlight is an important and free source of energy. The solar energy can be used not only for heating purpose, but also used for producing electricity. Solar heating systems in a building using solar energy heats water and the building. The use of photovoltaic cell converts light energy to electric energy. Application of photovoltaic cell in solar panel allows to convert sunlight into electrical energy.
- **Geothermal Power:** The main usage of geothermal energy is to cool down the temperature inside the building in summer, and an increase in winter without using fossil fuel. The geothermal power

systems move the heat through a pipe from the deep ground to the building in case of heating and vice versa in case of cooling. It uses the liquid antifreeze as a carrier which moves down through the pipe carrying the heat.

- **Wind Energy:** Another important natural source of energy is wind energy. Wind energy consists of energy produced by the flowing wind. The flowing wind is utilized in a turbine or sail to produce electrical or mechanical energy. If the building can be constructed in a place where enough wind resources are available, then the electricity produced from this wind resource can be used as home-based renewable energy which reduces the pollution as well as energy cost. Wind is considered to be a cubic energy resource. It means if the wind speed increases, then the power will also be increased cubically. Proper installation of the turbine at a higher altitude is more beneficial. Appropriate design and planning can make it possible to use wind power as one of the important sources of renewable energy.

Green Building Lifecycle

A building typically has four phases which correspond to different activities of the lifecycle of the building as shown below:

- Pre-Building Phase
    - Design
- Building Phase
    - Construction
- Post-Building Phase
    - Operation and maintenance
    - Demolition and disposal

Figure 3 shows a conceptual model of the lifecycle of a green and sustainable building along with the major aspects of sustainable construction (i.e., energy, water, waste, and materials) connected with each phase (Hui S. C., 2007). Green buildings consider and assess the environmental impact of every phase of its lifecycle and attempt to minimize the side effects and the environmental hazards at these phases. Since the phases are associated to different stakeholders (e.g., building designers, contractors, end-users, etc.) from varying turfs and interests it is crucial to ensure the joint efforts of each of them towards maintaining the green and sustainability requirements.

Dimensions of Sustainability

Green building technologies address three dimensions of Sustainable development of the construction industry:

- **Environmental Impact:** Includes the nature and climatic benefits.
- **Social Impact:** Includes the comfort, satisfaction, productivity, efficiency issues related to society.

Figure 3. Lifecycle of a green building construction and management
(Hui S. C., 2001)

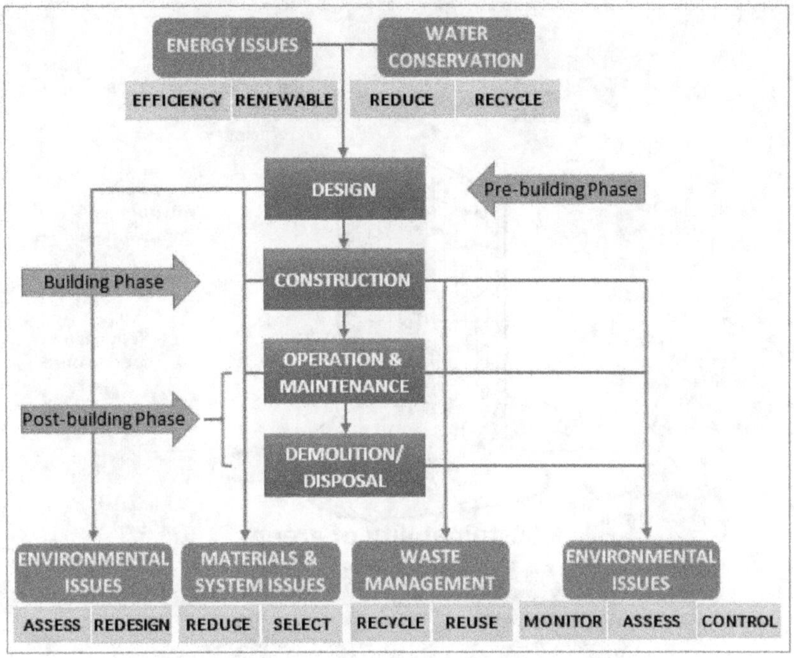

- **Economic Impact:** Includes both of the above because through the resource efficiency one nation's natural wealth reserve can be maintained and due to an increase in satisfaction and productivity national income can be increased.

Figure 4 shows how a green building addresses the different aspects of sustainability.

Benefits of Green Building

Some of the major benefits of green buildings are (Nationwide Construction, 2016):

- **Low Maintenance and Operation Cost:** Solitary design ensures the water and energy savings of the building.
- **Coherent Usage of Energy:** Usage of renewable energy, reduce the dependency on non-renewable resources like coal, etc. It reduces the cost as well as makes a pollution free environment.
- **Escalate the Indoor Environment Quality:** Installation of the operable window allows the sunlight to enter into the room sufficiently, which enhances the indoor air quality by removing the poisonous elements.
- **Water Efficiency:** Rainwater harvesting is one of the major components of green building architecture which reduces the wastage of water. Along with the recycling of wastewater with proper plumbing, the water can be used more efficiently.

Figure 4. Dimensions of sustainability of green buildings

- **Health Benefits:** Avoiding the plastic byproduct at the time of construction by applying the eco-friendly building material reduces the chances of cancer like diseases.
- **Material Efficiency:** Using long-lasting material reduces material waste.
- **Healthy Environment:** By avoiding the use of energy resources like coal, petrol which pollute the environment, green building can make a healthy environment.

Major Challenges in Green Building Adoption

Some of the major challenges faced in implementing green buildings are (GoSmartBricks, 2018):

- **Lack of Awareness:** A very few peoples of the world are aware of the concept of green building and its benefits. That few peoples perceive it to be an unachievable option.
- **Improper Government Policies and Procedures:** Though many countries have already thought about this green concept due to lack of, inadequate policies of the government and lack of synchronization of rules and regulations, green building implementation in large scale is a big challenge.
- **Extra Clearances and Approvals:** Due to the complicated procedure for the approval for green building, peoples are getting demotivated.

- **Lack of Incentives to Encourage Adoption:** There is very few, and a limited number of schemes are available to promote the green building concept in the developing countries. Though these schemes are available in the form of a rebate in property tax, are not sufficient to encourage the large scale adoption.
- **Lack of Expensive Equipment and Products:** Due to the huge cost of green building materials compare to other conventional building materials, developers are not getting motivated as it increases their budget.
- **Lack of Skilled Manpower:** Green building concept requires proper architectural design along with an efficient engineer, policy makers, contractors and manpower to implement this. Unfortunately, still, we lack that skilled manpower.

BASICS OF SMART BUILDING

Buildings play an important role in human life because the majority of the time a person spends inside a building which can be home, school, college, workplace, hotels or hospitals. Individual's physical and especially mental health often depends on the environment where he lives. So, a building has a significant role in setting morale, work efficiency, time efficiency, productivity, and overall satisfaction. Real estate holds one of the largest capital expenses for businesses. And, there is a constant demand from customers' and manufacturers' side to make building more flexible, efficient, sustainable, reliable, safer and more comfortable, thus, expressing in a single word the building would be 'smart'.

What Is Smart Building?

Smart building refers to a building that behaves smartly with the coordination of different heterogeneous system by incorporating smart technologies. In a smart home, various appliances such as lighting, heating, and air conditioning systems, security systems, parking systems, etc. communicate not only to the occupants but also with each other, directly or through a central control unit in order to provide the occupants a comfortable, secured, and energy efficient living. Smart buildings are sometimes referred to as 'automated buildings' or 'intelligent buildings',

The Information and Communication Technologies (ICT) and IoT are the two main pillars of innovative smart building solutions to meet the criteria of smartness by linking core systems such as lighting, power meters, water meters, pumps, heating, fire alarms, chiller plants, elevators, access systems and shading with sensors and control systems. A smart building uses hardware like sensors, actuators, and microchips connected to a network with intelligent and adaptable software to provide responsive, effective and supportive environments to make occupants' lifestyles easier and smoother. Hard sensing is done by numerous sensors deployed in different appliances, in different subsystems and with this vast amount of data are generated which are analyzed or soft sensed by cloud computing services using machine learning to give expert solutions. The global smart building market is expected to grow from \$5.37 billion to \$24.73 billion by the year 2021, according to research published by Markets and Markets, global research and consulting firm (Siemens AG, 2014).

The work (Lilis et al., 2017) suggests that compared to a building automation system (BAS), an IoT-enabled building performs in a better way because of the following reasons:

- Occupants' need and choice are key to decision making.
- Soft sensing is done on the basis of occupant's location and habits tracking.
- Low-cost sensors for monitoring and analytics are incorporated.
- Wireless and low bandwidth networks inherently supported.

Environmental Comfort Management System (ECMS) (Shaikh et al., 2014) incorporating intelligent control systems for buildings that uses computers, microprocessors, storage devices, and communication links (Zanella et al., 2014) is essentially incorporated in smart buildings to improve the indoor environment which includes thermal, humidity, indoor air quality, and illumination levels, etc. ECMS controls the indoor environment, analyzing the occupants' mindset, requirement, satisfaction level. For example, if the weather is cloudy the indoor lighting will be more than a proper sunny day or temperature in different bedrooms would be set with different values depending on the general user's preference level. Thus, ECMS also meets up energy efficiency and cost-cutting by this type of customization.

The building control system (BCS), also termed as building management system (BMS), is the general specifications of building monitoring and metering systems integrating with hardware-software networks and mechanical, electrical and plumbing (MEP) system controls.

An integrated building management system (IBMS) integrates a large number of software programs and applications along with different hardware to collect different data, analyze them to generate logic to perform in a customized manner for ensuring safe, secure, and energy-efficient operations (Siemens, 2018). For example, when one specific person drives into the parking lot, the building can automatically prepare heating and lighting conditions in his room.

Major Components of a Smart Building

Smart building emphasizes a multidisciplinary, cost-effective, energy-efficient effort to assimilate and optimize the building structures, systems, services and management create a productive, responsive, comfortable and environmentally approved atmosphere for the occupants.

The Physical Infrastructure

The physical infrastructure of a smart building includes (Shaikh et al., 2014):

- Traditional building automation equipment, such as hardware, controllers, cables, lighting, as well as fire, security, and other systems.
- Sensors.
- Actuators.
    - Heating, ventilation, and air conditioning (HVAC) systems for thermal comfort which includes control of its various parts:
        - Chillers.
        - Boilers.

        - Air handling units (AHUs).
        - Rooftop units (RTUs).
        - Fan coil units (FCUs).
        - Heat pump units (HPUs).
        - Variable air volume boxes (VAVs).
    - Visual comfort which includes artificial lighting technologies, dimming control, solar radiation blind control, etc.
    - Air quality control which includes window openings, air conditioning units, and fan regulators.
    - Humidity comfort which employs Dehumidifiers, desiccant flow rates, and certain humidifiers, etc.
    - Plug loads constitute kitchen appliances, laundry accessories, charging portables, etc.
- New and developing building systems that meet growing sustainability and efficiency requirements, such as:
    - Rainwater harvesting.
    - Sun-tracking system.
    - Exterior shading.
    - Water reclamation.
- Cabling plays a prominent role in intelligent infrastructure.
- Smart buildings with intelligent infrastructures require a consistent power supply to manage large amounts of data.

The Information (Data) Infrastructure

Smart buildings collect and communicate unprecedented amounts of data. The information infrastructure of a facility consists of four key components:

- **IBMS:** As mentioned in Section 3.1, IBMS is responsible to connect the building systems and management dashboards to the data analysis software.
- **Data Collection:** The intelligent field devices that are connected through IoT can record and collect data, on the basis of which different operational decision can be taken (Pramanik et al., 2018).
- **Data Analytics:** A true analytics platform which analyzes data, make knowledge base and turns knowledge into intelligence to improve building productivity as a whole.
- **Fault Detection and Diagnostics (FDD):** The primary role of FDD is to identify faults and errors, with the help of the collected data and based on the established hierarchical relationships and rules between the different equipment and processes. FDD is typically part of the data analytics platform.

The Cloud

The cloud computing provides a strong architecture to save, parallel processing, integration and security of big data generated by heterogeneous IoT and ICT technologies (Pramanik et al., 2018). Different cloud services deliver highly scalable and large computational infrastructure by which the heterogeneous data generated in the smart building system are stored, analyzed, computed, and applications or services are hosted responding every demand and overall management of the system.

How Can a Building Be Made Smart?

Smart Building is the concept with complete automation and communication technologies to create a smart, and dynamic environment. A building can be made smart by turning the devices in building into smarter ones, and by incorporating the automated building management system. IoT, which is an amalgamation of the internet and sensors allows devices to communicate with one another and services over the Internet to accomplish some objective. The appliance of IoT turns individual devices into smarter ones. The communication between the devices and their management entities across different buildings, the smart grid and cloud computing are can be facilitated through the Web of Things (WoT) (Mathew et al., 2013). Since a smart building contains many complex heterogenous subsystems, thus more computational power is needed. Smart Buildings could obtain more computational power by using cloud computing, fog computing, dew computing, and crowd sensing technologies.

Technological Aspects

IoT has the power to provide a ubiquitous network of connected devices and smart sensors for smart and connected communities (SCC). Boosted up by the power of mobile crowdsensing and cyber-physical cloud computing, IoT enhances SCC. Furthermore, the move from IoT to real-time control, considered necessary for SCC, can be realized by applying big data analytics.

An interoperable intelligent building architecture named "OpenBMS" has been proposed in (Lilis et al., 2017). This architecture has been deployed in a university building. Shang et al., (2014), proposed a secure data-centric Building Management System (BMS) architecture that uses Named Data Networking (NDN). The system performs the encryption of sensor data packets with a symmetric key which is sent to every authorized user. The ARM-compliant IoT security framework is another approach to handle the security and privacy threats in smart building (Hernández-Ramos et al., 2015).

An 'Intelligent Hospital System' is proposed in the paper (Plageras et al., 2017) where BMS is using NDN, replacing every IPv6 address with data names which consist of two data packet types (the Interest type and the Data type). Machine learning algorithms like neural network, random forest, and support vector machine can be used to predict the temperature of a conditioned space by inputting multiple variables of weather conditions (Paul et al., 2018).

Smart Building and Energy Efficiency

The world's predominant fossil resources are on the verge of exhaustion, and naturally, governments, researchers, policymakers, and scientists from all corners of the world are concerned with energy- cut objectives. According to a survey report published by the International Energy Agency (IEA) (Rode et al., 2011), by the end of 2050, the global energy consumption of HVAC components is going to be tripled which is almost equal to the today's total energy demand of the USA, EU, and Japan. Energy efficiency is thus becoming one vital criterion of smartness. Lightweight protocols, scheduling optimization, predictive models for energy consumption, a cloud service-based approach, low-power transceivers, and a cognitive management framework are some energy efficient solutions for IoT enabled smart buildings (Ejaz et al., 2015; Vlacheas et al., 2013; Qayyum et al., 2015). In addition, companies such as IBM, Cisco, Honeywell, Intel, and Schneider Electric are involved in various energy-efficient solutions for smart cities which are also applicable to the smart building framework.

The smart grid is a concept where all buildings, whether residential, tertiary, industrial, or public, can contribute to energy transition. Every building system generating power from renewable energy sources can distribute or share its excess power to another through this grid by connecting and communicating technologies of IoT, ICT, and cloud computing.

Other Associated Factor: Human Preference as a Sensor

Traditional work on smart building systems mainly focuses on issues of automatic control and energy efficiency, but omits the human-machine interaction problem. In a room, it often happens that one person is feeling cold at a certain temperature of AC where another person is sitting sweating. It is very difficult to regulate temperature by HVAC system at office premises because several factors are involved like a number of persons in that place, number and position of windows and the amount of sunlight reaching those windows at office time. It creates dissatisfaction of employees and thus hampers productivity. This problem can be solved if every occupant can be considered as one sensor of perceived comfort and thus every occupant will be included in the HVAC system. For example, occupants can make an individual request for warm or cold air based on their comfort needs via a smartphone or desktop/laptop computers, using the Intel-enabled Building Robotics' Comfy (Intel IoT Smart Building, 2015) which is connected to the cloud through an Intel-based gateway. Comfy enables integrating both human interaction and machine intelligence, in the cloud, to set the room temperature suitably. Comfy dynamically adjusts temperatures set points and ranges depending on occupants' comfort preferences rather than maintaining zone temperature at preconfigured levels. Comfy also reduces air conditioning when the room or building is unoccupied, resulting in improved energy efficiency. Comfy can be suitably customized to be used in a single area or floor within a building or may be expanded to an entire building or campus.

In (Lu et al., 2013) and (Chen et al., 2014) also, the authors address human activity recognition in the home environment. Conflict detection-based HVAC system is also proposed (Sun et al., 2015) where the conflict between the two services are resolved by proposing a model of User, Triggers, Environment entities, and Actuators that are involved and by decomposing a complex rule into multiple basic rules to design a rule conflict detection algorithm.

Benefits of Smart Building

Designing an intelligent infrastructure with the collaboration of hardware and software for a new or existing building ensures the following benefits:

- Lowering costs.
- The fully integrated and automated system.

- Increasing satisfaction, comfort and productivity.
- Energy efficiency.
- The automated fault detection system.
- Easy management.

Major Challenges

The major challenges in implementing and operating smart buildings are as follows:

- **Interconnectivity and Interoperability of Technologies, Including Hardware and Software:** Due to the widespread technological advancement of sensor-based devices, it is very much expected that IoT will penetrate in the everyday entities including furniture, food packages, dresses and many more. It is expected buildings to have contemporary IoT equipped "brain" so that they can behave smartly to ensure comfort, security, energy efficiency, and sustainability. Micro-location-based services which give a specific and accurate location of an entity, are widely used in smart building, but such different technologies like GPS, WLAN, Zigbee, radio frequency, ultrasound, infrared, various filtering techniques use different technologies. Integration of heterogeneous technology of heterogeneous devices and heterogeneous application software to make a system is thus a big challenge.
- **Efficient Connection and Coordination:** A continuous research process is required to come up with the solutions for efficient connection of the numerous devices of smart building to the Internet and other devices, requiring minimal bandwidth. These devices should be coordinated intelligently with big data analytics and edge/fog/cloud computing.
- **Security and Robustness of the System:** Smart building depends on the cloud-based approach and thus tightening of security, and the privacy mechanism is an utmost important and obvious point of the challenge.
- **Energy Efficiency:** To maintain sustainability with nature, it is important to control greenhouse gas emission. But more use of autonomous devices consumes vast energy; thus, it is also a big challenge to optimize energy consumption and which in turn should reduce the carbon emission.
- **Continuous Training:** Ultimately, it is very challenging to make every occupant of the smart building technologically aware and technically educated. A continuous training process should be carried out to make users 100% equipped with every recent technology used so that they can take part efficiently in the system.

GREEN SMART BUILDING

The Need to Merge Green Building and Smart Building

Smart building is an integrated system of all appliances and regulatory bodies which can communicate with one another with the help of IoT technologies to achieve maximum comfort and satisfaction by providing different utilities at anytime, anywhere, and in anyway. IoT, mobile sensing, and crowd sensing with the help of cloud computing, fog computing (Pramanik et al., 2018a) and crowd computing (Pramanik et al., 2018b) make the collaborative strategic decision-making process simpler thus increases operational efficiency. The data collection, integration of different electrical, electronic gadgets of different heterogeneous systems like HVAC, audio-video system, surveillance system, etc., and from that data analytics and respective decision making with full coordination of all heterogeneous systems present to make the management fully automated, pervasive and smart.

Due to the enormous growth of ICT and IoT, massive production of such supporting devices is needed, and this manufacturing process requires a large number of resources and energy especially electricity which ultimately increases carbon footprint on the earth. A considerable amount of electricity is also needed to operate this ICT and IoT related devices. Electrical energy consumption jumps from the double or triple level in the ICT sector to support dew, fog, and cloud computing. Another dangerous side effect is an electronic waste (e-waste) produced when this related hardware and software are trashed, and these pollutants are hazardous for the entire globe. From manufacturing to disposition, the world of connected devices is increasing environmental threat day-by-day.

To maintain the sustainability in the environment, to reduce carbon footprint, to cope with global warming, we have to pay enough attention to reducing the power consumption, the emission of greenhouse gases, and the e-waste. The smart building gives us maximum comfort, smooth coordination among heterogeneous systems present and hence promotes easy control, coordination, and management. From this point, we have to go forward to achieve energy-efficient and an anti-pollution model of the building that is known as a green building which maintains green IoT, green ICT, and green communication. So, GSB is the product of merging green with smart.

A smart building can be made 'green' mainly by using eco-friendly building materials, by deploying alternative source of energies, using energy in an optimized way, by reducing the emission of greenhouse gases and by putting proper steps to disperse wastes.

Passive Designing: A Key for Merging Smartness With Green

It plays an important role in making the building sustainable and eco-friendly. A good passive design is basically a smart design keeping the following points in mind and using low tech conservation-oriented approach:

- **The Orientation of the Building:** It is the key point to take advantage of solar power and the prevailing wind.
- **Building Shape:** It is also an important role to regulate heat.
- **Indoor Lighting:** Maximal use of sunlight by proper designing of building with the right place of windows, glass pane, etc. reduces energy consumption and thus cost-effective.
- **Indoor Air Control:** It is a very important aspect of maintaining a pollution free environment within a building. Pollutant gas emission and ventilation process are taken care of.
- **Insulation:** Suitable insulation techniques with pollution-free low-energy consuming insulating materials can be deployed. Poorly insulated walls, ceilings, windows, and doors generally have air and temperature leakage which causes more energy consumption for heating/cooling the room.

- **Use of Eco-Friendly, Biodegradable Materials:** Rammed earth, mycelium, wood, bamboo, recycled plastic, Timber Crete (an interesting building material made of sawdust and concrete mixed together), Ash Crete (made of fly ash instead of traditional cement) can be used as building materials. The Grass Crete is a flooring and walling method which allow grass or other flora to grow. The practice of reusing the recycled materials in new construction can help to promote environmental sustainability.
- **Use of a Renewable Source of Energy:** Due to overuse and over-exploitation non- renewable energy sources are on the verge of depletion. To encounter this problem, development and deployment of different renewable sources of energy are the primary criteria to go Green. Renewable energy sources which are available naturally and also are inexhaustible, include wind, solar, biomass, geothermal, and hydro are briefly discussed below (Yuan et al., 2013; Li et al., 2013):
    - Solar panels, also known as photovoltaic modules, consisting of a series of solar cells, which convert light from the sun into DC electricity. With pollution-free energy, having a low maintenance cost and unobtrusive, installing solar panels, solar hot water, solar pumps, solar inverters add smartness to the building.
    - A wind turbine is another example of affordable, sustainable, reliable, cost-effective, and pollution-free energy. It is affordable, clean, and sustainable.
    - Micro hydro systems can also be installed within a building territory if there is a continuous source of water flow available with high speed.
    - Geothermal heat pumps and heating-cooling systems save money, decrease demand on conventional power sources and reduce greenhouse gas emissions.
    - Dagdougui et al. (Dagdougui, Minciardi, Ouammi, Robba, & Sacile, 2012)developed a dynamic hybrid model, including biomass, wind, and solar PV.
- **Chemical and Nonchemical Health Impacts of Materials Used:** Toxic effects of the building materials and the related health hazards of the system also should be considered.
- **Disposal of Building's Waste:** The negative impacts of the construction, renovation, refurbishment and retrofitting activities of the building include the noise, dust, traffic congestion, water pollution, and waste disposal. This will involve the consumption of natural resources and energy and GHG emission as well. Proper waste control methods in the construction and demolition phase of a building can check carbon emission significantly.

Green IoT Technology: Another Key of Merging Smartness With Green

The green IoT is now an emerging field of study. Extensive research is going on in this direction. For instance, the authors proposed a novel self-adapting intelligent system (Byun & Park, 2011) that consists of the self-adapting intelligent gateway (SIG) and self-adapting intelligent sensor (SIS), which make consumer devices more energy efficient and intelligent with the help of energy-efficiency self-clustering sensor network (ESSN) and node type indicator-based routing (NTIR) to enhance service response time and network lifetime. Green ICT is the practice to use computing resources efficiently and effectively with a nominal impact on the environment. Green ICT applies technologies with minimum resource consumption like energy, water, etc. and harmful emissions like CO_2, toxic materials, e-waste, etc. to maintain sustainability. Similar to green IoT and green ICT, green cloud (Qiu et al., 2018) also gives emphasis on efficient power consumption and reduction of harmful wastes by the adoption of different power saving hardware and software solutions. Figure 5 symbolically represents the combination of green IoT, green ICT, and green cloud in making the GSB. Figure 6 shows the different factors of green IoT, green ICT, and green cloud that combiningly contribute to building a GSB.

The Vision of Green Smart Building

GSB guarantees full automation and thus enhances pervasive, demand-based control and efficiency and ultimately enhances productivity and satisfaction of occupants of a building.

ZEB, nZEB, and ZCB

GSB merges to the zero-energy building (ZEB) or nearly zero-energy building (nZEB) or zero-carbon building (ZCB). A building that consumes zero or nearly zero-energy on an annual net basis by reducing primary energy consumption in the building through enhanced energy efficiency performance and producing on-site renewable energy and contributing fewer greenhouse gases in the environment is termed as ZEB or nZEB or ZCB. Figure 7 displays the necessary components of a ZEB.

Figure 5. Ingredients of green smart building

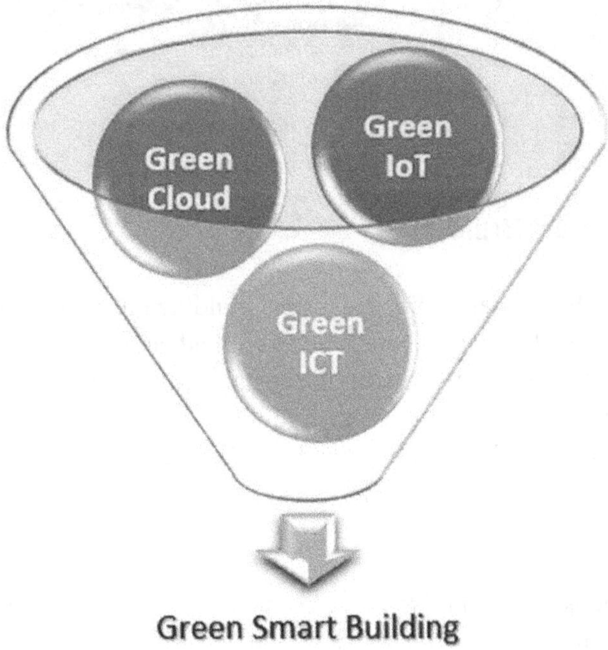

Figure 6. The combined contribution of green IoT, green ICT, and green cloud in green smart buildings

Benefits of Green Smart Building

"Go green" should not be the last keyword for making a building green, that's why we have invoked the concept of smartness in green building. There are plenty of advantages to a GSB. Figure 8 categorizes the advantages of GSB which goes in line with the sustainability dimensions of GSB as discussed in Section 2.5. Table 2 summarizes the positive impacts of GSB on the environment, society, and economy while enlisting the leading research articles that address these areas.

Environmental Benefits

Environmental benefits consist of natural resource conservation, reduction of waste materials generated from the building, enhancing the air and water quality by smart technology and by protecting the eco-system. Emission of greenhouse gas is one of the key pollutants emitted during the building construc-

Figure 7. Necessary components of a zero-energy building

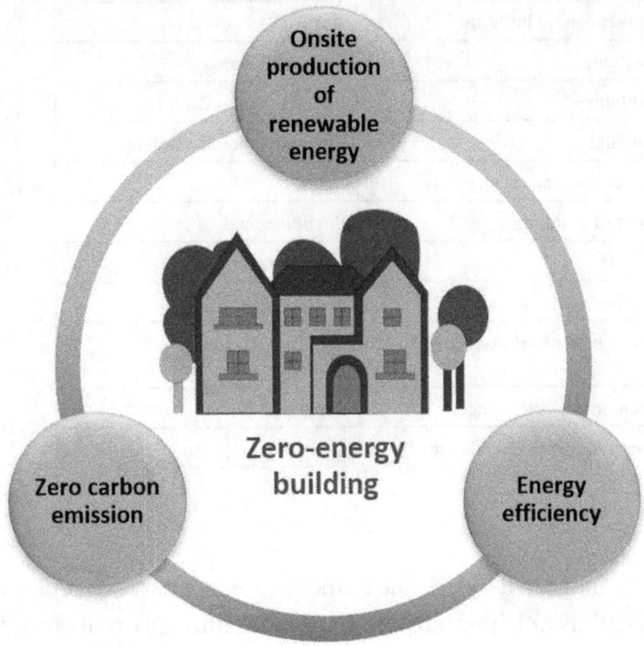

Figure 8. Advantages of green smart buildings

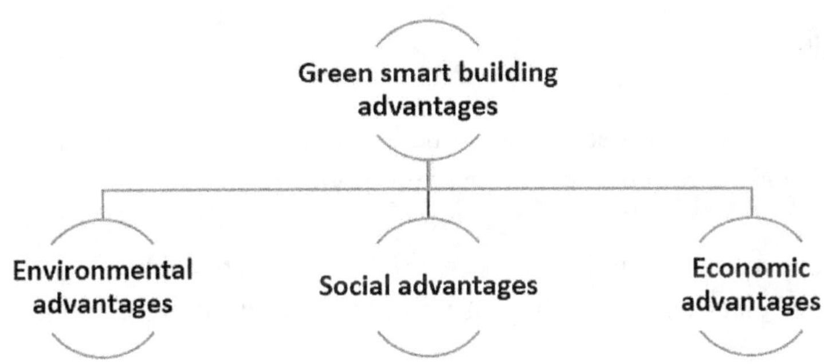

tion and operation. The major source of this greenhouse gas is the combustion of fossil fuel and coal. This combustion is required for producing electricity. So smart usage of electricity in green building reduces the usage of electricity by 30% compared to the non-green building. This green smart concept also decreased the impact of global warming (Turcotte et al., 2006).

Social Benefits

Smarter usage of green components in a building increased the productivity of workers and not only that, but also green building materials improved the health quality of the people. Increasing worker productivity and improved health are one of the major social effects of GSB. According to the Herman-Miller study,

Table 2. Impact of green smart building on socioeconomic and environmental issues

Positive effects of green smart building	Impact on	Sources
Reduce the lifecycle costs of buildings	Social, Economic	(Abidin & Powmya, 2014), (Ahmad, Thaheem, & Anwar, 2015), (Ahn, Pearce, Wang, & Wang, 2013), (Aktas & Ozorhon, 2015), (Annunziata, Testa, Iraldo, & Frey, 2016), (Qin, Mo, & Jing, 2016), (Zhang, 2015), (Love, Niedzweicki, Bullen, & Edwards, 2012), (Gou, Prasad, & Lau, 2014)
Greater energy-efficiency of buildings	Environmental, Economic	
Greater water-efficiency of buildings	Environmental, Economic	
Enhance occupants' health and comfort and satisfaction	Social	
Reduce the environmental impact of buildings	Environmental	
Better indoor environmental quality	Environmental, Social	
Reduce construction and demolition wastes	Environmental	
Increase overall productivity and a better workplace environment	Environmental, Social	
Better company image/reputation or marketing strategy	Economic, Social	
Preservation of natural resources and non-renewable fuels/ energy sources	Environmental, Economic	

workers' productivity was increased by 7% due to the go green smart concept. Many companies include indoor gardens, installed adjustable thermostats along with mirrors on the roof for proper utilization of sunlight. As a result, they observed that the rate of sick leaves had been decreased by 5% and 58% of the employees claimed to be the most productive employees. Proper monitoring of air quality also reduced the asthma rate among children from 8 to 18.

Economic Benefits

The concept of GSB is set to be accepted more due to its large range of economic benefits. Following are some important benefits from the economic point of view:

- Average maintenance and manufacturing cost reduced to noticeably.
- Efficient usage of electricity, recycling of waste water and different energy sources reduced the utility cost.
- Installing sensor-based IoT devices can monitor the air quality which reduced the medical cost.
- Improved health due to the comfort level inside the GSB causes extra productivity among employees, which led to more benefits for companies.

GREEN SMART BUILDING ARCHITECTURE

The architecture of any system describes the interconnection and functionalities of different components of the system. The architecture defines design, construction and operation, and maintenance of a building with respect to five issues, namely energy conservation, water conservation, waste management, material & system safety, and environmental safety. While designing a GSB architecture, the following perspectives should be considered:

- Using renewable resources like solar panel along with smart monitoring and IoT enabled controlling devices ensures energy conservation.
- Proper design and recycling methodology of the building also helps the occupants of the building to use the water efficiently which prevents the wastage of water.
- Waste management is another important aspect of green building which protects the building to get polluted by dangerous elements produced in daily waste.
- Using long-lasting and renewable building material compared to other conventional building material helps the building to make green.
- Construction should be made only taking concern about these five issues during design, which makes GSB different compared to other conventional building.
- Using smart technologies, the five GSB factors are controlled and coordinated for operation & management.

Figure 9 represents a model of GSB architecture. The architecture is divided into two sub-layers, i.e., operational system and construction. Each sub-system comprises certain components that carry respective functionality in resource management, distribution, and conservation.

Figure 9. A generalized model for green smart building

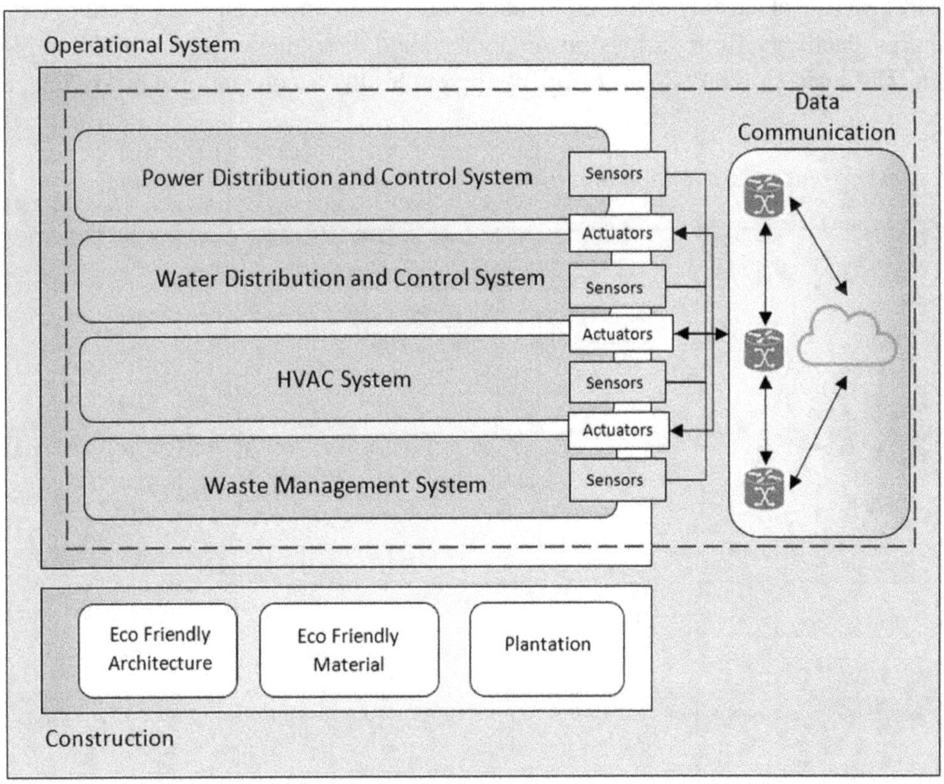

The Operational System of a Green Building Architecture

Some of the major components of the operational system of a GSB are discussed in the following. Besides these components, the operational system also comprises of Internet and cloud services. This subsystem comprises a combination of sensors and actuators that deals with the data. Sensors fetch the continuous data from each sub-system and for any change; the actuators take the corresponding actions. Data coming from all the different components of the operational system for aggregation and decision making is synced with the cloud system.

Power Distribution and Control System

This is the most important component in a GSB that is responsible for minimizing the carbon emission. It takes care of the electrical power distribution and consumption in the building and other subsystems. It also regulates and controls power related tasks. The detailed description of electric power distribution and control system component is depicted in Figure 10.

GSBs are connected to a complex external environment because weather influences a lot in the activity of such buildings. The energy consumption is a very vital point in the "green" aspect and the smartness of a building should ensure the optimization of power consumption and thus the reduction of greenhouse gas emission. GSBs can be used as a node of renewable power generation center and can share their excess of energy produced through micro-grids and smart grids.

The smart grid (the electricity distribution and control) is an important component of the subsystem which procures electricity from various power sources and distributes it in the building based on the requirement. The smart grid takes power from many sources like a conventional power line, using natu-

Figure 10. Electric power distribution and control system of a green smart building

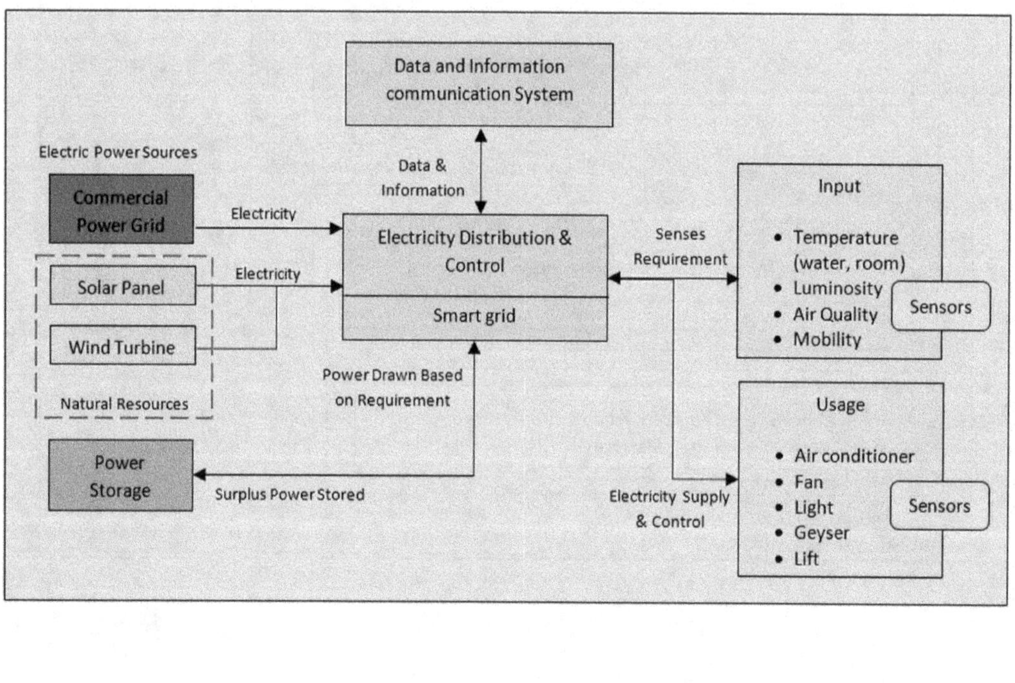

ral resources like wind and solar energy. The smart grid balances between the powers fetched from the different power sources. Electric power is mainly fetched from wind turbines and solar panel, and the deficient power is fetched from the conventional power grid. The power storage component is responsible for storing the excessive which can be used latter based on requirements. The smart grid sends and receives the information and data to the cloud services at regular intervals. The sensors fitted in and around the building senses different input parameters like changes in temperature, quality of air, mobility, luminosity, etc. Based on the input data decisions are made on which device and how long the various equipment's like for running fans, air conditioners, light, geysers, lift, etc. will be supplied electricity.

Water Distribution and Control System

This component of a GSB helps in water distribution throughout the building. This system controls the flow of water within the building. Figure 11 illustrates the detailed view of the water distribution and control system. This system fetches the water from different sources like underground water, rainwater and drainage water. Use of sensors helps in distinguishing the type of water and the incoming source. The collected water is further treated for purification by water purification section. Water treatment performs certain actions that are predefined in this system and based on these actions, it separates the water into drinking water and non-drinking water and is stored according to their respective test results. This stored water is then supplied for further use accordingly by the water distribution and control system. Drinking water may be used for drinking, bathing, and various kitchen work. Non-drinking water may be used in toilets, gardening purposes, for cleaning clothes or utensils, for cooling purposes, etc.

The system has involved various sensors to avoid the unnecessary use of the water. Water leakage and control system within this system detects water leakage in the pipeline system and elsewhere by using

Figure 11. Water distribution and control system of a green smart building

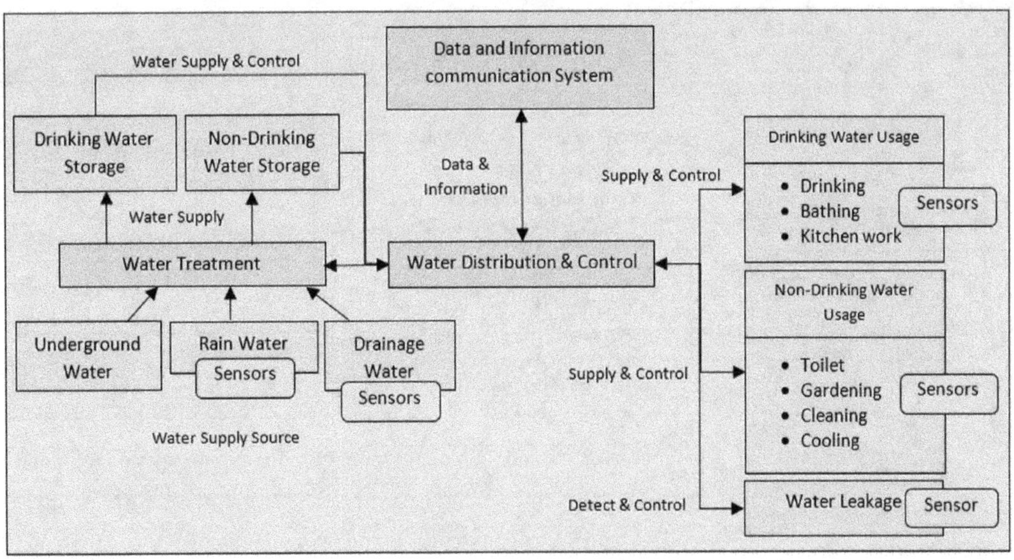

various sensors. The entire sensor data from this subsystem are synchronized with the cloud. Based on the inferred information, the water distribution and control system takes decision for water management in the building.

Heating, Ventilation, and Cooling (HVAC) System

Heating issues, proper ventilation, and facility to provide cooling within the building is managed and controlled by the component called the HVAC system. The HVAC system is responsible for heating, cooling and air flow in the building. The detailed description of an HVAC system component is depicted in Figure 12. An HVAC system has two subcomponents as discussed below.

Heating and Cooling Control System

This subcomponent is responsible for maintaining all heating and cooling related issues in the building. The system uses temperature sensing sensors for continuously assessing the changes in the room and water temperatures. The sensor data are aggregated and synced with the cloud. Based on the information inferred for any changes in temperature, the heating and cooling system take appropriate heating and cooling action at the required places. For the heating purpose, both conventional heating and natural heating mechanism are followed. Sunlight is used here as a natural source of heating for both rooms

Figure 12. Heating, ventilation, and cooling system of a green smart building

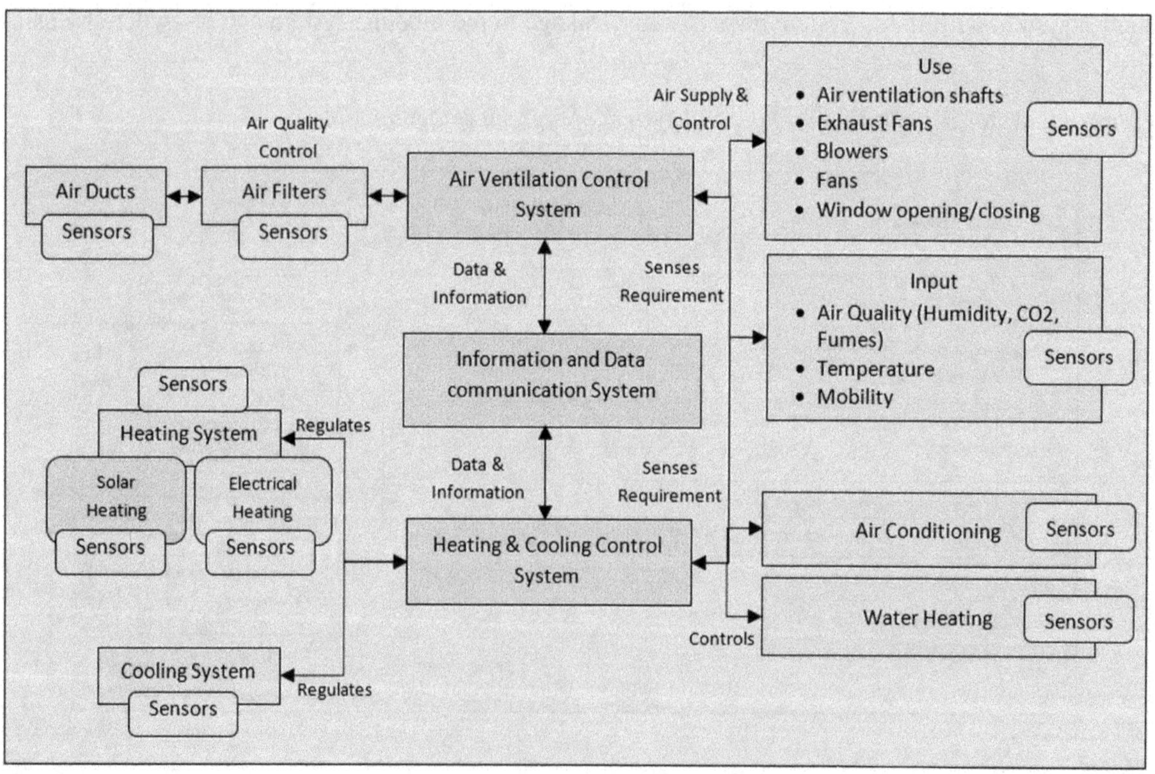

and water. It is both cost and energy efficient way of heating. The heating and cooling systems are continually sensed by various sensors, based on the sensor data, the heating, and cooling control system regulate their operation to avoid overheating or cooling. Thus, the heating and cooling control systems are responsible for the overall heating and cooling issues within the system.

Air Ventilation Control System

This subcomponent maintains the flow of air and proper ventilation in the building. The air ventilation control system controls the air flow within the building, and it comprises of air ducts which maintain the air flow in and out of the building, air filters and sensors. Each of the components is connected through sensors. Using sensors, air quality, temperature, and a person's mobility is assessed. The sensor data is synced with the cloud and based on the knowledge inferred the ventilation controls ventilation shafts, exhaust fans, blowers, fans, etc.

Waste Management System

The waste management system deals with the waste materials that are produced in the building. Figure 13 represents the elaborative view of the waste management system. The figure shows the working layout of the entire waste collection and disposal system. This waste collection component collects and manages the biowaste and non-bio waste of the building. The associated sensors help in understanding the bio and non-biowaste. Based on the type of waste material disposed of, they are collected correspondingly by the waste collection system. The degradable biowaste decomposing system uses the decomposing mechanism for decomposing the biowastes. While the non-biodegradable wastes are collected as building trash for recycling. Waste collection system sends all the information and data to the cloud at regular intervals. This helps to analyze the nature and behavior of the waste collection system.

The Construction System

The construction sub-section specifies the specification for making of a green building. To make a building green, it is important to consider the eco-friendly construction, construction materials and other

Figure 13. Waste management system of a green smart building

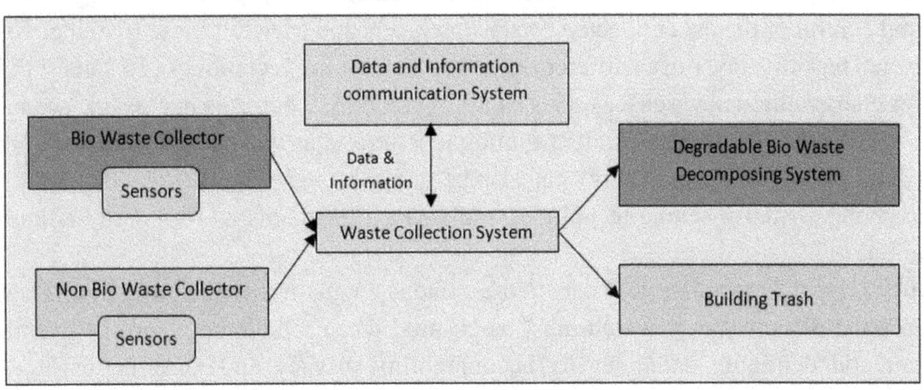

parameters which contribute to reducing carbon footprints and conserve energy. The building architecture plays a pivotal role in designing a green building. The architecture by considering the natural resources like sunlight, rain, wind and geographic position would help in designing a building which efficiently uses the natural resources for lighting, heating, cooling, cleaning, and air ventilation. This reduces the consumption of conventional energy resource considerably. Further, the eco-friendly material used in building construction considerably reduces negative environmental impact as well as the energy consumption in comparison to making conventional building materials. Eco-friendly materials are renewable and are nontoxic and are thus friendly to the environment. The plantation is another important aspect in green building construction subsystem. Use of plantation and gardening reduces the environmental temperature considerably and reduces the air pollution surrounding the building. The different approaches to make a building green is elaborately discussed in Section 2.3.

COMMUNICATION ARCHITECTURE OF GREEN SMART BUILDING

The communication system is the backbone of a GSB design. This system allows connecting all components of a GSB especially the operational system by interlinking each digital system allowing sharing of the data and information among each component for integrated and smooth operations within the building. The communication system consists of IoT (sensors and actuators), LAN, Wi-Fi routers, WAN routers, and cloud. The data collected by different sensors is accumulated by the corresponding router and communicated to the cloud over the internet. In the cloud, the data are analyzed, and feedback is sent back to each component of the required action. Figure 14 shows data communication and service architecture of a GSB. The figure shows the different components and the services they provide. The information and data communication system truly incorporates smartness into the operations of a green building, allowing each component to work in an optimum way, thus reducing energy wastage, preserving more energy and thereby increasing energy efficiency. Figure 15 mentions some of the crucial energy efficiency measures typically adopted in GSBs.

Figure 16 portrays a layered architecture of the core data communication in GSB. The layers are discussed below:

- **Physical Layer:** It consists of the different sensors, local embedded computing, and processing nodes, which are the main building blocks of IoT, present physically in the system.
- **Data Link Layer:** It consists of connected nodes, which are responsible for aggregation, acquisition, and control of the data collected by sensing layers and transmit those to higher core networking layer. Thus, this layer uses different edge/fog computing techniques. To satisfy the green-IoT goal for energy efficiency, sensor nodes should work only when it is necessary, use radio optimization techniques and energy-efficient routing techniques, as well as use miniature high-capacity energy storage technologies. Energy can also be saved through edge/fog/cloud computing, namely power-saving virtual machine techniques and various mechanisms for energy-efficient resource allocation.
- **Network Layer:** Transmission of data to the cloud is the main responsibility of this layer. Another critical issue is the security which must be ensured into each element and the overall system to maintain confidentiality, data integrity, accountability, privacy, and consumer trust.

Figure 14. Data communication and service architecture of a green smart building

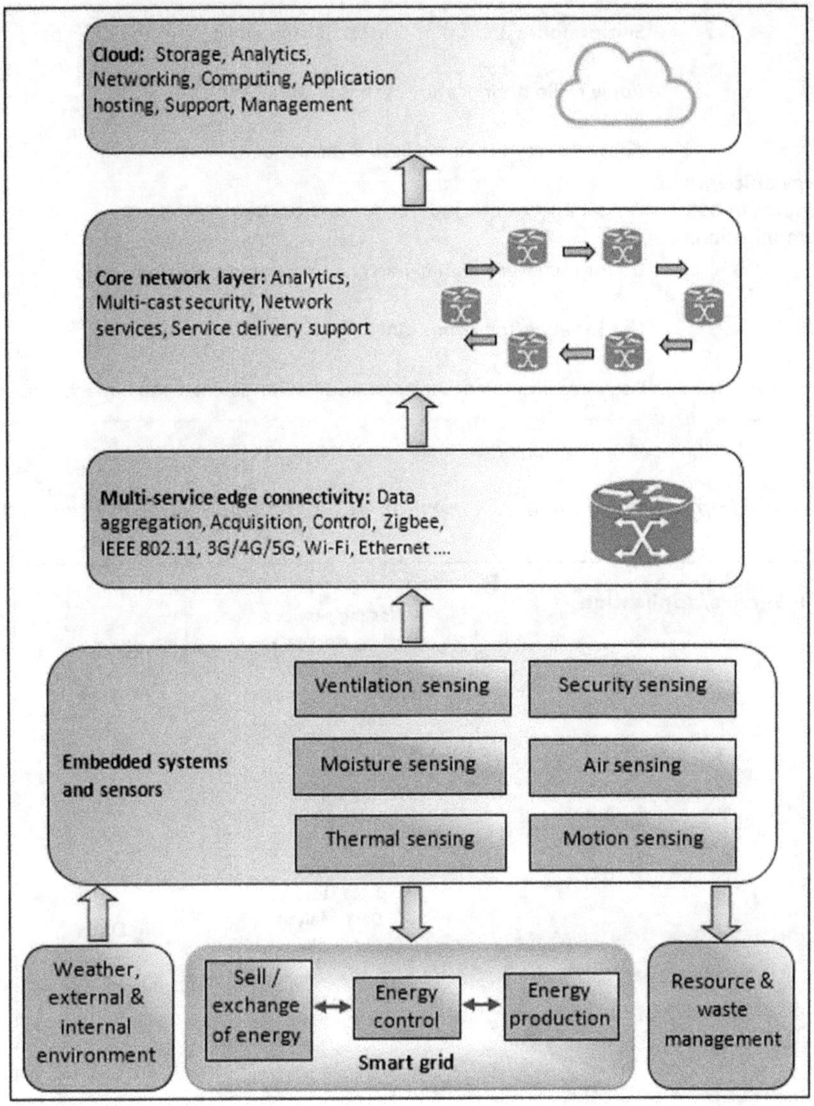

- **Application Layer:** Here, the data storage, management, and decision making are done by different applications with the help of analytical tools and cloud computing techniques, to be accessed ubiquitously and pervasively, thus offering possibilities of software coordination by data management and exchange. Several numbers of IoT/Green-IoT devices produce large quantities of data, and for this, it is characterized by volume, variety, velocity, value, and veracity. Appropriate big data mining techniques can play a big role in the extraction of useful information from these data. Additionally, other supporting technologies such as artificial intelligence, machine learning, computer vision, big data analytics, etc. are helpful in revealing hidden patterns, detecting anomalies, performing predictive modeling, and making actionable decisions.

Figure 15. Energy efficiency measures in green smart building's communications

Energy efficiency measures in GSB communications	
	Sensor nodes and other devices should work only when it is necessary.
	Apply radio optimization techniques.
	Adopt energy-efficient routing techniques.
	Exploit energy-efficient resource allocation mechanisms.
	Go for miniature high-capacity energy storage technologies.
	Opt for edge/fog computing.
	Use virtualization techniques and power-saving virtual machines.

Figure 16. The layered approach of the data communication system design in green smart buildings

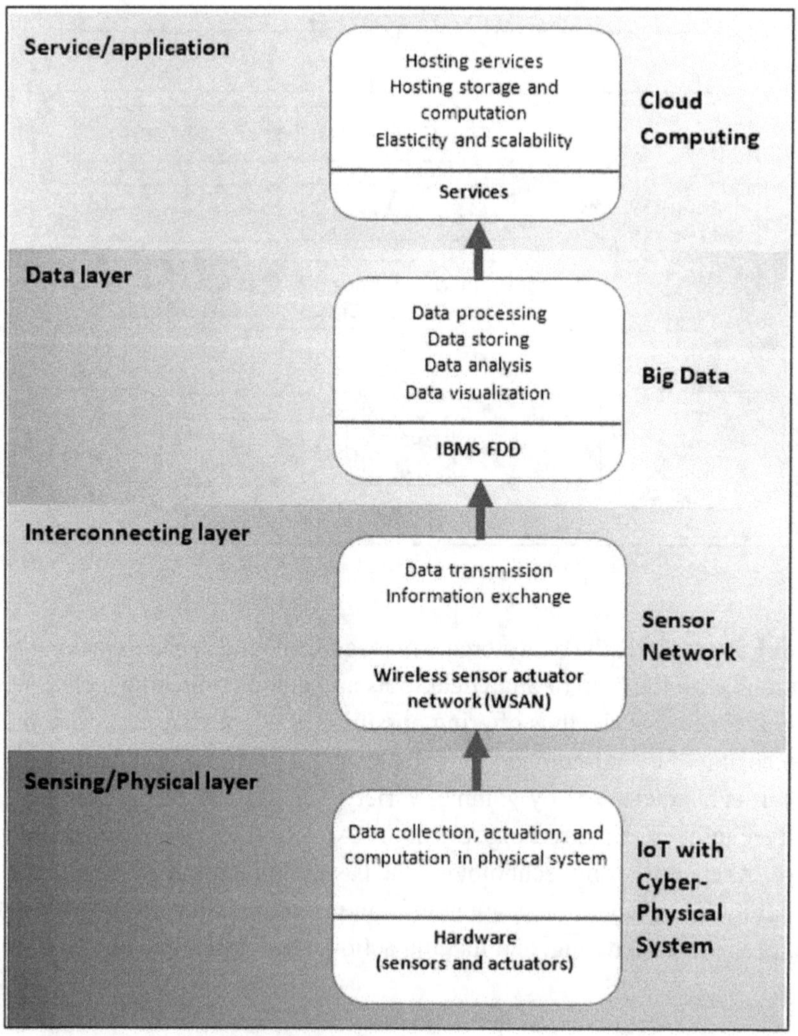

TECHNICAL CHALLENGES TO IMPLEMENT GREEN SMART BUILDING

Although many research efforts have been made on different aspects of GSB, still there are technical challenges that should be taken care of for successful implementation of GSB. Some of them are as follows:

- **Architectural Design Issues:** The architecture of GSB involves interactions and integration of hardware devices and software, in large scale, to manage, save, and analyze the big data received from the physical devices and also intra- and internetworking through heterogeneous systems with a high computing capability of edge, fog and cloud level. The architecture design still imposes many challenges, including the followings:
    - **Scalability:** It is tough to manage an enormous number of IoT devices, especially when they grow continuously in numbers.
    - **Interoperability and Heterogeneity:** Heterogeneous IoT devices to be connected through ICT to communicate, distribute, and collect vital information with other smart networks or applications. Designing a smooth and efficient interoperability structure between various devices working on different data formats and following different communication protocols is a challenging task.
    - **Security and Privacy:** Being an IoT-enabled and interconnected system, GSBs are volatile to security and privacy threats which need to be addressed to gain the confidence of the users.
    - **Energy Efficiency in Network Cloudification:** Spectrum and energy harvesting, efficient system is another challenging area. Energy efficient devices and protocols should be used. It is necessary to a trade-off between efficient dynamic spectrum sensing and efficient spectrum management.
    - **Lack of Standardization and Consistency:** Due to the varied nature of networks and devices, there is a lack of uniformity and standardization of both IoT systems and applications. International standards must be established for hardware interface, communication protocols, middleware, semantic rules, etc.
    - **Dynamicity of the System:** Ubiquitous computing (Abowd & Mynatt, 2000) or pervasive computing (Pramanik et al., 2018) is a very much challenging area to implement an IoT based system. Context-aware dynamic IoT is very much required to maintain a demand-based intelligent system.
- **Lack of Generosity:** The building management applications are designed and developed for various sustainability analyses such as energy performance, CO_2 emissions, solar radiation and lighting analysis, water conservation, indoor air improvement, ventilation, waste management, resource generation, etc. However, most of these applications are designed specifically for only a particular type of analysis and cannot address others.
- **Scalability in the Application:** It is very challenging to apply the and distributed computing techniques and algorithms with green-IoT system level architecture, extending the capacity of computing, networking, and storage from the cloud to the edge of the green-IoT network.
- **Smart Grid Issue:** The smart grid still faces a lack of sufficient infrastructure. The power requirement of GSB as well as the power generated from GSB (through various initiatives) should be well integrated to the smart power grids.

- **Maintenance of Renewable Energy System:** The lack of sufficient guidelines and infrastructure poses a great challenge in the maintenance and operation of different components of a GSB.
- **Continual Adoption and Implementation of Hardware with Application Software:** It is required to support dynamically changing environments of communication technologies, networks, and services.

OTHER ISSUES

- **Government Initiative:** The inadequacy of effective economic incentives that are needed to stimulate GSB innovation is a barrier. There is no adequate incentive policy for designers and developers for supporting them to develop knowledge and methods for GSB construction. The policymakers of a nation, state, and municipal areas should use their influence in encouraging building and buying GSB.
- **Economic Constraints:** The cost of establishing, maintenance, and operation of a GSB may mount significant challenges to implementing the technological innovations towards GSB. The absence of standardized policies is an obstructed in regulating the cost and price of the building materials as well as the constructed GSB. The builders should not superimpose the artificial costs by present only cosmetic facilities to the buyers. Also, the lack of accurate information about the cost and benefit of GSB can create confusion in clients' mind resulting in the inability to make decisions about the purchase of these buildings. Publication of good quality information about GSB is extremely important in order to create demand.
- **Need for Coordination:** The lack of collaborative working methods in the heterogeneous system is another barrier for GSB.
- **Need for Standard Methods for Comparison:** The lack of appropriate methods to compare different design approaches of different GSBs is another barrier.
- **Need for Standard Management Skill:** There are multiple managerial aspects of GSBs, i.e., project level, company level, and market level. Specific project management skills are required for managing GSBs at all the phases, i.e., design and planning, construction, and maintenance.
- **Lack of Standardization of Assessment Methods:** Standardized assessment method and tools are needed, which should enable users to assess and compare GSBs in terms of cost, energy-efficiency, comfort and liveability, sustainability and other GSB requirements.
- **Lack of Knowledge:** There is no adequate knowledge for considering potential new alternatives (for example, solutions for renewable energy and distributed systems of energy generation).
- **Social Awareness:** People are always tentative to embrace the changes - the same for adopting GSBs. General awareness programs should be promoted to the public, elucidating the socioeconomic and environmental benefits of GSBs.
- **Training for Awareness:** Handling any new system requires training and learning. Similarly, continuous education and training opportunities for all the stakeholders on different technical and other aspects of GSB are required.
- **Awareness About Risk Factors:** The proper information is needed regarding the risk that is caused by new technologies.

GREEN SMART BUILDINGS EXAMPLES

Adam Joseph Lewis Center for Environmental Studies, Ohio

The building of The Adam Joseph Lewis Center for Environmental Studies, Oberlin College, Ohio is described by The New York Times as 'the most remarkable of a new generation of college buildings' and as one of the 30 'milestone' buildings of the 20th century by the U.S. Department of Energy (David Orr, 2019). The building, shown in Figure 17, has following green smart features (Green Building Brain, 2009):

- The wastewater treatment facility.
- Energy efficiency by minimizing the use of air conditioning during the summer months.
- Use of a water-to-water heat pump instead of a less efficient electric boiler.
- Production of solar electricity and surplus is shared with the town of Oberlin.

Paryavaran Bhawan, New Delhi

The Paryavaran Bhawan, India's first on-site net-zero building (zero net energy consumption) constructed at Aliganj, Jor Bagh Road, New Delhi by the Ministry of Environment and Forests Department. The building, shown in Figure 18, is targeted to achieve LEED India Platinum Rating and GRIHA 5-star rating with the following green smart features (Indira Paryavaran Bhawan, 2011):

Figure 17. The Adam Joseph Lewis Center for Environmental Studies, Oberlin College, Ohio
(Green Building Brain, 2009)

Figure 18. Paryavaran Bhawan, New Delhi
(Indira Paryavaran Bhawan, 2011)

- A passive solar building design.
- Solar power plant.
- Sewage treatment facility.
- A geothermal heat exchange system.

The Geothermal Building of Boston University

Figure 19 shows the Boston University's geothermal building which uses the earth's geothermal energy to warm and cool the 95,000 square foot space without using fossil fuels (Lebovits, 2010). Under the building, a series of six wells, as depicted in Figure 20, are drilled up to 1500 feet deep. The water of the wells is kept warm at a constant temperature of about 55° F by using earth's thermal energy. The water is pumped out of the wells and put into a heat exchanger. Like a refrigerator, the heat pumps use a compressor to discharge the heat.

During the winter, the heat exchanger propagates geotherm to the heat pump. The heat pump then compresses the air to increase its temperature. This keeps the building warm.

In the summer, the heat exchanger absorbs the excess heat from the air. It then blows the afresh cooled air back into the building. The soaked-up heat is released into the earth.

Bee'ah Headquarter, Sharjah

The Middle East's environmental and waste management company Bee'ah commissioned Zaha Hadid Architects[1] to build headquarters in Sharjah. The design of the building, as shown in Figure 21, is inspired by the form of sand dunes. The building is architected to follow zero net energy consumption with the help of Tesla's Powerpack technology[2]that supports the building to be fully renewable energy run by

Figure 19. Boston University's geothermal building
(Lebovits, 2010)

storing the energy into a massive 1,890 kWh capacity battery along with passive solar power and wind power (Jr, 2019). The occupants and visitors will experience the AI-enabled services as the building integrates full-fledged AI technologies for smart building. This building ensures its green smart property by the following means (Lynch, 2017; Zaha Hadid Architects, 2019).

- This building's design is oriented to optimize prevailing winds.
- Recycled materials are used in construction.
- Production of renewable energy by integrated photovoltaic panels.
- Use of batteries to store the energy.
- Zero waste landfill.
- Zero net energy consumption building.
- Passive solar and wind-powered heating and cooling system.
- Native landscaping.

Figure 20. The ground-source heat pump technology of Boston University's geothermal building
(Lebovits, 2010)

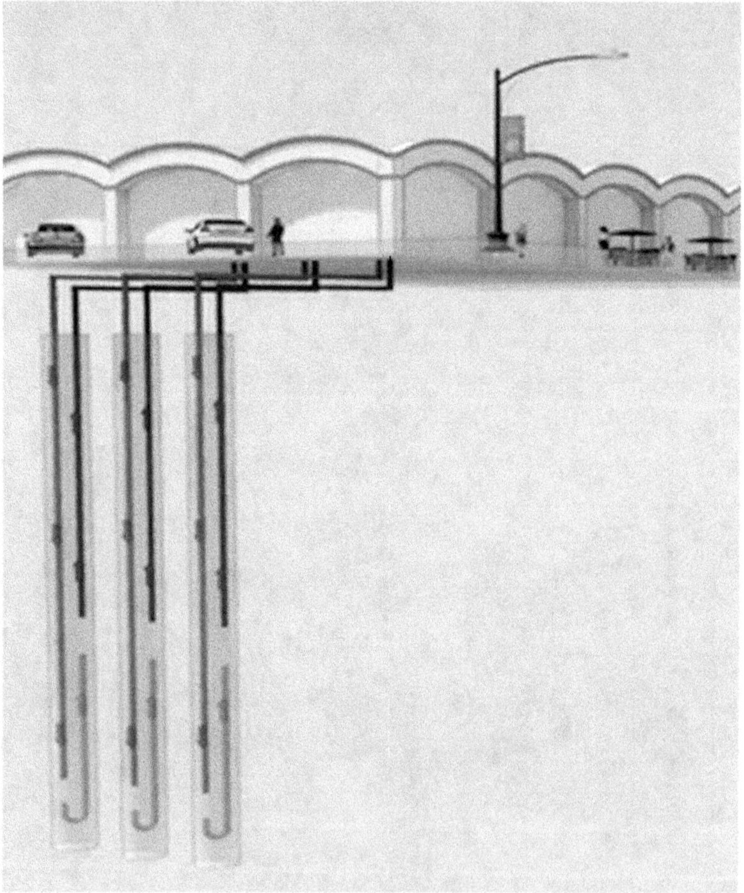

Figure 21. Zaha Hadid-designed Bee'ah Head Quarters in Sharjah
(Zaha Hadid Architects, 2019)

CH2 Building, Australia

Council House 2 (CH2) in Melbourne, Australia (Figure 22) was the first to be awarded the Six Green Star rating. The sustainable features of this building include (Australia's 10 Greenest Buildings, 2010; CH2: Australian's Greenest Building, 2008):

- Gas-fired cogeneration plants that reduce carbon emissions.
- Use of energy-saving lighting technology like T5.
- Water-mining plant in the basement.
- Wavy concrete ceilings which provide thermal mass.
- West façade of louvers made from recycled timber which are powered by photovoltaic cells.
- Shading screens.
- Photovoltaic array 3.5kW (tracks the sun).
- Chilled water-cooling system.
- Co-generation plant.
- Solar hot water collectors.
- Multiuse water treatment plant.
- Roof landscaping.
- Chilled ceiling panels/beams.

Figure 22. CH2 Building, Australia
(CH2: Australian's Greenest Building, 2008)

- Along with the mechanical ventilation the building opts for natural ventilation also, depending on the suitability at different times of the day (especially at the night). The natural ventilation scheme, as shown in Figure 23, uses a combination of top-down inlet chimneys and exhaust stacks, heated by solar gains and topped with wind cowls.
- CH2's north façade comprises ten dark colored air extraction ducts that absorb heat from the sun which helps the stale air inside the building to rise up and vent out of the building.
- The south façade comprises light-colored ducts that draw in fresh air from the roof and distribute it down through the building.
- Occupants can control the fresh air flow to their workspaces by floor vents.

Shanghai Tower, China

The Shanghai Tower, China's tallest and most sustainable super high-rise building (Figure 24), is considered highly energy-efficient. It is the first green building in China that has been certified with both LEED Gold and Three-Star Green Building Design Label (Inhabitat, 2011). It has a total of 43 different sustainable technological features which include:

Figure 23. Natural ventilation scheme in CH2 Building
(CH2: Australian's Greenest Building, 2008)

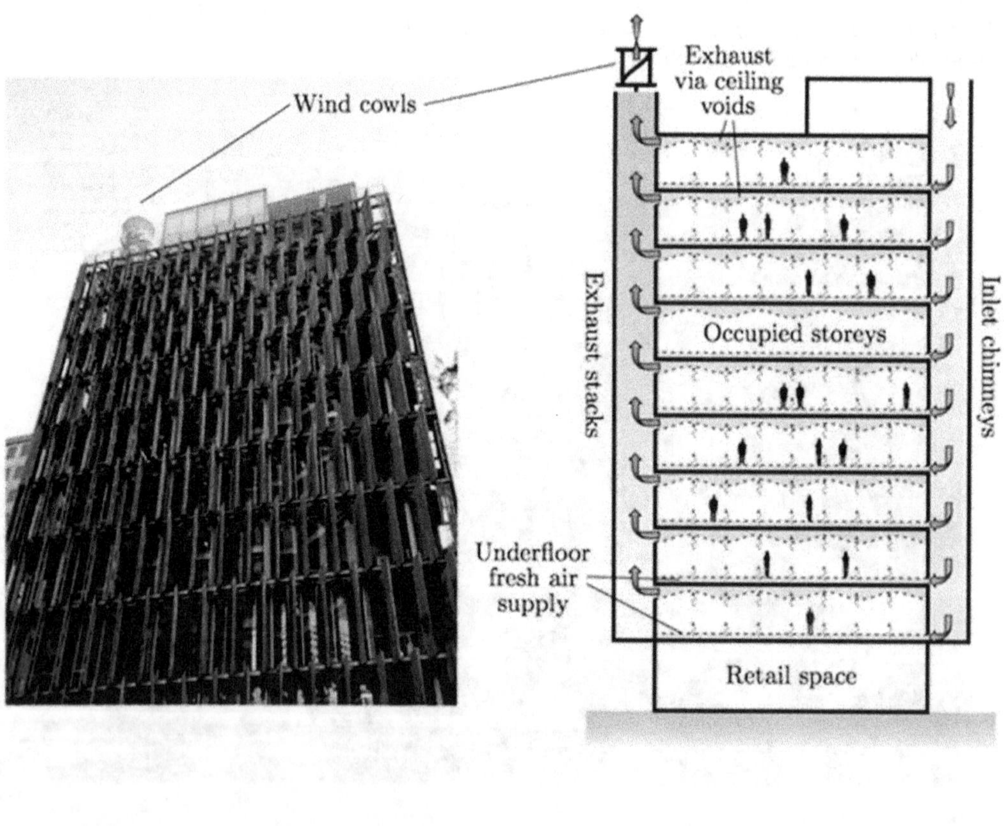

Figure 24. Shanghai Tower, China
(Inhabitat, 2011)

- The building's tapered, the spiral architectural design minimizes wind loads by as much as 24 percent to make the building resistant of the typhoon-force winds common in Shanghai.
- Renewable power generation by solar and wind.
- Extensive landscaping to help cool the building.

Infinity Benchmark, Kolkata

The Infinity Benchmark (Figure 25), a 20 storied building in Kolkata, India is the seventh construction in the world, second outside the USA and the first in eastern India to become a LEED Platinum certified. U.S. Green Building Council (USGBC) certified this building under LEED Core & Shell ver. 2.0. The building has the following green smart features (Infinity, 2016):

- Rainwater harvesting.
- CO_2 monitoring sensors.
- Humidification controls and wastewater recycle system.
- The roof is further covered in polyurethane for improved insulation.

Figure 25. Infinity Benchmark Building, Kolkata
(Infinity, 2016)

Skyshade

Skyshade[3], the first daylighting company in India, is launching different new products & services for daylight transportation and energy efficiency in buildings. This tubular daylighting system can capture ambient sunlight when mounted on a rooftop or side wall of the buildings. The collected light can be transported through pipes to a long distance up to 12 meters. It has the ability to light up basements also. Figure 26 shows the working principle of daylighting through a light pipe as described below (Skyshade Daylights, 2018):

Figure 26. Day lighting through light pipe

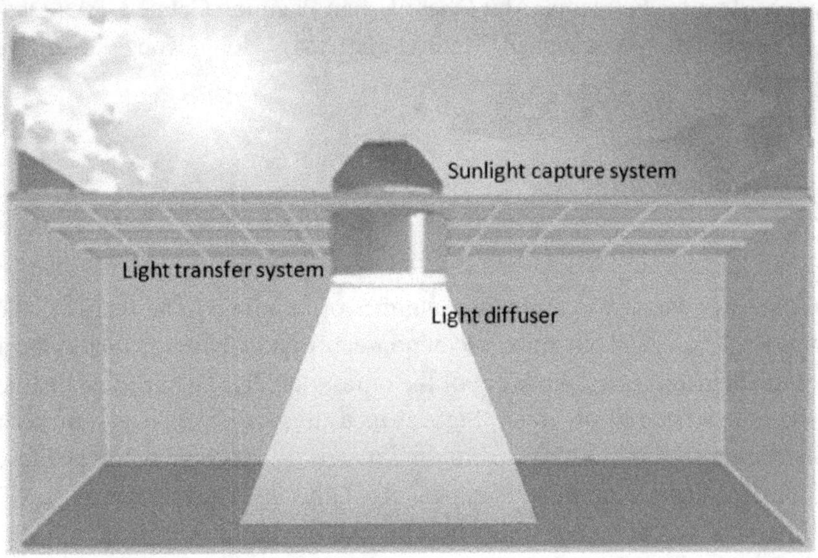

- **Sunlight Capture System or Light Collector:** The job of this component is to accumulate sunlight from all directions and distribute the light effectively throughout the day even. It is UV stable and works at low sun angles as well.
- **Reflective System:** The specially designed reflective system provides high luminance with minimal light loss and without color shift, even on cloudy days.
- **Light Diffuser:** The reflector system with silver metal oxide coating inside the pipe reflects the light and uniformly distribute the light through the whole area. It can control the light according to the glare and distribution of sunlight.

BUILDING ASSESSMENT TOOLS AND STANDARDS

There are different green building assessment tools and systems available in different countries, and they function depending on the environmental and climatic conditions and requirements of the locals. Thus, different countries put different weight on a specific criterion to evaluate a building for its complying with green and smart standards. Some of the important standards are:

- Leadership in Energy and Environmental Design (LEED), United States[4]
- BRE Environmental Assessment Method (BREEAM), United Kingdom[5]
- Green Building Council of Australia Green Star (GBCA), Australia[6]
- BCA Green Mark Scheme, Singapore[7]
- DGNB – German Sustainable Building Council, Germany[8]

- Comprehensive Assessment System for Built Environment Efficiency (CASBEE), Japan[9]
- Pearl Rating System for Estidama, Abu Dhabi Urban Planning Council, Abu Dhabi[10]
- Building Environmental Assessment Method (BEAM), Hong Kong Green Building Council Limited, Hong Kong[11]
- Green Building Index (GBI), Malaysia[12]

CONCLUSION

The growing urbanization has led to a growing number of buildings. The ineffective energy management, non-sustainable design, and planning, use of non-ecofriendly building materials, inefficient waste control, etc. have threatened the sustainability of the cities globally. To have sustainable cities and eco-friendly living, buildings need to be green. Several measures can be taken to make buildings green. The primary solutions are using sustainable building materials, reusing and recycling waste materials, energy-efficient design, using renewable energies, etc. Other measures such as rainwater harvesting, rooftop agriculture, wall gardening, etc. also help in making buildings green. On the other side, the IoT-enabled smart buildings not only have eased the livings of the residents, but also have technically assisted towards green buildings. But making the buildings smartonly does not solve the problem of sustainability. The excessive use of electronic devices and to transmit and process the huge amount of data generated out of these devices eats up a massive amount of energy. This high energy consumption of the smart buildings negates the green goal of the modern buildings. Hence, it is suggested that green technologies should be further stringently applied to every component of the smart buildings. Though these green smart buildings (GSBs) will bolster in achieving sustainability truly the achievement is not challenged-free. For successful design and implementation of GSBs, several challenges and issues, be it technical, administrative, or societal, are needed to be addressed. Expectedly, people have already started addressing these issues, and several such GSBs are coming into existence globally. Hopefully, the GSBs are going to offer us a smart, eco-friendly living in a sustainable green city.

REFERENCES

Abidin, N. Z., & Powmya, A. (2014). Perceptions on Motivating Factors and Future Prospects of Green Construction in Oman. *Journal of Sustainable Development*, *7*(5), 231–239. doi:10.5539/jsd.v7n5p231

Abowd, G., & Mynatt, E. D. (2000). Charting Past, Present and Future Research in Ubiquitous Computing. *ACM Transactions on Computer-Human Interaction*, *7*(1), 29–58. doi:10.1145/344949.344988

U. SAdministration. (2011). *The Benefits and Challenges of Green Roofs on Public and Commercial Buildings*. Tucson, AZ: The University of Arizona.

Ahmad, T., Thaheem, M. J., & Anwar, A. (2015). Developing a green-building design approach by selective use of systems and techniques. *Architectural Engineering and Design Management*, *12*(1), 29–50. doi:10.1080/17452007.2015.1095709

Ahn, Y. H., Pearce, A. R., Wang, Y., & Wang, G. (2013). Drivers and barriers of sustainable design and construction: The perception of green building experience. *International Journal of Sustainable Building Technology and Urban Development*, *4*(1), 35–45. doi:10.1080/2093761X.2012.759887

Aktas, B., & Ozorhon, B. (2015). Green Building Certification Process of Existing Buildings in Developing Countries: Cases from Turkey. *Journal of Management Engineering*, *31*(6), 05015002. doi:10.1061/(ASCE)ME.1943-5479.0000358

Annunziata, E., Testa, F., Iraldo, F., & Frey, M. (2016). Environmental responsibility in building design: An Italian regional study. *Journal of Cleaner Production*, *112*(1), 639–648. doi:10.1016/j.jclepro.2015.07.137

Australia's 10 Greenest Buildings. (2010, Jan. 25). Retrieved from https://sydneyarchitecturearchive.wordpress.com/2010/01/25/australias-10-greenest-buildings/

BossControls. (2019). *How Buildings Impact the Environment.* Retrieved Feb. 14 from https://bosscontrols.com/buildings-impact-environment/

Byun, J., & Park, S. (2011). Development of a self-adapting intelligent system for building energy saving and context-aware smart services. *IEEE Transactions on Consumer Electronics*, *57*(1), 90–98. doi:10.1109/TCE.2011.5735486

CH2: Australian's Greenest Building. (2008, June 28). Retrieved from https://sydneyarchitecturearchive.wordpress.com/2008/06/28/ch2-australias-greenest-building/

Chen, L., Nugent, C., & Okeyo, G. (2014). An Ontology-Based Hybrid Approach to Activity Modeling for Smart Homes. *IEEE Transactions on Human-Machine Systems*, *44*(1), 92–105. doi:10.1109/THMS.2013.2293714

CityMetric. (2015, April 15). *9 building materials made entirely from waste products.* Retrieved from https://www.citymetric.com/skylines/9-building-materials-made-entirely-waste-products-932

Dagdougui, H., Minciardi, R., Ouammi, A., & et al, . (2012). Modeling and optimization of a hybrid system for the energy supply. *Energy Conversion and Management*, *64*, 351–363. doi:10.1016/j.enconman.2012.05.017

David Orr. (2019). *Adam Joseph Lewis Center for Environmental Studies, Oberlin College*. Retrieved from http://www.mcdonoughpartners.com/projects/adam-joseph-lewis-center-for-environmental-studies-oberlin-college/

Ejaz, W., Shah, G. A., Hasan, N., & Kim, H. S. (2015). Energy and throughput efficient cooperative spectrum sensing in cognitive radio sensor networks. *Emerging Telecommunications Technologies*, *26*(7), 1019–1030. doi:10.1002/ett.2803

Go Smart Bricks. (2017, April 12). *Rain Water Harvesting – Why, Benefits, Techniques & More*. Retrieved from http://gosmartbricks.com/rain-water-harvesting-methods-in-india/

GoSmartBricks. (2018, April 13). *Top 6 Challenges Hindering Rapid Adoption of Green Building Practices In India*. Retrieved from http://gosmartbricks.com/top-6-challenges-hindering-rapid-adoption-of-green-building-practices-in-india/

Gou, Z., Prasad, D., & Lau, S. S.-Y. (2014). Impacts of green certifications, ventilation and office types on occupant satisfaction with indoor environmental quality. *Architectural Science Review*, *57*(3), 196–206. doi:10.1080/00038628.2014.908113

Grainger. (2019). *Focus on Three Components for Effective Sustainable Building Design*. Retrieved from: https://www.grainger.com/content/supplylink-sustainable-building-design

Green Building Brain. (2009). *Oberlin College Lewis Center*. Retrieved from https://greenbuildingbrain.org/buildings/oberlin_college_lewis_center

Hernández-Ramos, J. L., Moreno, M. V., Bernabé, J. B., & et al, . (2015). SAFIR: Secure access framework for IoT-enabled services on smart buildings. *Journal of Computer and System Sciences*, *81*(8), 1452–1463. doi:10.1016/j.jcss.2014.12.021

Howe, J. C. (2010). Overview of Green Buildings. *National Wetlands Newsletter*, *33*(1), 3–14.

Hui, S. C. (2001). *HVAC Design and Operation for Green Buildings*. Shaanxi-Hong Kong Refrigeration and HVAC Seminar, Xian, China.

Hui, S. C. (2007, September). *Sustainable building technologies for hot and humid climates*. Hangzhou, China: Joint Hong Kong and Hangzhou Seminar for Sustainable Building.

Indira Paryavaran Bhawan. (2011). Retrieved from http://www.moef.nic.in/ipb/home.htm

Infinity. (2016). *Infinity Benchmark*. Retrieved from http://www.infinityitpark.com/project/infinity-benchmark/

Inhabitat. (2011, Dec. 15). *Interview: Gensler's Chris Chan on the Sustainable Shanghai Tower, Asia's Tallest Skyscraper*. Retrieved from https://inhabitat.com/interview-genslers-chris-chan-on-the-sustainable-shanghai-tower-asias-tallest-skyscraper/shanghai-tower-7/

Intel IoT Smart Building. (2015). *Controlling Office Temperature with Building Robotics*. Intel Corporation. Retrieved from https://www.intel.in/content/www/in/en/internet-of-things/white-papers/iot-smart-office-building-robotics-brief.html

JrB. D. (2019, Jan. 15). *Arabian Business*. Retrieved from https://www.arabianbusiness.com/construction/411344-zaha-hadid-designed-beeah-hq-to-be-ready-by-summer

Lebovits, S. C. (2010). *Geothermal Building*. Retrieved from: http://www.bu.edu/sustainability/what-were-doing/green-buildings/geothermal/

Li, D. H., Yang, L., & Lam, J. C. (2013). Zero energy buildings and sustainable development implications – A review. *Energy*, *54*, 1–10. doi:10.1016/j.energy.2013.01.070

Lilis, G., Conus, G., Asadi, N., & Kayal, M. (2017). Towards the next generation of intelligent building: An assessment study of current automation and future IoT based systems with a proposal for transitional design. *Sustainable Cities and Society*, *28*, 473–481. doi:10.1016/j.scs.2016.08.019

Love, P. E., Niedzweicki, M., Bullen, P. A., & Edwards, D. J. (2012). Achieving the Green Building Council of Australia's World Leadership Rating in an Office Building in Perth. *Journal of Construction Engineering and Management*, *138*(5), 652–660. doi:10.1061/(ASCE)CO.1943-7862.0000461

Lu, C.-H., Ho, Y.-C., Chen, Y.-H., & Fu, L.-C. (2013). Hybrid User-Assisted Incremental Model Adaptation for Activity Recognition in a Dynamic Smart-Home Environment. *IEEE Transactions on Human-Machine Systems*, *43*(5), 421–436. doi:10.1109/TSMC.2013.2281586

Lynch, P. (2017, Dec. 19). *Zaha Hadid Architects' Bee'ah Headquarters Tops Out in UAE*. Retrieved from https://www.archdaily.com/885744/zaha-hadid-architects-beeah-headquarters-tops-out-in-uae

Mathew, S. S., Atif, Y., Sheng, Q. Z., & Maamar, Z. (2013). The Web of Things - Challenges and Enabling Technologies. In *Internet of Things and Inter-cooperative Computational Technologies for Collective Intelligence* (pp. 1–23). Berlin, Germany: Springer. doi:10.1007/978-3-642-34952-2_1

Nationwide Construction. (2016, Feb. 26). *Eco-friendly Construction: 8 Advantages of Green Building*. Retrieved from https://nationwideconstruction.us/eco-friendly-construction-8-advantages-of-green-building/

Padmanabhan, G. (2018). *Methods of Rainwater Harvesting - Components, Transport and Storage*. Retrieved from https://theconstructor.org/water-resources/methods-of-rainwater-harvesting/5420/

Paul, D., Chakraborty, T., Datta, S. K., & Paul, D. (2018). IoT and Machine Learning Based Prediction of Smart Building Indoor Temperature. In *4th International Conference on Computer and Information Sciences (ICCOINS)*. Kuala Lumpur, Malaysia: IEEE. 10.1109/ICCOINS.2018.8510597

Plageras, A., Psannis, K. E., Gupta, B., et al. (2017). Solutions for Inter-connectivity and Security in a Smart Hospital Building. In *IEEE 15th International Conference on Industrial Informatics (INDIN)*. Emden, Germany: IEEE.

Plant Connection. (2019). *Benefits of G-O2® Living Walls*. Retrieved from http://myplantconnection.com/green-wall-benefits.php

Pramanik, P. K. D., & Choudhury, P. (2018). IoT Data Processing: The Different Archetypes and their Security & Privacy Assessments. In S. K. Shandilya, S. A. Chun, S. Shandilya, & E. Weippl (Eds.), Internet of Things (IoT) Security: Fundamentals, Techniques and Applications (pp. 37–54). Gistrup, Denmark: River Publishers.

Pramanik, P. K. D., Pal, S., Brahmachari, A., & Choudhury, P. (2018). Processing IoT Data: From Cloud to Fog. It's Time to be Down-to-Earth. In P. Karthikeyan, & M. Thangavel (Eds.), Applications of Security, Mobile, Analytic and Cloud (SMAC) Technologies for Effective Information Processing and Management (pp. 124-148). Hershey, PA: IGI Global. doi:10.4018/978-1-5225-4044-1.ch007

Pramanik, P. K. D., Pal, S., & Choudhury, P. (2018). Beyond Automation: The Cognitive IoT. Artificial Intelligence Brings Sense to the Internet of Things. In A. K. Sangaiah, A. Thangavelu, & V. M. Sundaram (Eds.), Cognitive Computing for Big Data Systems Over IoT: Frameworks, Tools and Application (pp. 1–37). New York, NY: Springer. doi:10.1007/978-3-319-70688-7_1

Pramanik, P. K. D., Pal, S., & Choudhury, P. (2019). Green and Sustainable High-Performance Computing with Smartphone Crowd Computing: Benefits, Enablers, and Challenges. Scalable Computing. *Practice and Experience*, *20*(2), 259–283.

Pramanik, P. K. D., Pal, S., Pareek, G., Dutta, S., & Choudhury, et al. (2018). Crowd Computing: The Computing Revolution. In R. Lenart-Gansiniec (Ed.), Crowdsourcing and Knowledge Management in Contemporary Business Environments (pp. 166–198). Hershey, PA: IGI Global.

Pramanik, P. K. D., Upadhyaya, B., Pal, S., & Pal, T. (2018). Internet of Things, Smart Sensors, and Pervasive Systems: Enabling the Connected and Pervasive Health Care. In N. Dey, A. Ashour, S. J. Fong, & C. Bhatt (Eds.), *Healthcare Data Analytics and Management* (pp. 1–58). Cambridge, MA: Academic Press.

Qayyum, F. A., Naeem, M., Khwaja, A. S., & et al, . (2015). Appliance Scheduling Optimization in Smart Home Networks. *IEEE Access: Practical Innovations, Open Solutions*, *3*, 2176–2190. doi:10.1109/ACCESS.2015.2496117

Qin, X., Mo, Y., & Jing, L. (2016). Risk perceptions of the life-cycle of green buildings in China. *Journal of Cleaner Production*, *126*, 148–158. doi:10.1016/j.jclepro.2016.03.103

Qiu, C., Shen, H., & Chen, L. (2018). Towards green cloud computing: Demand allocation and pricing policies for cloud service brokerage. *IEEE Transactions on Big Data*, 1-1.

Rode, P., Burdett, R., & Gonçalves, J. C. (2011). Buildings Investing in energy and resource efficiency. In Towards a Green Economy: Pathways to Sustainable Development and Poverty Eradication (pp. 331-373). United Nations Environment Programme.

Shaikh, P. H., Nor, N. B., Nallagownden, P., & et al, . (2014). A review on optimized control systems for building energy and comfort management of smart sustainable buildings. *Renewable & Sustainable Energy Reviews*, *34*, 409–429. doi:10.1016/j.rser.2014.03.027

Shang, W., Ding, Q., Marianantoni, A., & et al. (2014). Securing building management systems using named data networking. *IEEE Network*, *28*(3), 50–56. doi:10.1109/MNET.2014.6843232

Siemens, A. G. (2014). *Our future depends on intelligent infrastructures.* Siemens Industry. Retrieved from https://www.siemens.com/content/dam/internet/siemens-com/us/home/company/topic-areas/intelligent-infrastructure/buildings/documents/bt-cpp-intel-infrstrctr-wp.pdf

Siemens. (2018, Nov. 6). *Unlock your city's and building's potential.* Retrieved from https://new.siemens.com/global/en/company/topic-areas/intelligent-infrastructure/buildings/digital-buildings.html

Skyshade Daylights. (2018). *Solar light - Pipe.* Retrieved from http://www.skyshadedaylights.com/products-solar-light-pipe-energy-solar-tube/

State of Green. (2019). *Energy efficicency in buildings.* Retrieved from https://stateofgreen.com/en/sectors/buildings/energy-efficicency-in-buildings/

State of Green. (2019). *Green buildings as key elements of future sustainable cities.* Retrieved from https://stateofgreen.com/en/creating-smart-green-liveable-cities/green-buildings/

Sun, Y., Wang, X., Luo, H., & Li, X. (2015). Conflict Detection Scheme Based on Formal Rule Model for Smart Building Systems. *IEEE Transactions on Human-Machine Systems, 45*(2), 215–227. doi:10.1109/THMS.2014.2364613

Tuballa, M. L., & Abundo, M. L. (2016). A review of the development of Smart Grid technologies. *Renewable & Sustainable Energy Reviews, 59*, 710–725. doi:10.1016/j.rser.2016.01.011

Turcotte, D., Villareal, J., & Bermingham, C. (2006). *The Benefits of Building Green.* Lowell, MA: UMass Lowell.

Venkata, G., Bhargav, S., Choudhary, S., & Sharma, B. (2015). Use of Renewable Energy Sources in Construction of Green Building. *International Journal for Research in Applied Science and Engineering Technology, 3*(VII), 108–113.

Vlacheas, P., Giaffreda, R., & Stavroulaki, V. (2013). Enabling smart cities through a cognitive management framework for the internet of things. *IEEE Communications Magazine, 51*(6), 102–111. doi:10.1109/MCOM.2013.6525602

Vlacheas, P., Giaffreda, R., Stavroulaki, V., Kelaidonis, D., Foteinos, V., Poulios, G., ... Moessner, K. (2013). Enabling smart cities through a cognitive management framework for the internet of things. *IEEE Communications Magazine, 51*(6), 102–111. doi:10.1109/MCOM.2013.6525602

Williams. (2012). *Passive Solar Design.* Retrieved February 11, 2019, from https://sustainability.williams.edu/green-building-basics/passive-solar-design

Williams. (2012). *Passive Solar Design.* Retrieved from https://sustainability.williams.edu/green-building-basics/passive-solar-design

Yuan, X., Wang, X., & Zuo, J. (2013). Renewable energy in buildings in China - A review. *Renewable & Sustainable Energy Reviews, 24*, 1–8. doi:10.1016/j.rser.2013.03.022

Zaha Hadid Architects. (2019). *Zaha Hadid Architects.* Retrieved from http://www.zaha-hadid.com/architecture/bee'ah-headquarters-sharjah-uae/

Zanella, A., Bui, N., & Castellani, A. (2014). Internet of Things for Smart Cities. *IEEE Internet of Things Journal, 1*(1), 22–32. doi:10.1109/JIOT.2014.2306328

Zhang, X. (2015). Green real estate development in China: State of art and prospect agenda – A review. *Renewable & Sustainable Energy Reviews, 47*, 1–13. doi:10.1016/j.rser.2015.03.012

ENDNOTES

1 http://www.zaha-hadid.com/architecture/bee%E2%80%99ah-headquarters-sharjah-uae/
2 https://www.tesla.com/powerpack
3 http://www.skyshadedaylights.com/
4 https://new.usgbc.org/leed
5 https://www.breeam.com/
6 https://new.gbca.org.au/
7 https://www.bca.gov.sg/greenmark/green_mark_buildings.html
8 https://www.dgnb.de/en/index.php
9 http://www.ibec.or.jp/CASBEE/english/
10 https://www.upc.gov.ae/en/estidama/pearl-rating-system/pearl-building-rating-system
11 https://www.hkgbc.org.hk/eng/BEAMPlus.aspx
12 http://new.greenbuildingindex.org/

Chapter 2
Role of Renewable Energy Techniques to Design and Develop Sustainable Green Building

Pradeep Tomar
Gautam Buddha University, India

ABSTRACT

This chapter presents the role of renewable energy techniques to design and develop a sustainable framework for green building. The first viewpoint is identified with the earlier structure and the low encapsulated energy building materials for the design and development of a framework for green building. The primary perspective is to manage energy protection using renewable energy techniques in the green building. Green building interchangeably can be used with the term's sustainable construction or green construction. So, durable construction means using environmentally responsible and resource-efficient procedures in development to be ensured of sustainability throughout the lifetime of the building. This chapter also presents the combination of renewable, energy-based technology for green building construction and sustainability with the economics of renewable energy.

INTRODUCTION

A green building refers to a structure which is developed, designed, built, operated, re-used or renovated in a resource efficient and ecological way that is also called as a sustainable building. Generally, in environmental terminology, a sustainable system can be described as a living system which continues as a result that the resources are not used up faster than what can be replenished naturally. A sustainable economic policy in terms of financial terminology can be defined as one in which the expenditures are less or at least equal to the income. A sustainable social system in terms of common terminology can be described as one in which empowerment of members is done to collect; as a result, a synergistic whole is created. It is progressively clear that decisions that are made on an individual level have an effect on

DOI: 10.4018/978-1-5225-9754-4.ch002

the global scale and vice-versa. Through energy efficiency, energy preservation of the green building has gained important significance everywhere in the world. The four primary viewpoints for renewable energy production in a green building incorporate as a matter of first importance the about zero energy uninvolved building plan before genuine design and development, also the utilization of building materials that are of low energy amid its growth, and utilization of energy efficient types of gear for the low operational energy necessity and the ultimate mix of sustainable power source advances for different applications. These perspectives have been talked about alongside their financial matters and natural effects for future generation. During the current period, the habitat and dwelling and are linked invariably, making buildings as convenient and comfortable possible all across the world. By investing 30 to 40 percent of total essential resources of the world, the building sector is moving at an exponential pace. In the current period, after industry and agriculture, the most significant consumer of energy from fossil fuels is buildings. The European Commission has initiated the Asia-Link program to spread and promote the knowledge with the close approach of zero energy on a sustainable built environment. In this program of the sustainable built environment, there is the integration of the building for various applications such as cooling or heating, water heating and electricity production with the proven technologies of renewable energy. All over the world, in the building, the use of operational power insignificant. Hence, for sustainable development, all over the world, assessment of sustainability framework is required especially in the building sector. Reduction in depletion of resources that are critical such as water, raw materials, and energy is the main aims and targets of sustainable design; environmental degradation to be prevented which is caused by infrastructure and benefits in the whole life cycle; and develop environments are to be created that are effective, productive and safe utilization of the solar and water energy.

GREEN BUILDING

Two additional imperatives must be satisfied with the construction of a natural green building: the renewable of natural materials, alongside the measurable data and from the landscape to its adaptation of the architecture. According to original data the selection of the site must be made. Architectural forms must not use artificial color rather than take inspiration from nature. The material provided by the environment must be used in natural building techniques. The use of natural material is highly recommended as these materials replace synthetic products as their production consumes energy in huge quantity. The raw materials are covered under the term "natural materials" that can be used either accordingly to modern techniques or traditional craft methods. The construction standard refers to the expression "passive building" that can be accomplished using a different variety of materials used for construction. It can be interpreted as a green building construction which claims the climate of the interior is as comfortable as without the need for a conventional heating system in winter as it is summer. For obtaining low energy consumption, this is one of the conditions. The following requirements must be fulfilled to build a passive building: to capture passive solar energy building's orientation; triple-glazed windows of high quality. Such as, by taking into consideration the direction of a structure and change its windows, to catch the passive solar heat, architects and promoters must keep in mind that energy consumption should be less than create their designs, enhance the aspect of sustainable development and daylight penetration improvement, which increases the productivity of employee without incurring additional cost of construction.

ROLE OF RENEWABLE ENERGY RESOURCES AND TECHNIQUES FOR GREEN BUILDING

Renewable energy technologies have a lesser impact on the environment than conventional energy technologies as they are a clean source of energy. Greenhouse gases are not produced, and thus it is a boon to the environment. Renewable energy is everlasting. Non-renewable sources of energy are limited and will be exhausted someday. An increase inthe economy will open an avenue for new jobs. A necessary component of green building is energy efficiency, and it is a significant concern. It is one of the significant factors in its success. A green building must be suited with different types of solutions that offer a contribution to supplying quality energy, reduction in consumption and better management of electrical power. The advantage of renewable energy is that its sources are available in unlimited quantities. Their use not only conserves the environment, but also is a way of meeting all of our energy requirements. Sources of renewable energy (https://www.conserve-energy-future.com) are hydraulic power, wind power, biomass power, solar power, geothermal power, etc.

Solar Power

As a sustainable construction technology, solar power has been exploited exponentially. The utilization of active solar power is appropriate for green construction. The use of functional systems from solar energy that absorb the radiation of the sun for supplying electricity and for heating is called active solar power. The usage of gas or electricity for daily needs is reduced. The installation and setting up expenseare not economical, but it saves decent on energy bills in the long-term and a reduction in the emissions of greenhouse gases like fossil fuels from non-renewable energy sources. And the other type is passive solar power; it is a design and implementation in which the rays of the sun are used to warm homes by using surfaces that absorb heat by the calculated position of the windows. During winters, this reduces the need for house warming as the windows let in temperature and the energy.

Biodegradable Materials

To make the construction sustainable, the eco-friendly means is to use such as biodegradable materials. Most of the traditional methods of construction cause build-up of toxic chemicals and waste products, most of them decay in hundreds or even thousands of years. And after the decomposition, it harms and contaminates the ecosystem. To restrict the harmful impact on the ecosystem, biodegradable materials, for example, organic paints, should be used as they readily decompose without releasing harmful toxins. Sustainable construction technologies include the usage of biodegradable materials for walls, insulators and building foundations.

Green Insulation

One of the serious concerns among the construction of homes and buildings is of insulation. Insulations need not be made from highly finished and costly materials; they are merely wall filters. It can be concluded that the application of green insulation is a sustainable construction technology as it ultimately gets rid of the requirement of high-end products constructed from non-renewable materials. Used and old documents such as newspaper or denim can be used as a solution for green insulation.

Smart Appliances

Majority of the consumption of energy is used by the commercial buildings and houses and on this ground; it has become a necessity to use sustainable construction technologies and to use smart appliances made from them. Installation of efficient and energy-saving devices is the focus of sustainable construction technologies. Examples of sustainable technologies are refrigerators, dishwashers, washing machines, and the smart grid. The establishment of zero-energy commercial buildings and homes is the orientation of this technology.

Cool Roofs

One of the sustainable green design technologies is cool roofs whose purpose is reflecting sunlight and heat away. It helps in maintaining building and homes at the desired room temperatures by adjusting thermal emittance and heat absorption. The design is madeusing individual tiles and reflective paints which soaka small amount of heat, and most of the solar radiation is reflected away. During summers, more the 50° Celsius temperature is reduced by cool roofs. Dependence on systems like air conditioning is minimized and thus the use of energy is reduced which results in low emission of cumulative greenhouse gases from power plants.

Sustainable Resource Sourcing

One of the primary examples of sustainable construction technology is a sustainable resource sourcing because it emphasizes on the adaptation and usage of construction materials which are created and designed for products that were recycled and are friendly to the environment. In the majority of the cases, agricultural by-products or wastes are used to manufacture the construction materials. As a whole, the materials are recyclable, recycled, reproduced and fetched from sustainable sources.

Low-Energy and Zero-Energy Building Design

Mechanisms to lower the consumption of energy are included in sustainable construction technologies. Sustainable construction technology, for instance, provides wood for construction of buildings as it has lesser embodied energy in consideration to those constructed using concrete or steel. Sustainable green construction in addition-built designs that let air to flow freely and reduce air leakage while in the meantime using high action insulation techniques and windows. Dependence on central heating and air conditioning are reduced using these techniques. The windows are placed strategically in such a way that supports day-lighting and hence reduce the reliance on electric lighting during the day. Low-energy and zero-energy building design includes the use of renewable energy, for example, solar energy for water heating and lighting. The implementation costs of installation of zero-energy buildings may be costly, but in the long-term, they pay off. In winter conditions, the pre-heating of the air is done by using the preheated capacity of the water. The heat exchanger is combined with Water Air Heat Exchanger (WAHE) for this purpose. Then further heating occurs as a result of the exchange of heat between the preheated fresh ventilation air at the Air-Air Heat Exchanger (AAHE) and the exhaust room air. Ventilation air heating occurs due to the heating coil which is joined with the solar water heater. Proper circulation of this ventilation air to the different locations of the house with separate flow rates as set by the person in the day and night time along with the efficient controlled working operation of underground AWHE pump and AAHE with smart control sensors. Such homes are kept airtight as they are combined with smart air handling units to achieve proper heating of the room's air at Belgium (Arvind et al., 2015).

Electrochromic Smart Glass

In sustainable construction, it also includes technologies such as Electronic Smart Glass. It is a new technology. In summer periods, to slam the solar radiation's harsh heat, the electronic smart glass is used. The intensity of solar radiation reflection is changed by the smart glass which uses tiny electric signals to charge the windows to a small degree. The control systems in the buildings are incorporated with it, hence the users are allowed to define the degree to block the solar radiations. Commercial buildings and homes can benefit a lot by saving on air conditioning, ventilating, and heating costs with the help of this technology. Soon the smart glass will be fully used as it is still in the research and will be applied as a smart energy-saving technology in sustainable construction.

Water Efficiency Technologies

Sustainable construction technologies include many water efficient technologies. The water is adequately recycled, managed and used for non-drinking purposes like flushing toilets and washing cars, is ensured by these methods. For example, dual plumbing increases the potential to recycle water on-site and decreases sewer traffic. And conversely, rainwater harvesting provides water to be stored for future and multi-purpose usage. The water efficiency technologies of sustainable construction help in conservation of water and lower the cost of water usage. Up to 15% of water wastage is reduced in urban areas and assists in freshwater shortages.

Sustainable Indoor Environment Technologies

During the establishment of any home or building, the safety and health of the residents of the building is a must and fundamental. In green construction the sustainable indoor technologies are mandatory. To ensure green safety standards, the materials used include moisture resistance, low volatile emissions, non-toxic materials, and hazardous free elements. For example, materials like bamboo, wood, and cork are sourced from nature and do not contain any toxic, carcinogenic or irritating ingredients. Low VOCs material usage also intensifies IAQ and restricts radiations to a chemical which are harmful to health such as vinyl, lead, and phenol-formaldehyde.

Self-Powered Buildings

The construction of self-powered buildings is the advancement and can be achieved by sustainable construction techniques. It is concerning the realization of the zero-energy construction of self-powered buildings. The self-powered buildings are constructed in such a way that they can produce power to be able to support their needs for energy and also direct back excess energy to the power grid. The use of wind power technology in self-powered buildings, in many skyscrapers, rooftops are mounted with wind turbines. The turbine blade is propelled with the help of heavy and constant and air currents at higher altitudes which in return produces the requirement of power for the building.

Rammed Earth Brick

Ancient construction technology has been re-introduced called a rammed earth brick, for the fulfillment of the demands of environmental sustainability. Raw materials that are sourced sustainably is used in this technique. Due to advancements in the technologies, the procedure of constructing a rammed-earth structure has become, but the ancient preparation process is still followed. Hard substances like clay or gravel and moist earth mixture are mixed with stabilizing elements such as compacted and concrete and to create hard, dense walls. It is ideal for sustainable construction because of the formation and sourcing procedure of rammed-earth bricks as the temperature of the building can be equally stabilized by the material, and it reduces environmental impacts. Rammed-earth structures ensure the buildings remain warm in the winter and chill in the summer and also contribute to secondary emissions.

GREEN ENERGY CONSERVATION IN BUILDING

Green building interchangeably can be used with the terms' sustainable construction or green construction. So, sustainable construction means using environmentally responsible and resource-efficient procedures in development to be ensured of sustainability throughout the lifetime of the building. Mainly, the reference of sustainability to the structure implies site design, building operations, repair, demolition and maintenance with the minimum impact onnature. The procedure requires active collaboration with the architects, the client, and the construction engineers, in the whole project of construction. The goal is to check that the construction and building methods reduce, durable and cost-effective the effects as a whole on the human health and environment with a significant aim of using resources, energy, improved occupational health, water preservation, wastage and reducing pollution efficiency, and minimizing wastage and corruption. To reduce the building's consumption of energy, there are following ways which subsequently results in reducing CO_2 emissions with the help of energy conservation.

Artificial light usage can be reduced during daytime by applying proper design of daylighting and hence reducing the consumption of energy by the building. The utilization of low embodied energy is the second important aspect and using materials available locally for construction of a building to prevent primary energy supply in development and hence a reduction in the emission of CO_2. To reduce energy obtained from fossil fuels is the third aspect of making for operation. Zero-emission or highly energy efficient green building is called when all the power of the building is produced by renewable sources of energy. For the acceptance of renewable source energy systems, the economics are given in comparison with the conventional sources of energy.

Maximum conservation of energy is the most sustainable energy technique. The design of the building is directly related to the usage of power, and hence the composition of the passive solar building design can help in conservation of energy. Passive solar building designs in buildings use natural energy such as sun's energy for free of charge heating, daylighting and cooling. This provides a comfortable atmosphere, and the requirement to consume power from other energy sources is reduced. Passive solar design principles are compatible with an existing building and diverse architectural styles for zero energy usage. Passive solar design adds a combination of features of construction and eliminates or reduce or the requirement for mechanical heating and cooling and artificial lighting in the daytime. Special attention to the sun is given by the builders and designers to ensure minimize cooling and heating requirements. The design of the building does not require being complex, but the plan should involve window technology, local climate, and knowledge of solar geometry. Virtually, passive solar design can be integrated into any architecture if the proper site of the building is given (Aksamija, 2015).

In passive solar energy, the basic natural procedures are the flows of thermal energy linked with natural convection, conduction, and radiation. When a building is struck by sunlight, the material of the structure can transmit, reflect, or absorb ration of the sun. Furthermore, the sun's heat causes movement in the air and in designed spaces it can be predictable. Solar heat's primary responses lead to placements, material choices and design elements that can supply cooling and heating and change in the home. Passive solar energy systems do not employ mechanical means to utilize energy from the sun. Some thumb rules to be considered for the effective utilization of solar energy from passive solar systems. Applying the approximately zero energy building concepts the energy management of the building can be reached smartly (Angelis et al., 2015) and (Perlova et al., 2015).

Passive solar heating system's goal was to record the heat of the sun within the element of the building and when the sun is not brightly releasing that full heat. In the meantime, the weather is being absorbed by the parts of the structure or materials for later use, heat from the sun is available for the space to be comfortable. A combination of energy-efficient doors and windows, proper insulation, daylighting, ventilation, and shading will keep homes with low temperature and with less use of energy (http://www.eere.energy.gov). The approach comprises of, use of wing walls, thermal chimney, and operable windows. By providing vents in the upper or top level of building natural ventilation can be made to allow air to escape outside as the warm air rises up by convection. In the meantime, the vents in the lower or bottom level can draw cooler air in. The lower or bottom vent is provided beside the building where trees are planted which helps in providing shade for more cooling air present outside (http://en.wikipedia.org/wiki/Solar_chimney).

COMBINATION OF RENEWABLE ENERGY BASED TECHNOLOGY FOR GREEN BUILDING CONSTRUCTION

Renewable energy is obtained from sources that will be steadily renewed. Multiple types of renewable energies are derived from the light of the sun and from the heat energy that is produced deep inside the earth's surface. The electrical energy and heat energy gathered from wind, sun, ocean, water; geothermal sources are procured from these inexhaustible, renewable resources (http://www.iea.org). The technologies obtained from renewable energy resources comprise of the wind power, solar power, hydroelectric power, and several others (http://en.wikipedia.org/wiki/Renewable). The global status report of 2007, states that 18% out of the total worldwide energy usage in 2006 was from the renewable, inexhaustible resources, that included 13% obtained from biomass like burning of wood.

The second biggest inexhaustible source was hydropower that provided 3%, next was heating or hot water that contained 1.3%. Latest technologies, like the ocean, solar, wind, geothermal energy jointly gave about 0.8% out of the total energy utilization (http://www.ren21.net). Hence the scientific perspective for their usage is enormous, which surpasses all the other easily obtainable sources (http://www.undp.org). A large quantity of electricity and heat requirements of different buildings could be efficiently met by the extensive usage of thermal/solar photovoltaic and collector cells. The future years might find several other Renewable EnergySources (RES) like the water, biomass, wind and many more could also be used, reducing the usage of the traditional energy resources. Nuclear Energy and RES are considered to be the substitute energy sources for preventing the greenhouse effect of greenhouse gases. Between the above two energy sources, RES is eco-friendly, safe and harmonious with the biosphere and are evenly divided worldwide. Since they are renewable and inexhaustible, they can be used with a lot of ease by the customers with little ownership undertaking and market trust (http://www.inive.org).

Technologies procured from renewable energy have a lot of advantages like greater employment, sustainability, the lengthy lifecycle of energy sources and security of constant energy supply. Although the solar energy systems have a quite high price, it looks to comply with international and European policies. The sole cause is that the technology of the sun is eco-friendly in biosphere conditions for modern implementations and buildings. Such type of technology is very beneficial for the economic growth of several nations as they can substitute the costly conventional sources of energy such as nuclear fuel, gas, coal, and oil. The solar energy-based systems or implementations can be put up in a very harmonic manner on buildings to shield the cooling, lighting, heating and electricity needs. The inclined or the flat rooftops and the facades of homes and other buildings constitute an expanded surface for using photovoltaic panels and solar thermal collectors. The structure of the buildings could be constructed in terms with the bioclimatic architectural designs to reduce the energy requirements and their ecosystem impact by bringing in use, the new special glasses and heat-enclosing materials such as smart windows that minimize the heat losses effectively in the winters and energy utilization for cooling in summers. Due to this, the future savings of energy in the assembly of buildings could be roughly more than 50% of the total energy utilization of model buildings (http://www.inive.org). The installing of several solar energy units and devices is about their higher price and their consonance with the architecture of the building and as well as the ecosystem. Systems based on solar energy are also used in preference for aesthetic causes, to avoid the negative phenomena caused due to diesel engines. It becomes essential again to apply them if they are implemented in consonance into the current natural and local settings of the ecosystem through strategic planning and intelligent environmental analysis. The extensive usage of systems based on solar energy for developing applications that are sustainable and could lead to the production of creative building structures, supported with bioclimatic characteristics targeted to save energy consumption. By study, it was observed that the building sector was causing about 35% of the total energy utilization and 40% of the gaseous emissions and it is roughly proved that the stored percentage of energy will come at 60% when solar energy-based systems are utilized for cooling and heating applications. The European Commission (EC) has put up a new directive into force by the policies of saving energy in freshly constructed buildings. Hence the RES application to build structures with increased performance along with a beautiful combination will lead to the rise of the living standards (http://www.inive.org).

There are several integrated solar energy systems adopted globally for setting up green buildings. Such methods might also be used as individual units in series order with an enclosed insulated storage tank present inside the building for single-family houses. This happens in case of large size applications. Solar collectors have multiple color absorbers other than black color which could be a prominent solution for the broad applications of solar energy-based systems. Colored collectors possess lower absorption and therefore they work with less thermal efficiency as compared to the traditional collectors of black type. In terms of cost, an extra amount of 20% to increase the area of the collector, for overcoming their little thermal efficiency in comparison to the similar type of collectors with the black absorber. The usage of collectors of blue color in the island buildings of white color or reddish-brown type collectors in buildings that have inclined/tilt rooftops could support a more significant use collectors solar thermal (Kalogirou et al., 2005).

Several buildings are covered with flat roof due to which solar collectors' installation, in parallel row format can take place, positioned at an appropriate length, to avoid shading of collector during the winters. The gap between the multiple parallel rows might be utilized to provide extra sun radiations on the aperture surface of the collector by positioning the booster reflectors at the collector-base of the adjacent row from the collector-head top of a row. Such glasses can support to the rise in the output of thermal energy by 20–50% and this type of equipment installation from spring season to fall are the most beneficial for the operation of the collectorat high temperatures, adapting cooling needs of the ecosystem (Tripanagnostopoulos & Souliotis, 2005). The utilization of glass coatings and the optical annoyance coming from the reflected light, which diffuse the reflected light is considered the best-suited method. To prevent the difficulties due to the glare, it is appropriate to utilize unglazed solar collectors in lower temperature-based purposes such as swimming pools. Similarly, on the other hand, unglazed solar collectors might work as an alternative to current typical collectors that have glazing, for heating the water up to 35 °C. The uncovered collectors could be integrated with the colored surface to provide a more innovative integration of solar collectors in inclined roofs and facades of several buildings (Tripanagnostopoulos et al., 2000).

Thermal / Hybrid Photovoltaic (PV/T) Systems

The thermal/hybrid photovoltaic (PV/T) systems are the solar energy-based systems that provide heat and electricity. They consist of photovoltaic modules that are coupled to water or air heat extracting equipment targeting a large energy conversion efficiency of the absorbed solar radiation (Ji et al., 2007). Installing such systems can contribute to natural ventilation.

Using Fresnel Lenses for Building Atria

Optical devices such as the Fresnel lenses are used for concentrating sun radiations. The lenses are of smaller focal length, lower weight and volume and lower price in comparison to the conventional thick lenses. This property to segregate the direct sun radiation from the diffused sun radiations makes fresnel lenses more appropriate for illumination control of the interior space of the building, by giving light with appropriate intensity level and no sharp contrasts. This Fresnel lens-based system promotes solar dominance of the buildings to keep the interior temperatures and illumination at the easiest level (Tripanagnostopoulos et al., 2005).

Integration of Wind/Solar Systems

The building combination of wind and solar energy system is another innovative topic that has been observed and studied. Both of these systems look to be very interesting and prominent among other renewable energy sources for the built ecosystem. The horizontal or inclined rooftops, the facades of houses, hostels, and buildings of multiple types are suitable surfaces for the application or operation of solar energy converting systems as they include thermal collectors or photovoltaic panels for accumulating the heat and electricity demands. Other than this building roof can be mounted with small size Wind Turbines (WT), mostly at locations where wind velocity potential is satisfactory. The usage of small size turbines in mesh-connected residential building areas, hotels, etc. is considered to provide for the significant supply of electricity, and also, they are appropriate for decentralized applications.

Additionally, they can be integrated with thermal collectors or photovoltaic cells (Tripanagnostopoulos & Tselepis, 2003). Due to these multiple systems are developed, like the wind turbines of the vertical (VAWT) axis or horizontal axis (HAWT) wind turbines and wind concentrators. Hybrid diesel systems or photovoltaic-wind systems can provide significant support in the generation of energy based on sun and wind energy. The regions or localities where sunlight and wind conditions are superb, such as the Greek islands, then the integrated use of wind turbines and photovoltaic cell systems have good outcomes for most days and nights. This happens for a very long time period in the year. Hence the thermal collectors meeting our thermal requirements are needed for the whole year. The combo of photovoltaic/thermal (PV/T) and Wind Turbine systems has been put up as a brand-new idea, and the different energy conversion systems are the mix of solar (electric and thermal) or wind (electronic) systems (i.e., PV/T or WT). They are regarded as very suitable in rural areas where electricity supply comes from stand-alone, single units or small-grid connection. PV/T or WT based systems can find use in grid-connected works (Tripanagnostopoulos et al., 2004). For such stand-alone PV/WT systems, diesel generators are utilized as the wind, and solar energies are not enough to meet the electrical load.

ECONOMICS OF RENEWABLE ENERGY BASED TECHNOLOGIES

The capacity utilization and capital cost affect the per unit price of electricity produced from WT power generation systems. Or researchers and practitioners can say that the wind electricity price can be minimized by reducing the capital price and increasing the energy results occurred due to improved aerofoil, bigger turbines and the launch of more productive operating strategies. The capital price can also be minimized as an outcome of innovations that might comprise the usage of lightweight design materials amounting to lesser manufacturing prices. Different traditional methods are used for obtaining the required thermal energy. The unit cost of fuel and efficiency of fuel consumption that is used for heating of water has a more significant effect on the unit price of thermal energy which is given by a solar hot water system. It is observed that the unit price of consumable thermal energy generated critically relies on the number of total days when the water heated by solar is either used by or delivered to the investor. The unit price of a solar water heater compared to the conventional cheapest methods of water heating for obtaining greater values of utility-supported with a lower rate of discount. The simplest devices are still the traditional solar that provide distilled water in the remote areas which have the least rainfall and small freshwater sources. The traditional distillation that uses fuels from fossils is more complicated than distillation systems based on solar energy. The old systems need cost for operation, maintenance, and capital amortization. The traditional distillation system produces distilled water whose unit cost will depend on the above aspects besides the price of fuel used. The distillation systems that are solar based are similar to the traditional desalination systems in all features except for the associated solar energy collecting small sub-systems and energy source. Hence distillation systems assisted by solar energy are more involved in working than conventional plants because of its dependency on different sources of input. The wind, biomass, ocean and thermal energy can also be used for the distillation process as other alternatives of renewable energy (Kandpal & Garg, 2003).

SUSTAINABILITY OF GREEN BUILDINGS USING RENEWABLE ENERGY TECHNOLOGIES

The prominent and safe sources of renewable energy generation for an eco-friendly, sustainable and green buildings are wind, hydroelectric (for rural areas near a river), geothermal (Heat of the Earth) and sun. The use of such renewable energy resources for green buildings critically helps in minimizing our carbon footprint. These resources have also led to the conservation of energy at various levels. Also, the renewable energy-based systems will provide a greater share of the building's gross energy requirement for the same price as compared to a structure that is less efficient. The rapid technological advancement has provided for a ten-time increase in the wind turbine's size. It has changed the units from 50 kW to 5 MW in the last 25 years and also a decrease in cost more significant than 50% in the previous 15 years. Photovoltaic-based systems of energy production are one among the readily growing industries of energy because of the average yearly growth rate of 35% in the previous ten years (Liberali, 2007).

CONCLUSION

It is highly essential to analyze that the ecological aesthetics of a building are taken as an extra price. Due to which, the establishment of green, safe buildings becomes more expensive, inevitably as it uses highly efficient good quality materials and a more complicated work process. These days, researchers, scientists, architects, designers, etc. observe the activity that is oriented to safe, sustainable, andeco-friendly development that may assist others in finding out alternative methods that would bring global and social benefits without any additional price. The construction of green buildings helps to eliminate expenditure with renewable energy technique. Incorporating such technology based on renewable energy in building construction will eventually minimize our requirements of fossil fuels and reduce environmental dangers due to extensive consumption of natural resources and energy. Hence, I can sum up to the fact that the usage of renewable energy technologies for green buildings has proven to be a safety initiative, speeding up the journey of the biosphere towards safety and sustainability.

REFERENCES

Angelis, E. D., Ciribini, A. L. C., & Tagliabue, L. C. (2015). Smart campus demonstrator: Renovation towards a zero energy classroom building. *Procedia Engineering*, *118*, 735–743. doi:10.1016/j.proeng.2015.08.508

Chel, A., Janssens, A., & Michel, D. (2015). Thermal performance of a nearly zero energy passive house integrated with the air-air heat exchanger and the earth–water heat exchanger. *Energy and Building*, *96*(2), 53–63. doi:10.1016/j.enbuild.2015.02.058

Conserve Energy Future. (2019). Top 10 sustainable construction technologies used in green construction. Retrieved from https://www.conserve-energy-future.com/top-sustainable-construction-technologies-used-green-construction.php

International Energy Association. (n.d.). Retrieved from http://www.iea.org/textbase/nppdf/free/2000/Renew_main2003.pdf

Ji, J., Yi, H., He, W., & Pei, G. (2007). PV-Trombe wall design for buildings in composite climates. *Journal of Solar Energy Engineering*, *129*(4), 431–437. doi:10.1115/1.2770751

Kalogirou, S., Tripanagnostopoulos, Y., & Souliotis, M. (2005). Performance of solar systems employing collectors with colored absorber. *Energy and Building*, *37*(8), 824–835. doi:10.1016/j.enbuild.2004.10.011

Kandpal, T. C., & Garg, H. P. (2003). Financial evaluation of renewable energy technologies. Macmillan India Ltd.

Liberali, R. (2007). Renewable Energy Technologies: Long Term Research in the 6th Framework Programme 2002–2006. Project Synopses, Community Research, European Commission.

Perlova, E., Platonova, M., Gorshkov, A., & Rakova, X. (2015). Concept project of zero energy building. *Procedia Engineering*, *100*, 1505–1514. doi:10.1016/j.proeng.2015.01.522

REN21. (2008). Renewables 2007 Global Status Report. Retrieved from http://www.ren21.net/pdf/RE2007_Global_Status_Report.pdf

Regenerative design of existing buildings for net-zero energy use. (2015). *Procedia Engineering*, 72–80.

Tripanagnostopoulos, Y. (2006, July). Energy and aesthetics of building integrated RES. In *Proceedings of International Workshop on Energy Performance and Environmental Quality of Buildings. Milos, Greece*. Greece: Department of Physics, University of Patras; Retrieved from http://www.inive.org/members_area/medias/pdf/Inive%5CMilos2006%5C46_Tripanagnostopoulos_6P.pdf

Tripanagnostopoulos, Y., & Tselepis, S. (2003). Hybrid solar/wind PVT/WT building integrated system. *Proceedings of 2nd International European PV-Hybrid and Mini-Grid Conference,* 329-333.

Tripanagnostopoulos, Y., Nousia, T., & Souliotis, M. (2000). Low-cost improvements to building integrated air-cooled hybrid PV-Thermal systems. *Proceedings of 16th European PV Solar Energy Conference,* 1874-1877.

Tripanagnostopoulos, Y., Siabekou, Ch., & Tonui, J. K. (2005). The Fresnel lens concept for solar control of buildings. *Proceedings of International Conference PALENC-2005*, 977-982.

Tripanagnostopoulos, Y., & Souliotis, M. (2005, May). Booster reflector contribution to performance improvement of solar collectors *Proceedings of the WREC*, 63-68.

Tripanagnostopoulos, Y., Tselepis, S., & Souliotis, M. (2004). Practical aspects for small wind turbine applications. *Proceedings of International EWEC 2004 Conference, Track 7; Autonomous and Hybrid Systems*.

United Nations Development Programme. (n.d.). Retrieved from http://www.undp.org/energy/activities/wea/drafts-frame.html

U.S. Department of Energy. (n.d.). Retrieved from http://www.eere.energy.gov/consumer/your_home/space_heating_cooling/index.cfm/mytopic=12350

Wikipedia. (n.d.a). Renewable energy. Retrieved from http://en.wikipedia.org/wiki/Renewable_energy#cite_note-REN21-2007-0

Wikipedia. (n.d.b). Solar chimney. Retrieved from http://en.wikipedia.org/wiki/Solar_chimney

Chapter 3
Improved Design of Vertical Cavity Surface Emitting Laser for 3D Sensing in Internet of Things Applications

Priyanka Goyal
Gautam Buddha University, India

Prachi Singh
Gautam Buddha University, India

ABSTRACT

Internet of things (IOT) and Vertical Cavity Surface Emitting Laser (VCSEL) future can be seen together, since VCSEL technology-based 3D sensors are introduced for IoT applications. The improved VCSEL structure design with fixed wavelength using a thermally actuated cantilever structure is presented. This improved structure of VCSEL will help us in realizing athermal VCSEL. In athermal VCSEL the dependency of VCSEL on temperature will be much less because it will not require temperature controllers. Realizing fully temperature-independent VCSEL (i.e., athermal VCSEL) is still a challenge but we can reduce it to some extent. In this chapter, recent diversification of application of VCSEL technology from data communication to sensing has been discussed. This proposed VCSEL structure may give us an opportunity to improve the VCSEL technology. Therefore, smart 3D sensors based on VCSEL will help in making internet of things applications more reliable and will directly or indirectly serve the concept of smart homes and smart cities.

DOI: 10.4018/978-1-5225-9754-4.ch003

INTRODUCTION

Kenichi Iga invented Vertical Cavity Surface Emitting Laser (VCSEL) in 1977. From that point of time VCSEL come up a long way and diversified their applications from data communications to upcoming 3D sensing technology for Internet of Things (IoT) purposes. Earlier VCSEL was used as an optical source for data communication in optical fibers, as infrared illuminators for military or surveillance. VCSEL also became popular in medical science. VCSEL based printers are also very common. Nowadays VCSEL based 3D sensors are hitting the market. By 2023, VCSEL market is expected to reach United States Dollar (USD) 3.89 billion, which is USD 1.78 billion in present scenario given by Report buyer (2018).

In 1999, Kevin Ashton formulated the term 'Internet of Things', in a presentation to Proctor and Gamble. FurtherArik Gabbai (2015) observed IoT concept in between 2008-2009. IoT is all about Big data, analytics, cloud computing, software, and sensors. The sensor is one of the crucial components for realizing the concept of IoT in reality. Sensors play a very pivotal role in various applications of IoT. In IoT, devices totally rely on sensors because sensors sense everything in the environment, collect data which is further processed for continuing the work or for making decisions manually or automatically by the application itself. In order to fulfill the requirement of IoT, every device is made smart such as smart T.V., smart mobiles, smart tablets, smart computers or any portable device with internet access. Further development and innovation of smart devices will contribute to the development of IoT only. One important characteristic of smart device is its strong sensing ability and object recognition. These smart devices with the strong sensing capability can be employed anywhere in vehicles, airports, railway station, libraries, buildings, shopping malls, schools, colleges, roads, etc. Installing smart devices wherever possible will contribute to the intelligent applications of IoT. IoT applications use a plethora of sensors among which VCSEL based 3D sensors in the future can replace multiple sensors.

Devices are made smart when the sensors are made smart. Keeping in view of smart sensors, 3D sensors are realized by using VCSEL as optical source by various smartphone brands. According to Evangeline H (2018), Apple released the latest iPhone X smartphone with 3D sensing technology based on sensors. These 3D sensors are not used for just face recognition but also for sensing the environment. iPhone X used VCSEL technology for proximity sensing and face identification. Now other smartphone brands are planning to integrate 3D sensors based on VCSEL in their upcoming smartphone release. In 2018 Xiaomi launched Xiaomi Mi8, Oppo launched Oppo Find X phone. By 2019, other smartphone brands like Huawei, Vivo, and Samsung are believed to integrate VCSEL for making more efficient 3D sensors in mobile phones.

Rising adoption of VCSEL technology for 3D sensing is not just limited to mobile phones only, but can also be realized in tablets, PCs and notebooks, gaming, automotive, drones and above all for the application of smart homes and smart cities in IoT. Jabil (2019) gives that for IoT applications like smart home, 3D sensors are employed for safety and security purposes, as 3D sensors can sense 2D or 3D gesture. For realizing smart cities in IoT, VCSEL based sensors can be employed in parking and museum or in the retail industry for safety and security purpose. The concept of smart home and smart cities can be envisioned with the innovation of VCSEL based smart sensors employed in every possible device of home or city. The idea of smart home can be visualized when every electronic or non-electronic object like T.V, refrigerator, A.C., lighting, microwave, oven, or flowering pots etc. can communicate with each other and can be controlled manually or automatically.

Now days VCSEL based 3D sensors are integrated with smart devices for scanning the external environment. In future these sensors are set to integrate with each and every device possible to give the extended view of the environment. This will enable huge security and safety in homes and cities. In IoT applications, 3D sensors will reduce the burden of multiple sensors to very few 3D sensors in order to work efficiently according to Powell (2016).

In this chapter, the improved VCSEL structure for 3D sensing is designed and studied. Various principles and technologies i.e. time of flight and structured light for 3D sensing using VCSEL are introduced in detail. Realization of a thermal VCSEL and the challenges are discussed.

LITERATURE REVIEW

VCSEL was first commercialized by Honeywell in 1996, to design high-speed data communication over multi-mode optical fiber. Philip Moser et.al (2013) proposed the energy efficient oxide confined VCSEL which works on the 850 nm wavelength. They were designed for applications in data centers and optical interconnects. VCSEL can work in the range up to 100 m or less. To make the VCSEL efficient the aperture diameter of cavity was changed from 2.5 μm to 9 μm. The results after varying aperture came out to be error-free data transmission at 25 GB/s. Record of low heat energy dissipation was observed, which was only 56 fJ/bit. The VCSEL works efficiently using a single mode optical fiber. Another framework of VCSEL at 980 nm wavelength was proposed by Hui Li et al., (2014) in which VCSEL was well suited for short reach communication, high-performance computers, board to board or chip to chip communication where optical reach was less than one meter for optical interconnects. The VCSEL was highly temperature stable for data communication with high bit rate and minimum error-free transmission at 38 GB/s. Also, the high speed of VCSEL was analyzed. Error-free data transmission was observed between 25-degree centigrade and 85-degree centigrade. The designed VCSEL consumed very low power. This error free operation was possible by applying various techniques like numerically investigating the temperature dependence over gain in quantum well, by making certain changes in the active region like increasing the decreasing number of quantum wells. Then temperature stable and energy efficient VCSEL was again designed by Hui Li et al., (2015) at 980 nm wavelength, but improved the reach of VCSEL from less than one meter to less than two meters where the application was same for optical interconnects in board to board, chip to chip communication and high- performance computers. To make the VCSEL temperature stable, cavity oxide aperture diameter less than 5 μm was used and if the oxide aperture was increased up to 7 μm, error-free data transmission of 42 GB/s at 25-degree centigrade was obtained. After varying oxide aperture between 5-7 μm author found low energy dissipation at 3 μm and to improve the temperature stability, made a gain to etalon (fixed separation between cavities) offset value to -15 nm. Hence error-free data transmission of 38 GB/s observed up to 85-degree centigrade and concluded that VCSEL with small oxide aperture was more efficient in terms of energy as compared to high oxide aperture. Another oxide confined VCSEL was designed by Rohan Bajaj et al. (2016) but instead of operating the VCSEL at short wavelength, the author designed the VCSEL using a long wavelength, i.e. 1310 nm. Temperature insensitivity, power dissipation were the main problems occurring in VCSEL and the solutions for temperature stability and power reduction in the application of on-chip optical interconnects were also given by the author. The emerging need of long wavelength VCSEL due to fiber dispersion and absorption at a short wavelength for high-speed LAN was observed to be inefficient. VCSEL with large oxide aperture was capable of providing high modulation bandwidths and meeting future demands of large bandwidth, but these large apertures were found out to be inefficient at high temperature whereas small oxide aperture of VCSEL was capable of meeting highly efficient VCSEL. Very low bias current was observed along with a modulation bandwidth of 22 GHz and the thermal behavior was found out to be effective with a range of 283 K-323 K. Another large bandwidth VCSEL with small current density and temperature stability has been proposed, but the wavelength chosen by the author Philip Moser et al. (2017) was 980 nm and the reason for not using long wavelength was that the glass fibers made in the market is not optimized for this long wavelength and also the sensitive and fast photodiodes are also not available to compensate and work efficiently with long wavelength. Hence, it was observed that lifetime of photons can be controlled by altering the DBR mirror reflectivity and error-free data transmission was also observed with a modulation bandwidth of 50 GB/s. High bit rate was achieved due to small oxide aperture diameter and the reach of VCSEL was nearly 2 km.

From Table 1, it can be concluded that VCSEL is very commonly used in data communication and the researchers are aggressively working on VCSEL to make it more efficient. Due to the high reliability, efficiency, low cost, symmetrical, low divergent optical beam, and unique compact structure of VCSEL gives tough competition to other optical sources like Edge Emitting Laser (EEL) diodes. Hence now VCSEL is used not only for traditional applications, but also for sensing the environment. Hence huge diversification of application of VCSEL from optical interconnects to VCSEL based 3D sensors can be observed.

Table 1. Conclusion of VCSEL technology in optical interconnects

Application	Technique	Wavelength	Optical Reach (Distance)	Result
Optical Interconnects	Change aperture diameter of the cavity from 2.5 to 9 µm	850 nm	Up to 100 m	Error-free data transmission at 25 GB/s and low heat dissipation of 56 fJ/bit.
Chip to chip communication	-15 nm quantum well gain to etalon offset	980 nm	Up to 1 m	High temperature stable, high bit rate, error-free transmission of 38 GB/s-42 GB/s at 25°C-85°C.
Optical interconnects, high performance computers	Oxide aperture <5 µm, -15 nm quantum well gain to etalon offset	980 nm	Up to 2 m	Temperature stable, error-free transmission of 38 GB/s-42 GB/s at 25°C-85°C range increased by 1m.
Optical interconnects	Small oxide aperture i.e. 3 µm	1310 nm	Up to 1 m	Energy efficient, modulation bandwidth 22 GHz.
Short reach Optical interconnect	Mirror reflectivity adjusted, cavity length reduced by λ/2	980 nm	2 km	Error free transmission of 50 GB/s-42 GB/s at 25°C-75°C

DIVERSIFICATION OF VCSEL APPLICATIONS

The incremental pressure sensor based on optical feedback in VCSEL was presented by Van Hoe (2012). The growing demand for pressure sensor in various sectors like, health sector, automotive industry etc. leads the author to design a miniaturized VCSEL based pressure sensor. VCSEL based sensors provide immunity from electromagnetic interference and also work efficiently in harsh environmental conditions by giving accurate results. The technique used was self-mixing interference in which the external cavity of 100 μm has been designed and based on optical feedback pressure was sensed by Bram Van Hoe et al. (2012). Holger Moench et al., (2016) represented VCSEL based sensors for measuring distance and velocity in 3-dimensional space for IoT applications. The wide view of VCSEL based sensors for scanning 3D system for developing various IoT applications has been discussed. With the humongous growth of 3D sensors in the market, the function of 3D sensors in IoT applications like for transportation in vehicles, in mobile phones and in sensing 3D environment with a 3D sensor has increased. The VCSEL operates at 980 nm and used time of flight technique. VCSEL arrays were used along with micro lenses for better illumination where the emission of high power VCSEL shows the thermal shift of 0.07 nm/K. The wavelength shift was plus-minus 10 nm for all operating temperatures of VCSEL and the velocity of the object was measured by observing (Doppler frequency) periodic variation of feedback light due to the movement of the object. The changing interference signal due to moving object when modulated with Doppler shift helped in deriving velocity. The demand of VCSEL will be 10-100 times more for 3D sensing applications rather than the traditional applications. A short review of various applications of VCSEL like illumination, 3D photography and video with a brief introduction to the important technique of 3D sensing like time of flight has been given by Ebeling (2018). VCSEL as gas sensor for oxygen gas sensing in the combustion process, efficient inverted grating AlGaAs-GaAs VCSEL is used. For 3D photography and 3D video, 2D array VCSEL of 8-micrometer cavity length was used in which no optical damage was found. To determine the phase and depth of object for each pixel of the sensor, Photon Mixing Device (PMD) image sensor was used by Karl Joachim Ebeling et al. (2018). The discussion about the superiority of VCSEL over other light sources for sensing applications has been done by Kent D. Choquette & Chun Lei (2017). If the array of VCSEL is used, then they are more efficient for sensing the 3D environment and indoor navigation. The main advantage of using VCSEL based sensor is for security requirements in autonomously moving vehicles. The main emphasis is on VCSEL array because of their brilliant focused illumination property which has the potential to replace alternatives like radar, which is bulky and their ultrasounds which are less accurate. Though the time of flight method is used for distance measurement of objects in the environment still there is a limitation of artificial light or sunlight, which can hinder the process of distance measurement of the particular object. But results showed the thermal shift of 0.06 nm/K and wavelength shift comes out to be plus-minus 5 nm at 80-degree centigrade for a single array, whereas for VCSEL array wavelength shift comes out to be plus-minus 10 nm given by K.J. Ebeling and R. Michalzik, (2017). Then Marciniak (2018) designed a new and compact optical sensing system. The proposed VCSEL design used Monolithic subwavelength High index Contrast Grating (MHCG) VCSEL for gas and liquid sensing. Another framework for VCSEL based sensing was introduced where the VCSEL technology applications with a wider view of the sensing and 3D imaging has been stated. The review of VCSEL based upcoming technologies and developments have been discussed by Karl Joachim Ebeling et al. (2018). The technology used to get a depth image was the time of flight and used Photon Mixing Device (PMD) image sensor, but the depth of the image was not visualized in 3D. Hence, along with PMD, CMOS active pixel image sensor, the thin multi-quantum well, Electro-absorption Modulator (EAM) has been used to obtain a more depth image. VCSEL was operated at 862-864 nm and could get the depth image of the object kept up to 4 m distance.

Therefore, on the whole, VCSEL has the humongous capability for growth due to the development of the internet of things and 3D sensing based on VCSEL technology. From Table 2 it can be concluded that VCSEL is in huge demand for meeting various sensing applications, but at the same time if results are observed each year there will be a thermal shift of 0.07 nm/K. Hence, in order to make the VCSEL more efficient, there is a need to make the VCSEL which depends less on the temperature and contribute to the IoT applications in an efficient way without thermal expansion at higher temperature.

Since IoT and sensing go parallel, therefore need of smart sensors in every IoT application is mandatory. The smart sensors are important in IoT applications like industrial automation, home automation, traffic management, medical and health care etc.

Internet of things is a platform where all the devices with built-in smart sensors are connected to the internet to perform efficiently. The IoT platform, then collects data from the sensors embedded in various devices and objects. This collected data is then processed to bring important information which can be useful for any specific IoT application. And finally, this information can be used for decision making either manually or by the application itself. Main building blocks of IoT are sensors, smart devices, IoT cloud and data analysis. The sensors connected to different devices continuously collect the useful data or sense any physical change in the environment and send it to IoT cloud. The IoT cloud is the platform which collects the data, process it efficiently and at last stores and manages the data for easy access. The data can further be accessed whenever required for making important decisions.

Table 2. Conclusion of VCSEL technology based on sensor applications

VCSEL Based Application	Techniques Used for VCSEL	Wavelength/Distance	Results
Pressure sensor, Automotive industry	External cavity made outside VCSEL of 100 μm	850 nm	Responsivity dropped to 50%
LIDAR for driving assistance smart phone, IOT	Calculate distance and velocity using time of flight.	980 nm, distance up to 100 m	Thermal shift 0.07 nm/k
Monitoring oxygen in combustion process, 3D photography and 3D video.	Inverted grating AlGaAs-GaAs VCSEL Used synchronous time selective in PMD image sensor	763 nm	Ideal for gas sensing, Phase and depth of objects determined for each pixel of the sensor
3D imaging	Time of flight, VCSEL array	980 nm	Thermal shift 0.1nm/K
3D depth information of environment Security zones in industries and surveillance	Time of flight +PMD (photon mixing device) image sensor PMD+CMOS+thin Multi Quantum Well (MQW) +EAM (Electron Absorption Modulator)	862 nm-864 nm	300 mW power consumed, depth image realized. Efficient depth image.

From all the components of IoT, the important and crucial component for realizing IoT applications in various fields is only possible because of sensors. Sensors are like ear and eye of the IoT. Hence VCSEL technology-based sensors also contribute to the IoT by realizing various sensors like VCSEL based gas sensor, or VCSEL based distance measuring and 3D imaging system. VCSEL based sensors for 3D sensing applications also contributes to IoT applications.

It can be illustrated from figure 1 that for any IoT application to be possible it needs sensors. Hence, in the block diagram first application includes home automation and home security with the help of VCSEL based sensors employed at home. For example, to sense any harmful gas by VCSEL based gas sensor or by counting the number of the person in the room by gesture recognition. Second application consists of VCSEL based 3D sensors employed at roads for safety. 3D sensors employed in vehicles to get the wide view of the roads and hence safety from accidents. In third application, VCSEL sensors in mobiles and industries are shown. In IoT, all these different applications are connected through the internet and the continuous data collected by the sensors goes for data classification via internet. This whole data is collected in the database of the IoT infrastructure. Database can also be considered as IoT cloud and finally, analysis of data takes place from where decisions are made, for example, 3D sensors placed in vehicles scan the environment and indicates the driver about road obstruction by the traffic.

APPLICATIONS OF VCSEL IN 3D SENSING AND IOT

VCSEL technology is best suited for 3D sensing applications as it is the most reliable diode laser with high power efficiency. They also work efficiently at the extended temperatures. From communication to sensing brought great impact to hold VCSEL technology in future applications with great authenticity.

Various sectors like industrial, environmental security, building automation, health care and life services, public safety retail, consumer and home automation, transportation, and logistics etc., need IR sensors for IoT applications. There are various IoT applications which are based on VCSEL technology as VCSELs have the inherent quality of being highly reliable because of its innumerable advantages. Their adoption in various IoT applications is just escalating day by day. Therefore, various 3D sensing applications based on VCSEL are:

Figure 1. Block diagram of various IoT applications like smart city and smart homes possible by VCSEL technology-based sensors

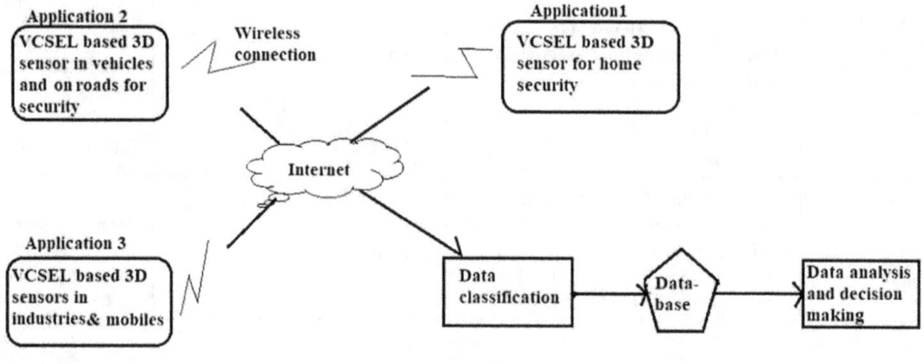

- For industrial automation, VCSEL based optical sensors are used for gate control, sensing the size and content of containers, checking seals, and inspecting PCBs etc.
- For home automation, the 3D image sensor can be used for security at homes by tracking, detecting and identifying individual motion.
- For traffic management VCSEL based 3D sensors can be used for tracking the number of vehicles. Also, VCSEL based 3D sensors can be deployed on roads to get the in-depth view of the traffic and security. 3D sensors deployed in vehicles can sense the distance of the object from the car.
- Due to increase of High Definition Multimedia Interface (HDMI) range, display port, interface range and bandwidth range over Multimode fiber, VCSEL has been adopted. Because VCSEL as optical source in cameras make feasible high resolutions, fast frame, and 3 Dimensional videos in diverse applications as given by Jim Tatum (2017).
- In IoT, various sensing applications run through electrical wiring which sometimes becomes unserviceable. Therefore, VCSEL based light system as a source is required to supply optical light energy to sensor without any wiring.
- Various IoT applications like illumination, night navigation, aerospace and line of sight communication require high power. Such high-power requirement applications can be attained by VCSEL arrays which can achieve one watt of continuous wave optical power.
- GPS applications require an intact and accurate position of data which is easy to realize by the source time information from the VCSEL based clock.
- Determining the presence of nearby objects in the 3D environment by transmitting infrared light and sensing the reflected light back is achieved with great accuracy by VCSEL technology.
- In any 3-dimensional environment like the living room, in order to detect the movement of people across the room, high power and accurate detection system is required which is possible through VCSEL based 3D sensor.
- In the entertainment and gaming sector, VCSEL based 3D sensors plays a very important role in which consumers can control games with their body gestures without any need for remote control.
- VCSEL based 3D sensors are also used to capture depth image information across the whole room hence contributing to the smart home.
- VCSEL based technology has also enhanced the capability of the camera which enables in-depth image or object detection with ease.
- According to Albert (2018), in IoT, wireless proximity sensors are widely used for monitoring machines and sensing objects. These sensor applications range from the industrial sector to medical, from food to the chemical industry, transport etc.
- Proximity sensors are highly reliable because there is a huge gap between the sensor and sensed object, they sense the object in the environment without any physical contact by emitting electromagnetic radiation like infrared light from the VCSEL based proximity sensor on the target. For installing best and reliable proximity sensors for IoT applications, VCSEL technology-based proximity sensors can be used.
- Due to the development of internet of things and cloud computing, which totally rely on the plethora of sensors among which VCSEL technology-based sensors or 3D sensors are very common.

ADVANTAGES OF VCSEL TECHNOLOGY IN 3D SENSING AND IOT

VCSEL technology brings various advantages that led market to rely on VCSEL for 3D sensing applications. VCSEL based 3D sensing gives advantages from the old concept of gesture recognition in mobile phones to the new concept of IoT which requires in-depth data to enhance any IoT application like smart cities, smart transportation system etc. Hence, various advantages of VCSEL are listed below:

- VCSEL is the first choice for 3D sensing applications because it is the unique optical source.
- VCSEL lasing wavelength is very stable, which operates in a single longitudinal mode.
- Fabrication of 2D array or 3D array which is mostly used for 3D sensing can be achieved easily with minimum cavity wavelength deviation (less than 2 nm deviation).
- VCSEL can also be operated up to 80-degree centigrade of temperature and delivers nearly 1200 Watt per centimeter square high-power output per unit area which in the future can go up to 2.4 kW/cm^2 as given by Boucart (2018).
- VCSEL is very efficient and can be coupled easily as it has circular beam.
- VCSEL is highly reliable because there is no optical damage.
- They are easily scalable too, i.e. 2D array of the VCSEL can be directly processed, unlike EEL (only one D array is possible).
- High power VCSEL like 2D array has very efficient heat sinking process because heat traverses only a few microns of semiconductor material like (AlGaAs).
- Low cost and high optical efficiency of VCSEL with focused and coherent circular light made it efficient to integrate with other optical elements, electronic devices and control logic within the same package for 3D sensing in IoT applications.

ISSUES AND CHALLENGES IN VCSEL TECHNOLOGY

After looking at the various applications and advantages of VCSEL in IoT and especially in 3D sensing, there are few problems that need to be addressed to make this VCSEL technology better for 3D sensing and IoT applications. Hence various challenges to realize efficient VCSEL for sensing purpose are:

- In IoT, there is a demand that the VCSEL based 3D sensors embedded in any smart device consumes less power as possible. Along with less power the sensor modules are light in weight and have less complexity. Whenever VCSEL technology is used, it is not used alone, it has heavy and costly thermal controllers which makes the sensor module too heavy and complex for various applications of IoT. Hence the elimination of these thermo-controllers can be only possible if VCSEL is thermally independent.
- There are various challenges to realize fully temperature independent VCSEL. Also manufacturing thermally independent and more efficient VCSEL is still a challenge.
- It is challenging to realize athermal VCSEL because of various reasons. VCSEL has an energy efficiency of around 25-35% only. The residual 70% of the energy supplied to the VCSEL is wasted

as heat energy inside the VCSEL structure. This heat dissipation inside VCSEL structure happens due to certain phenomena like Joule heating, internal optical absorption and the recombination of electron-hole to generate photons inside the VCSEL cavity.

TECHNOLOGIES ON WHICH 3D SENSING WORKS

Karl Joachim Ebeling et al., (2018), VCSEL technology-based Time of Flight (ToF) sensors are in huge demand because they are set to launch in robots, vehicles, household appliances etc. after smartphones for IoT applications. Every IoT smart Device needs to have sensors which can provide distance measurement of objects present in 3-dimensional space. Also, 3D imaging is very beneficial for all IoT and augmented reality applications.

The development of IoT for smart cities or smart homes is related to the development of technology which makes a device aware of its environment, position or distance of objects in the environment. Therefore 3-dimensional environment can be realized by three basic operating principles -structured light, stereovision and Time of flight (ToF).

The structured light method projects a multiple known dot kind of pattern over the object present in the 3-dimensional environment. The multiple dots pattern gets deformed according to the shape of the object which helps in observing the object in detail and gives in-depth information to calculate the distance and hence the position of an object. The limitation of bright sunlight or an artificial light source is not supported in the best way with this method.

The stereovision uses two cameras which are positioned accurately relative to each other. The images obtained from two cameras are correlated to generate the depth map. Its dependency on lightning condition, high processing cost, leading to strong heat dissipation puts this method back forth given by Time of Flight Basics, (2017).

Time of flight can be direct or indirect. In the direct method distance of an object present in the 3D environment can be measured by time of flight technique. In it, time is calculated when the photons emitted by VCSEL travels to the object and back to the detector. Single Photon Avalanche Diode (SPAD) is a sensitive photodetector which detects an innumerable amount of photons emitted by optical source. The time difference of light hitting the object and coming back is used to calculate the distance by Using equation 1:

$$\mathrm{D} = \frac{s \times t}{2} \tag{1}$$

where 's' is the speed of light, 't' is the time taken by light to hit the object and coming back, and 'D' is the distance of the object. The time divided by 2 indicates that photons emitted from the sensor travels to the object and then comes back to the photo-detector. Hence calculated time is twice, therefore, halving the time gives the accurate distance.

In the indirect method, the source light, for example, VCSEL illuminates the object in the environment and calculates the phase delay of the reflected light from the object. The measured phase delay is converted to distance by using a sampling technique called quadrature sampling technique. The advantage of the ToF 3D sensing technique is that this technique can be realized with low cost, low software complexity, compact size, medium depth accuracy, better low light performance, and good sunlight performance wherein range and field can be adapted.

As shown in Figure 2, the modulated signal is transmitted in the environment towards the object. The reflected light gets back to the photodetector, the correlation block correlates the signal and forward it to Analog to Digital Converter (ADC) for the further sampling of the signal. The sampled signal is then used for calculating the phase delay. The derivation to recover phase and amplitude is beyond the scope of this chapter, only conclusions are given here. After the whole derivation in the correlation function is sampled at equal steps over one period, i.e. illumination phase is changed in steps of 90 degrees. As shown in Figure 3, the different phase shift is used to calculate the distance and amplitude of the signal.

Figure 2. Indirect time of flight

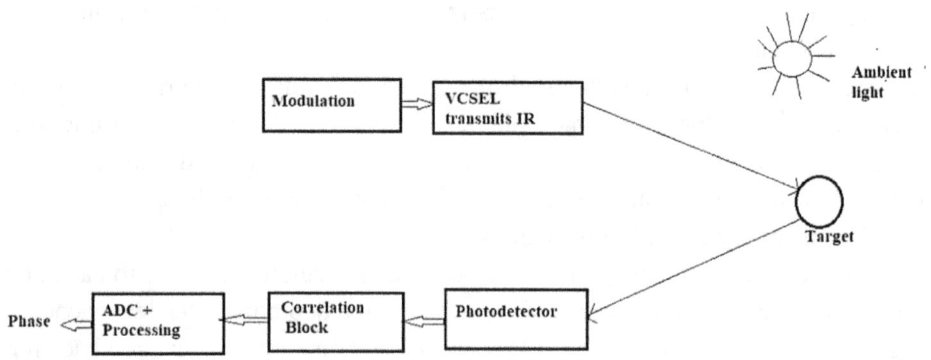

Figure 3. Different phase shifts are used to calculate distance and signal amplitude, then after calculating all phase shifts for each pixel stepwise gives in depth image

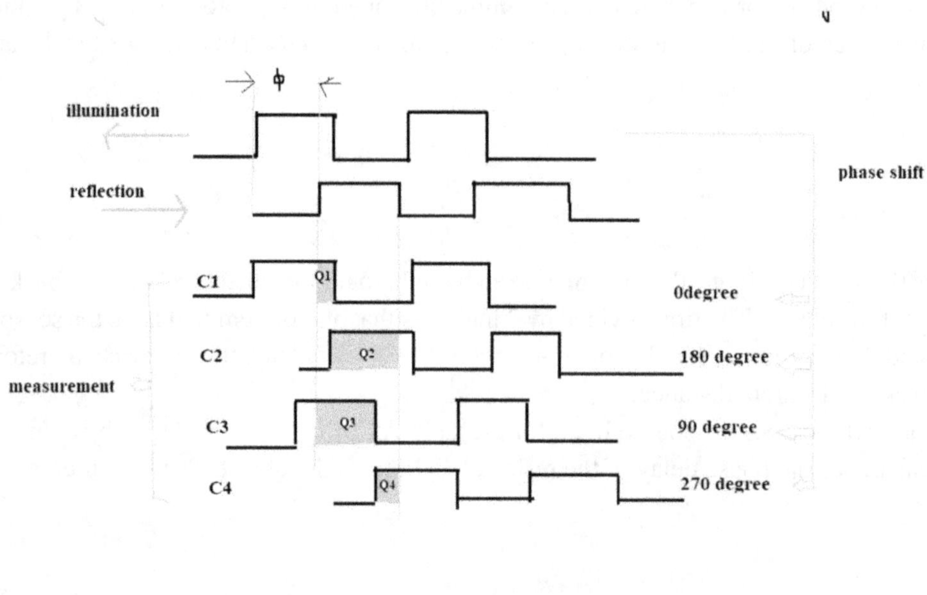

Changing phases for each pixel stepwise will help in realizing the in-depth image of the object, then for different terms of correlation function are found to be, C (t_0), C (t_{90}), C (t_{180}), C (t $_{270}$) as shown in Equations 2, 3, 4 and 5. After combining all correlation terms phase ϕ comes out to be

$$C\,(t_0) = \frac{a}{2}\cos\,(\phi) \tag{2}$$

$$C\,(t_{90}) = \frac{a}{2}\sin\,(\phi) \tag{3}$$

$$C\,(t_{180}) = \frac{-a}{2}\cos\,(\phi) \tag{4}$$

$$C\,(t_{270}) = \frac{-a}{2}\sin\,(\phi) \tag{5}$$

After combining these correlation terms, phase **ϕ** can be found out as shown in Equation 6 and amplitude can be found by using Equation 7:

$$\phi = \tan^{-1}\left[\frac{C(t_{90}) - C(t_{270})}{C(t_0) - C(t_{180})}\right] \tag{6}$$

$$\text{Amplitude} = \frac{\sqrt{(C(t_{90}) - C(t_{270}))^2 + (C(t_0) - C(t_{180}))^2}}{2} \tag{7}$$

With the help of phase and amplitude, the distance can be calculated as shown in Equation 8, where 'f_{mod}' is the modulating frequency of signal 'c' is the speed of light, and '**ϕ**' is the phase delay.

$$\text{Distance} = \frac{C}{2f_{mod}} \times \frac{\phi}{2\P} \tag{8}$$

The example to calculate amplitude and phase in order to find the distance of the object is shown in Figure 3. When the illumination is done over the object in the environment using VCSEL then the reflected light shows phase delay ϕ, this phase is varied in steps of 90-degrees in order to get amplitude of the object as shown in Equation 9, where Q is the quantity of change in phase at different phase shifts from 0 degree to 270 degrees as shown in Figure 3. The amplitude due to change in phase at every step is shown in Equation 9:

$$\text{Amplitude} = \sqrt{(Q_1 - Q_2)^2 + (Q_3 - Q_4)^2} \quad (9)$$

As shown in Figure 3, the total phase delay can be found by using the Equation 10:

$$\phi = \tan^{-1} \frac{Q_3 - Q_4}{Q_1 - Q_2} \quad (10)$$

The modulated signal in time of flight distance measurement is achieved by infrared illumination like VCSEL. The block diagram for 3D sensing application is shown in Figure 4.

The VCSEL based 3D sensor is sensitive to the ambient light like the sun or artificial light at the same time it is sensitive to reflected light coming from the modulated signal. The reflected light is the only light which carries distance information, but high ambient light like sun generates photon shot noise which eventually lowers the signal to noise ratio.

Therefore, to gather the reflected light from an object in the environment, the lens plays an important part. The lens, as shown in block diagram of Figure 4 will determine the field of vision of the sensor. The lens will also calculate the amount of modulated light gathered by the lens in order to create a depth image. Lens aperture of the lens or low focal length of lens leads to better signal to noise ratio. Because of this imaging lens, in the lens field of vision, every pixel coordinates (xi, yi) of the image will correspond to a certain angle. This information is used to create some 3D point images by using depth image. Hence, the time of flight is considered as the best technique to calculate the distance of the object in the environment due to its simplicity and accuracy in ambient light too.

Figure 4. Block diagram of VCSEL based sensor

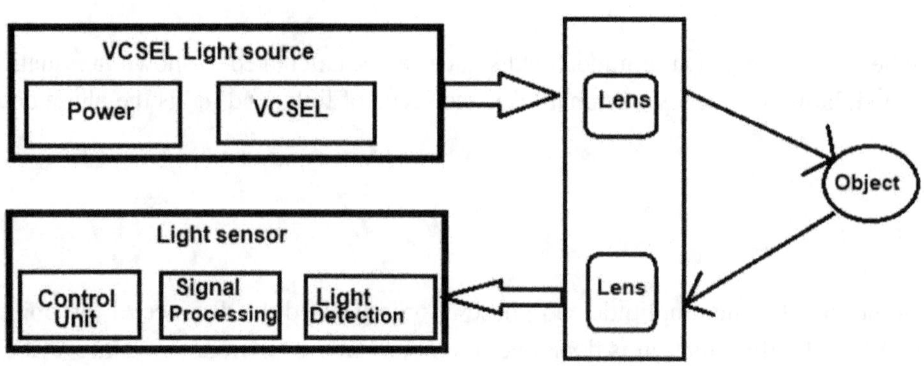

WORKING OF VCSEL TECHNOLOGY BASED ON 3D SENSOR

The block diagram of the VCSEL based 3D sensor is shown in Figure 5. An infrared light using VCSEL as optical source, present in the optical component of VCSEL based 3D sensor is spread in the environment either using time of flight method (a sheet of light) or structured method (structured pattern). The whole system of 3D sensor is capable of capturing the depth information across an entire environment by measuring the reflected light reflected off the objects.

Application-Specific Integrated Circuit (ASIC) is a microchip designed for a specific purpose or application. ASIC interfaced with 3D VCSEL module decides kind of transmission protocol by sending control instructions to the sensor module and receiving data for processing to get the in-depth image. ASIC intelligently empowers the IoT system because of its small size and efficient power consumption. ASIC interfaced with a 3D sensor makes the sensor more reliable and of high performance. Hence ASIC platform with VCSEL technology enhance the 3D sensor and hence enhance the IoT devices given by Chouhan (2018). Since IoT is made up of sensor-based devices, therefore, deploying ASIC platform in IoT sensor device enhances design flexibility and functionality of 3D sensor. This permits consumers to control games with gestures without using any remote control. Gesture recognition using 3D sensor can also boost touchscreen capabilities by controlling touchscreen without touching the screen of the device with just gestures far away from the device.

VCSEL DESIGN

VCSEL is the specialized semiconductor laser whose cavity is formed in the direction perpendicular to the active layers. Hence emits coherent light perpendicularly coming out from the horizontal layers.

Figure 5. The block diagram of VCSEL based 3D sensor

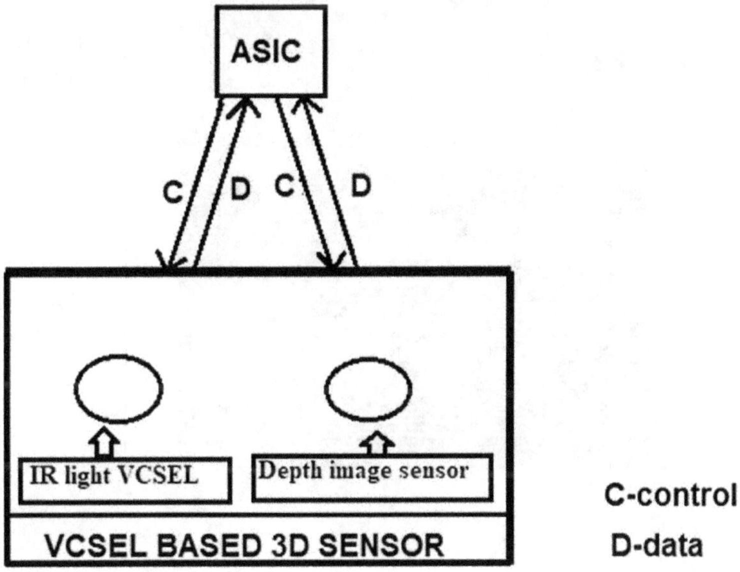

VCSELs are surface emitting because the cavity is vertical, i.e. VCSEL diodes are typically designed where multiple layers of VCSEL are stacked horizontally and the output coming out of the VCSEL emerge vertically which is circular in shape and coherent in property as shown in Figure 6. VCSEL is manufactured by using various semiconductor materials which include Aluminium Gallium Arsenide (AlGaAs), Gallium Arsenide (GaAs), and Indium Gallium Arsenide Phosphide (InGaAsP). VCSELs are constructed in such a way that they emit light in the near-infrared region. The mirrors of VCSEL laser are specially designed multiple layers periodically refractive varying Bragg's stack. The top and bottom mirrors of the VCSEL are formed by depositing Bragg stacks, i.e. in order to grow vertical cavity, deposit the Bragg stacks on the surface for reflection then deposit active medium and again Bragg stack for reflection.

In 1962, the first lasing action through Edge Emitting Laser (EEL) has been achieved. Then various improvements and research is carried out in order to enhance the performance of EEL. Unlike VCSEL, EEL is in-planed cavity laser in which light propagates parallel to the wafer surface and the mirror of EEL is formed by cleaving the ends of the semiconductor. But the VCSEL has superiority over EEL because it is cheap, easy to test and much more efficient. VCSEL is the topmost optical source because the optical beam of a VCSEL is coherent in nature, circular in shape and shows less optical divergence therefore, chosen over other optical sources like LED, EEL etc.

In 1977, a different structure of the laser which has a vertical cavity perpendicular to the active layers of the laser has been proposed. This vertical cavity leads to emit light from the surface and hence optical output is perpendicular to the surface. A semiconductor VCSELs core parts are lower n-type Bragg

Figure 6. VCSEL simulation diagram depicting major parts of VCSEL responsible for lasing

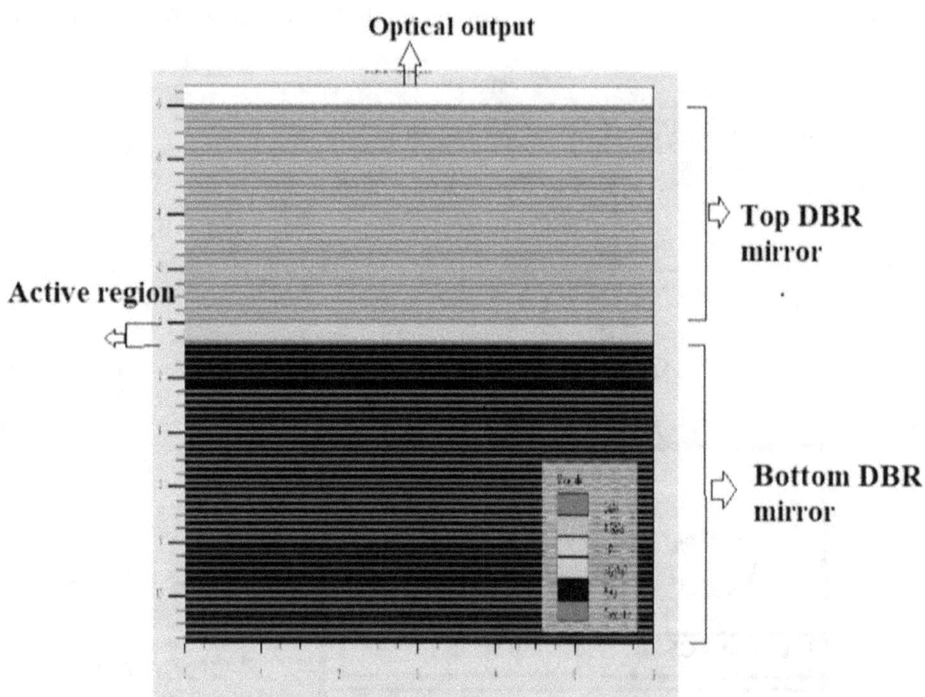

reflector (bottom mirror), an active layer made of quantum wells and upper p-type Bragg reflector (top mirror) as shown in Figure 6. The active layer is sandwiched between the top and bottom Distributed Bragg Reflector (DBR) which are highly reflective. DBRs are thick layers of an alternating high and low reflective index. The reflectivity of these DBR mirrors is in between 99.5% to 99.9%.

VCSEL works on the basic principle that is the creation of light in the cavity and amplification of light by stimulated emission. Stimulated emission engages coherent light emission with a goal of positive optical net gain and large optical power. Philip Moser et al., (2015) shows that the basis of VCSEL is the creation of light and then the amplification of light with the help of stimulated emission

In stimulated emission, as shown in Figure 7 a photon of energy E12, energize an electron placed at higher energy level E2, which reaches to a lower energy level E1 by emitting a photon, with the help of Equation 11, it can be explained as:

$$E12=E2-E1=h\upsilon \tag{11}$$

The emitted photon basically has the same direction, same polarization as well as same wavelength which make initial (incident) and final (emitted) photon coherent.

The Rate of stimulated emission R_{st} is given by the Philip Moser (2015) in Equation 12:

$$R_{st}=P_{st}\cdot N_1\cdot \phi \tag{12}$$

where P_{st} is proportionality factor of stimulated emission, N_1 is electron concentration at energy level E_1 and ϕ is the light intensity of the laser. In semiconductor laser like VCSEL stimulated emission is used to create emission of a coherent photon and for efficient light amplification, stimulated emission has to be very large.

An electron can also reach to higher energy E_2 by absorbing a photon of energy E_{12}, where rate of absorption R_{abs} is given by the Equation 13:

$$R_{abs}=P_{abs}\times N_2\times \phi \tag{13}$$

where P_{abs} is absorption proportional factor also known as Einstein coefficient. N_2 is the concentration of electrons at energy level E_2 and ϕ is light intensity of laser diode. Since absorption eradicate photons and the phenomena of stimulated emission create photon therefore to maintain the balance rate of absorption is equal to the rate of stimulated emission which is given by Equation 14:

Figure 7. Stimulated emission of photon by electron of energy E12

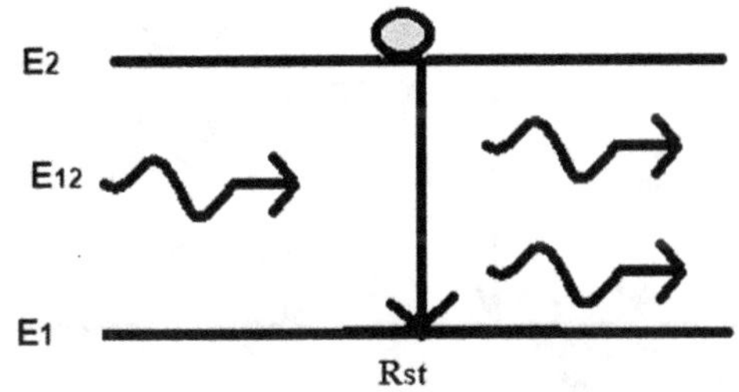

$$R_{abs}=R_{st} \tag{14}$$

Therefore, for all temperatures, absorption proportionality constant is equal to proportionality factor of stimulated emission which can be expressed as P in Equation 15. Hence:

$$P_{abs}=P_{st}=P \tag{15}$$

Net stimulated emission rate in semiconductor laser diode VCSEL is the difference between stimulated emission and absorption rate. Therefore:

$$R_{st}-R_{abs}= (N_1-N_2)\times P \tag{16}$$

The condition for net light amplification can be achieved if the rate of stimulated emission is greater than the rate of absorption i.e. $R_{st}>R_{abs}$ and $N_2>N_1$ which means concentration of electrons at higher energy level is more than the concentration at the low energy level.

This condition of greater stimulated emission rate and high concentration at high energy level is called population inversion as shown in Figure 8 which is the important aspect for lasing in Figure 9.

Population inversion in VCSEL is achieved by forward biasing p-n junction formed by a top DBR mirror (p junction) and bottom DBR mirror (n junction). In order to enhance the rate of stimulated emis-

Figure 8. Phenomena of population inversion

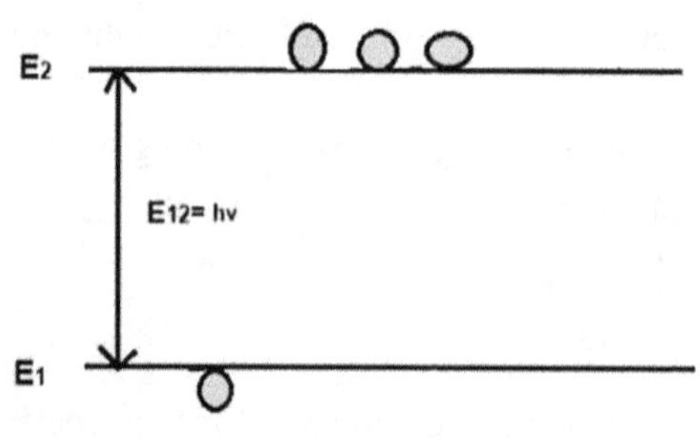

Figure 9. Lasing action after population inversion

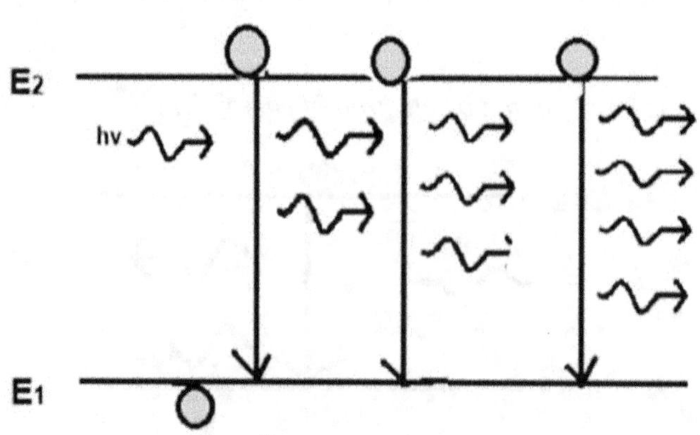

sion, DBR mirrors are used which confine the emitted photons in a VCSEL cavity in such a way that the confined photons pass multiple times through active region before leaving VCSEL lasers. Higher the power reflectivity of the DBR mirror, higher will be the light intensity.

Because of VCSEL great light amplification principle via stimulated emission, they are adopted in various IoT applications for sensing purposes among which VCSEL based 3D sensors is in great demand. Hence integration of VCSEL with sensing module makes the sensors efficient and more reliable for 3D sensing applications.

IMPROVED VCSEL DESIGN

The traditional VCSEL device structure has top mirror composed of DBR, the bottom mirror also composed of DBR and the active region of multiple quantum wells as can be seen in Figure 6. According to Fumio Koyama (2009), VCSEL has more additional components, for example; cantilever which is thermally actuated and also elastic in nature with strain control layer. Both the additional parts make the VCSEL unique in structure and in the application too (athermal). In this unique structure of VCSEL, the top DBR mirror is supported at one end only which is also known as cantilever mirror. This mirror is composed of DBR and Strain Control Layer (SCL) as shown in Figure 10. Hence, when there is thermal expansion of ΔT in-one end supported mirror, this top cantilever mirror gets deflected downwards to compensate the shift in wavelength due to change in temperature. The sacrificial layer over here is providing the air gap so that the top cantilever mirror gets deflected without touching the bottom mirror. The parameters and values of VCSEL are shown in Table 3.

In VCSEL top DBR mirror is composed of AlGaAs semiconductor material. The thermal expansion coefficient begins to be smaller as the content of Al is increased. But the use of strain control layer instead of regular DBR layer further has a smaller thermal expansion coefficient of Al content as its composition is increased. Hence, whenever the temperature is increased further, the cantilever structure deflects downwards hence compensating the shift in wavelength.

A unique design structure of VCSEL, which is a micro-electro-mechanical system based VCSEL i.e. MEMS VCSEL has been developed by Fumio Koyama & Masanori Nakahama (2015) for tuning the wavelength of the VCSEL. But along with MEMS VCSEL structure thermally actuated cantilever structure has been proposed in order to realize athermal VCSEL. This VCSEL works on the fixed wavelength and is self-compensated whenever temperature changes occur. Though the MEMS system in VCSEL is used to tune the VCSEL but the cantilever structure has been used to make the VCSEL thermally inde-

Figure 10. Thermally actuated cantilever structure of VCSEL, which compensates at thermal expansion of ΔT

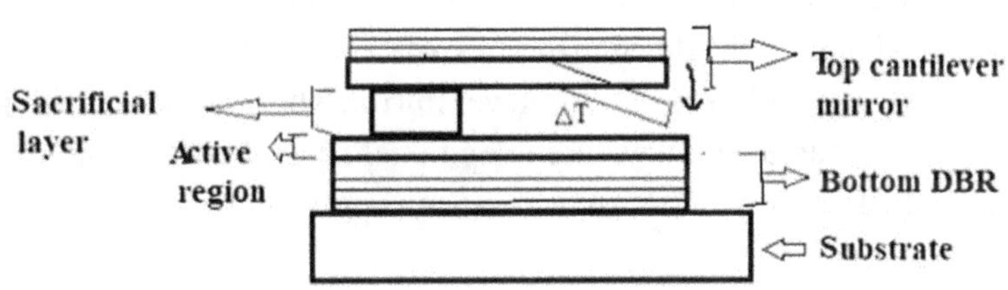

Table 3. Parameters and values of VCSEL

Parameter	Value
Emission wavelength	1.54 μm
Top mirror material	GaAs, AlGaAs
Number of top layers	56
Laser cavity material	InGaAsP, InGaAsP
Number of laser cavity layers	6
Bottom mirror material	GaAs, AlGaAs
Number of bottom layers	60
Reflectivity of top mirror	0.866

pendent that is athermal VCSEL. The athermal VCSEL has been realized at the wavelength of 1550 nm and 850 nm. The VCSEL has the advantage of the vertical cavity which has an enormous free spectral range. The structure and material used are as follows. Bottom mirror of the VCSEL is made by 40 pairs of AlGaAs/InP based DBR. After that the active region is formed by quantum wells of AlGaInAs. The tunnel junction of InP has been formed on the top of quantum wells. Then a sacrificial layer is formed above an active region only by growing GaInAs layer. Then after that, the additional part of a VCSEL is its cantilever structure in which 5.5 pairs of GaInAsP/InP top DBR (top mirror) are added to it. Also, Si/SiO2 dielectric mirror is grown on the top of the cantilever mirror. Then the Strain Control Layer (SCL) is added at the top and bottom of the cantilever mirror where a $\lambda/2$ thickness of material InP thermal SCL is added on the top of the cantilever and λ thickness of InPisadded at the bottom of the cantilever. These InP layers on top and bottom are added to realize the asymmetric structure of the cantilever. This asymmetric structure could have been realized by using GaInAsP but due to its large thermal coefficient, InP is used which has less thermal expansion coefficient as compared to GaInAsP.

The sacrificial layer added in the cantilever structure of VCSEL is responsible for the air gap between the top mirror and the active layer in order to notice athermal operation. When the wavelength is proportional to air gap thickness, the change in the thickness of the air gap is x, λ is the wavelength shift, then the change in wavelength along with temperature variation 'T' can be expressed by Equation 17:

$$\frac{\Delta\lambda}{\Delta T} = \frac{1}{n}\left(\frac{\Delta n}{\Delta T}\right)\lambda + \frac{1}{L}\left(\frac{\Delta x}{\Delta T}\right)\lambda \tag{17}$$

where η is the refractive index and L is effective cavity length. From Equation 17 it is clear that VCSELs temperature coefficient depends on the refractive index of the semiconductor material. The equation also shows that the temperature coefficient depends on the length of the cantilever structure. The wavelength shift can be controlled by changing the design structure of cantilever, if there is any shift in wavelength of the VCSEL due to thermal actuation of the cantilever mirror. The thermal actuation is denoted by 'T_A', cantilever length by 'CL' and radius of curvature in epitaxial layers by 'R'. At the end of the cantilever structure of VCSEL, the total amount of thermal actuation is given by:

$$\Delta T_A = \frac{CL^2}{2R} \tag{18}$$

From the equation, it can be stated that the thermal actuation is directly proportional to the square of the cantilever length and inversely proportional to the radius of curvature in epitaxial layers. Also, thermal actuation increases linearly with the linear increase in temperature. The thermal actuation also depends on other structural parameters like cantilever length CL, the thermal expansion coefficient of each layer, etc.

Various characteristics like current vs voltage or current vs light output has been observed. At continuous wave operation of VCSEL under room temperature the threshold current comes out to be 1.3 mA and maximum power 0.3 mW. The VCSEL structure was composed of a tunnel junction with a radius of 5 µm and the length of cantilever mirror was 75 µm. In order to avoid self-heating of VCSEL structure, it has been operated at the constant current of 4 mA. By counting the amount of thermal actuation of the cantilever structure using Equation 18, the calculation of dependence of resonating wavelength of VCSEL at different values of cantilever length is given in Koyama & Nakahama, (2015). Results stated that at different temperature changes in the cantilever mirror VCSEL had a wavelength shift of only 0.03 nm which is acceptable. Especially with the temperature change of 19 K wavelength of 0.03 nm was affected at the cantilever length of 95 µm. With different cantilever lengths, at a particular length (95 µm) there was a thermal shift of only 0.0016 nm/K i.e. when the cantilever was taken to be of 95 µm length then the temperature dependence of the VCSEL is very less which is 0.0016 nm.

The heat dissipation inside VCSEL structure is due to Joule heating, internal optical absorption and the recombination of electron-hole to generate photons inside the VCSEL cavity. Hence all the above factors when merged together expand the internal temperature of the VCSEL device as shown in Figure 11 and Figure 12. When this expansion of temperature inside the VCSEL device becomes out of

Figure 11. Temperature vs anode voltage graph, as the anode voltage is increased further the temperature increases rapidly

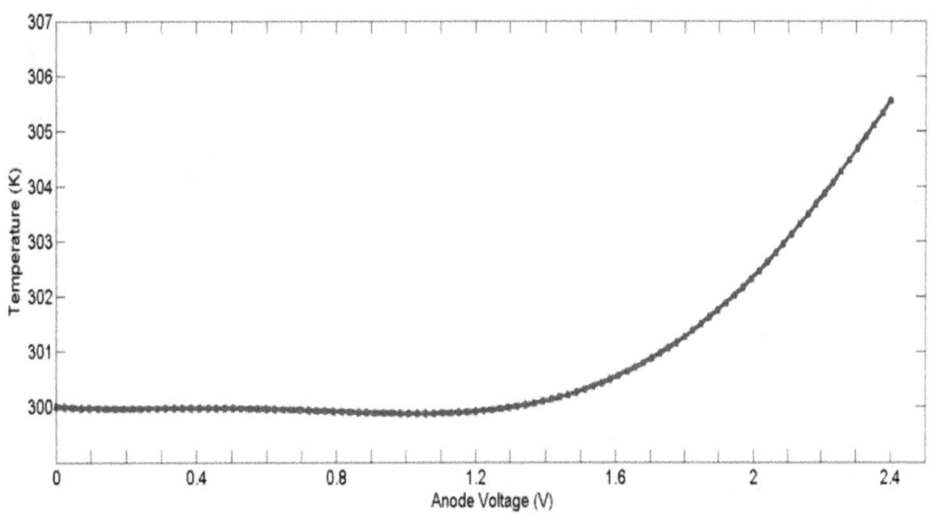

Figure 12. Temperature vs anode voltage graph gives increment in the temperature of the VCSEL device due to leakage of carriers in the active region

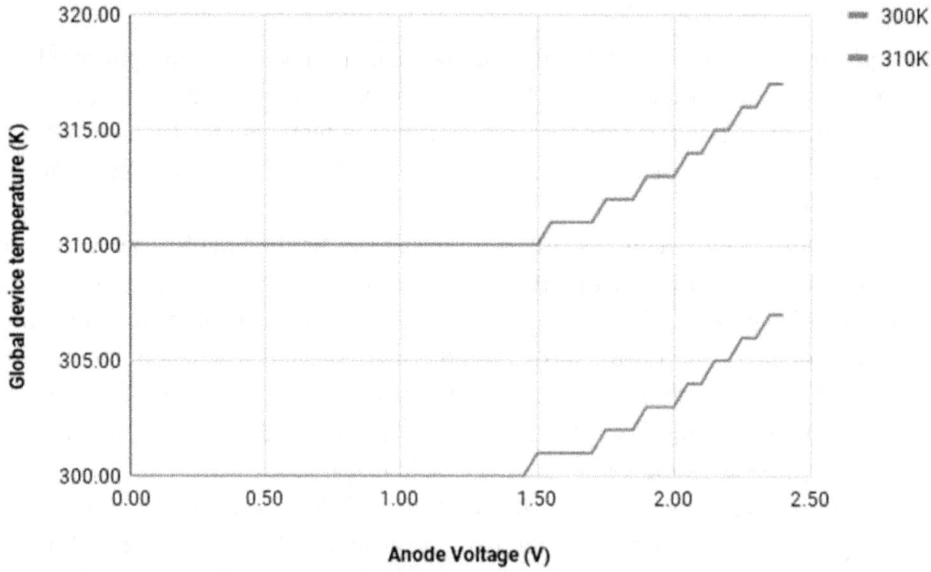

**For a more accurate representation see the electronic version.*

control, this very high temperature leads to decrease in the quantum efficiency of VCSEL laser. That is photons which are responsible for the very important phenomena of lasing called population inversion are reduced in number.

In Figure 13, as the thermal expansion takes place from 290 K to 310 K within the laser, the photon density reduces per cm^{-3}. The numbers of pump photons due to which the lower level photons go to higher

Figure 13. Photon density of VCSEL at different temperatures

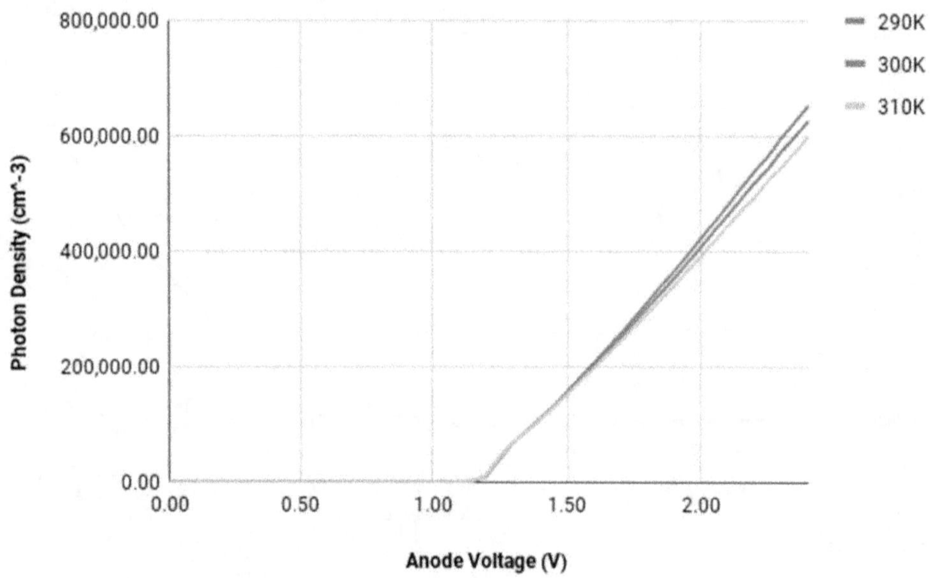

**For a more accurate representation see the electronic version.*

level are decreased. This reduction of pump photons occurs due to leakage of carriers in the quantum wells in the active region, which results in an increment in the temperature of the VCSEL device. The decrease in lasing action (gain) and quantum efficiency inside the cavity of the VCSEL laser results in saturation of photon density as well as output power is shown in Figure 14.

As shown in Figure 15, the total power emitted is reduced as the temperature of the laser is increased further from 290 K to 310 K whereas the gain and mirror loss shows incomparable deflection at different

Figure 14. Lasing action (gain) in VCSEL laser reaching to saturation point at different temperatures

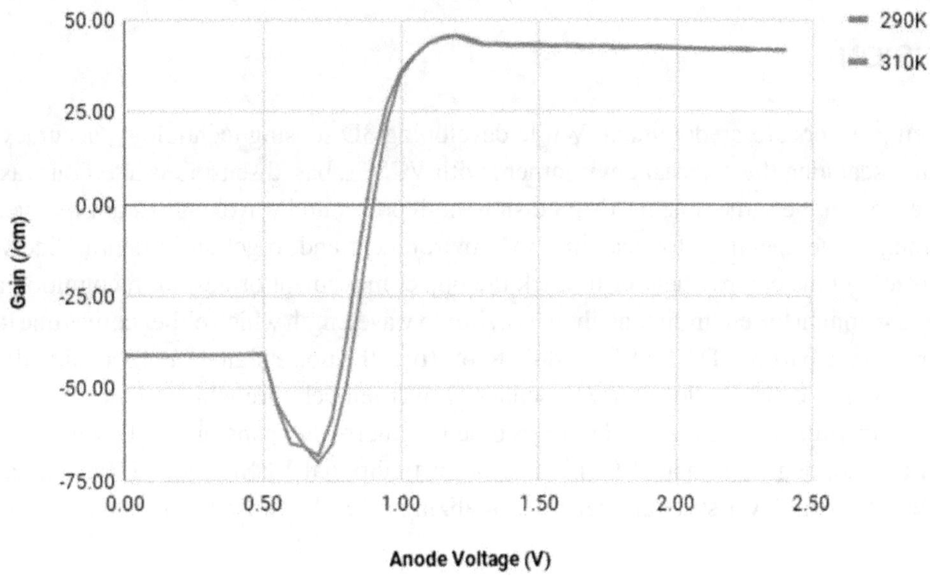

**For a more accurate representation see the electronic version.*

Figure15. Total power emitted vs anode voltage as the temperature increased from 290 K to 310 K

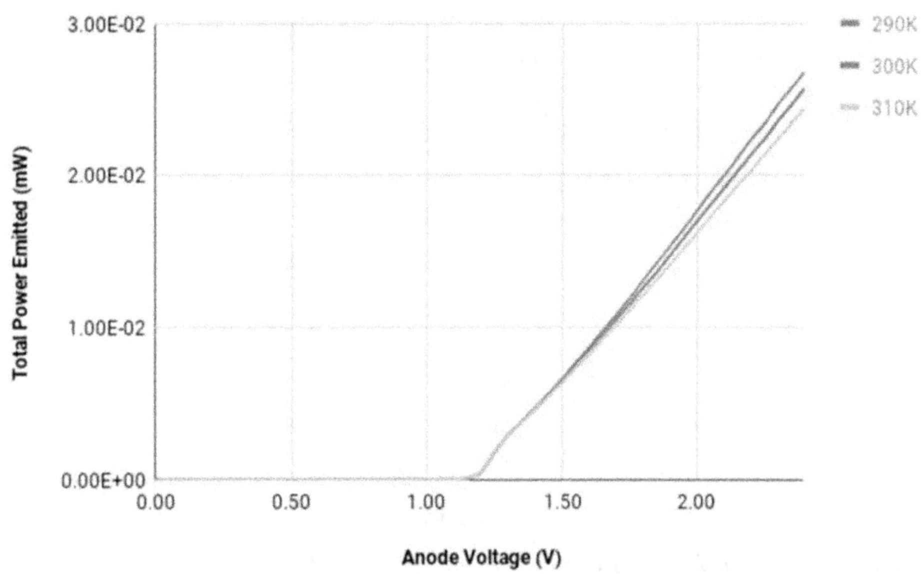

**For a more accurate representation see the electronic version.*

temperatures as in Figure 12 and Figure 16. VCSEL works efficiently only up to 85-degree centigrade temperature because VCSEL starts heating at very high temperature.

The typical value of dependence of single-mode semiconductor lasers like VCSEL over temperature is 0.1 nm/K. This dependency of VCSEL over temperature leads to thermal shift that is whenever there is a change in temperature there will be plus-minus change in wavelength. If this temperature dependence is reduced further even up to few decimals, the efficiency of the VCSEL can be improved. This reduction in temperature dependence will further help in realizing low power consumption and even smaller packaging of the VCSEL.

CONCLUSION

In IoT system, sensors are predominant. While developing 3D sensing technology, accuracy is the key point because scanning the external environment with VCSEL based sensors is a tedious task. VCSEL based 3D sensors at the same time needs precision, high power and wavelength stability over the entire operating range of temperature for scanning the environment and to get an in-depth wide view of everything present in the environment or to track the object movement or gesture recognition at 5 feet or above. While scanning the environment, there is a shift in wavelength which either comes due to the filters applied at receiving part of 3D VCSEL module to obstruct the noise signal (ambient signal) or most of the time due to the VCSEL performance dependency over temperature which hinders the performance of 3D sensing module. Hence the need to overcome this thermal expansion in 3D sensing module can be realized by adopting 'athermal' VCSEL instead of traditional VCSEL for 3D sensing application. Athermal VCSEL cantilever structure helped in realizing VCSEL whose thermal shift comes out to be

Figure 16. Mirror loss vs anode voltage graph at 290 K, 300 K and 310 K, shows the loss is negative but becomes positive after cavity formation in VCSEL at 1V

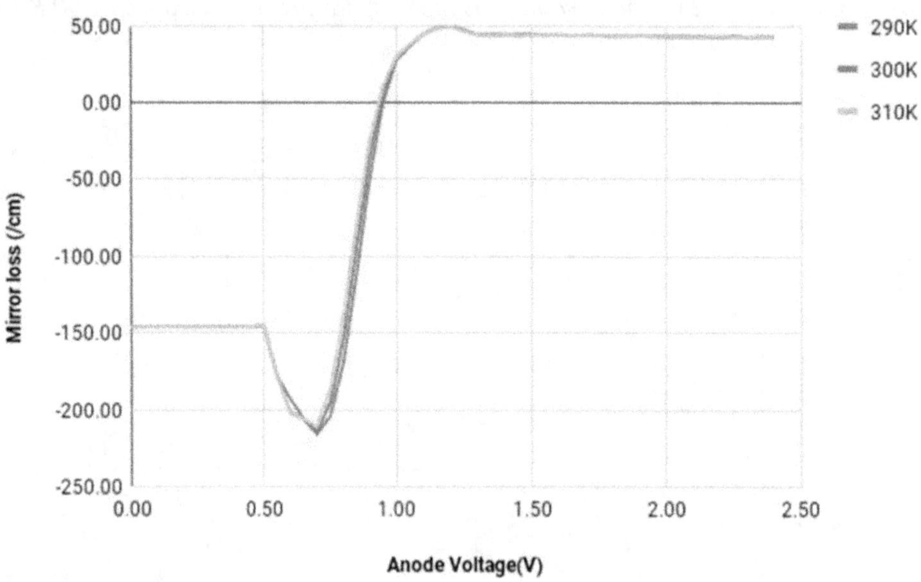

*For a more accurate representation see the electronic version.

very low and the traditional VCSEL show wavelength shift of plus-minus 5-10 nm. The total power emitted starts increasing at 1.2 V and with the increase in temperature it starts decreasing. The mirror loss remains same with the increase or decrease in temperature. Thermal VCSEL shows temperature dependence and global device temperature is expanding as the temperature increased from 290 K to 310 K.

Hence, more efficient VCSEL with least temperature dependency will reduce the problem of thermal expansion problem up to much extent. Therefore, more efficient, more effective, more productive VCSEL based 3D sensors can be realized not only for scanning the environment in a few applications, but also in more wider applications of Internet of Things in order to meet the concept of smart home, smart cities in a more upgraded way in future.

REFERENCES

Albert. (2018). *IOT Devices Proximity Sensor.* Retrieved from https://www.infiniteinformation.com

Bajaj, R., Mishra, H. K., & Goyal, P. (2017) Design of Oxide Confined and Temperature Stable Long Wavelength Vertical Cavity Surface Emitting Laser for Optical Interconnects. *Optik, 131,* 506-514.

Boucart, J. (2018). *Laser Diodes for (3D) Sensing, Swiss Photonics Meeting, Industrial 3D Vision.* Retrieved from https://www.swissphotonics.net

Ebeling, K. J., & Michalzik, R. (2017). VCSEL technology for imaging and sensor systems applications. *Proceedings of the 22nd Microoptics Conference (MOC2017),* 19 – 22. 10.23919/MOC.2017.8244477

Ebeling, K. J., Michalzik, R., & Moench, H. (2018). Vertical-cavity surface-emitting laser technology applications with focus on sensors and three-dimensional imaging. *Journal of Applied Physics, 57*(8S2), 08PA02. doi:10.7567/JJAP.57.08PA02

Evangeline, H. (2018, Jan. 4). *3D Sensing: Present and Future.* Retrieved from https://www.ledinside.com/news/2018/1/3d_sensing_present_and_future

Moench, H., Frey, M., & Grabherr, M. (2017, February). VCSELs as light source for time-of-flight sensors. In Vertical-Cavity Surface-Emitting Lasers. International Society for Optics and Photonics.

Gabbai, A. (2015, Jan. 2). Kevin Ashton describes 'the Internet of Things': The innovator weighs in on what human life will be like a century from now. *Smithsonian Magazine,* 8-15. Retrieved from https://www.smithsonianmag.com/innovation/kevin-ashton-describes-the-internet-of-things-180953749/

Gesture recognition and 3D sensing. (2019). Retrieved from https://www.jabil.com/capabilities/3d-sensing-gesture-recognition.html

Haglund, E. (2015). VCSELs for high-speed, long-reach, and wavelength-multiplexed optical interconnects. Gothenburg, Sweden: Chalmers University of Technology.

Komal, C. (2018). *4 Reasons to Use Custom ASICs in the IoT Era.* Retrieved from https://www.dzone.com

Koyama, F. (2009). VCSEL Photonics-advances and new challenges. *IEICE Electronics Express, 6*(11), 651–672. doi:10.1587/elex.6.651

Koyama, F., & Nakahama, M. (2015). Micromachined Vertical Cavity Surface Emitting Lasers athermalization, tuning, and multi wavelength Integration. IEEE Journal of Selected Topics in Quant. Electron., 21(4).

Li, H., Wolf, P., & Moser, P. (2014). Impact of the quantum well gain-to-cavity etalon wavelength offset on the high temperature performance of high bit rate 980-nm VCSEL. *IEEE Journal of Selected Topics in Quantum Electronics, 50*(8), 613-612. Retrieved from ieeexplore.ieee.org/xpl/RecentIssue.jsp?punumber=2944

Li, H., Wolf, P., & Moser, P. (2015). Temperature-Stable, Energy-Efficient, and High-Bit rate Oxide-Confined 980-nm VCSELs for Optical interconnects. *IEEE Journal of Selected Topics in Quantum Electronics, 21*(6). Retrieved from ieeexplore.ieee.org/xpl/RecentIssue.jsp?punumber=2944

Li, H., Wolf, P., & Moser, P. (2017). Large bandwidth, Small Current Density, and Temperature stable 980-nm VCSELs. *IEEE Journal of Selected Topics in Quantum Electronics, 53*(6), 2400908. Retrieved from ieeexplore.ieee.org/xpl/RecentIssue.jsp?punumber=2944

Marciniak, M., Piskorski, L., & Gebski, M. (2018). The Vertical-Cavity Surface-Emitting Laser as a Sensing Device. *Journal of Lightwave Technology, 36*(16), 3185–3192. doi:10.1109/JLT.2018.2834620

Melexis. (2017). *Time of Flight Basics.* Retrieved from https://www.melexis.com/

Moench, H., Carpaij, M., & Gerlach, P. (2016, March). VCSEL-based sensors for distance and velocity. In *Vertical-Cavity Surface-Emitting Lasers*. International Society for Optics and Photonics.

Moser, P. (2015). *Energy-Efficient Oxide-Confined VCSELs for Optical Interconnects in Data centers and Supercomputers*. Berlin, Germany: Academic Press.

Moser, P., Lott, J.A., & Bimberg, D. (2013). Energy efficiency of directly modulated oxide-confined high bit rate 850-nm VCSELs for optical interconnects. *IEEE Journal of Selected Topics in Quantum Electronics, 19*(4). Retrieved from ieeexplore.ieee.org/xpl/RecentIssue.jsp?punumber=2944Journal

Powell, J. K. (2016). *Sensors to make your home smarter and safer.* Retrieved from https://www.forbes.com/sites/jenniferhicks/2016/11/20/3d-sensors-to-make-your-home-smarter-and-safer/#2

Reportbuyer. (2018). *The VCSEL market is expected to reach USD 3.89 billion by 2023 from USD 1.78 billion in 2018, at a CAGR of 16.9% from 2018 to 2023.* Retrieved from https://www.prnewswire.com

Tatum, J. (2017). *VCSEL in Various Sensor Applications.* Retrieved from https://www.finisar.com

Van Hoe, B., & Bosman, E., & Missinne, J. (2012). Photonic Incremental Pressure Sensor Based on Optical Feedback in a Polymer Embedded VCSEL. IEEE Photonics Technology Letters, 24(13).

Chapter 4
Environmental Analysis of Construction Materials:
Material Specifications for Green Built Environment

Aysem Berrin Cakmakli
Middle East Technical University, Turkey

ABSTRACT

There is a growing universal awareness of protecting the living and non-living environment and making enlightened decisions to achieve a sustainable development without destruction of the natural resources. In this point of view, selecting building materials according to their energy and health performances gains importance in sustainable design. 3Rs (reducing, reusing, recycling), and supplying a healthy, non-hazardous indoor air for building occupants are two important parameters of environmental life-cycle assessment for materials. Information on exposure to gases and vapors from synthetic materials made from petrochemicals, to heavy metals and pesticides, and to some combustion pollutants that cause acid rain should be determined by analyzing environmental product declarations or material specifications. After studying on building materials individually, they are analyzed in the form of tables for four different stages; manufacturing, application, usage, demolition phase. Consequently, this chapter can guide the designer and engineer to think on the elements of design and construction activity.

INTRODUCTION

Buildings are considerable exporters of pollution. Polluted wastewater, indoor air, smoke and vehicle exhaust gases, rubbish, garden pesticides and nitrate fertilizers are all emitted from the buildings and gardens. (Quadrennial Technology Review, 2015) Sustainable design methods offer the architects the opportunity to integrate design skills that overcome damaging consequences. The effects of climate on the energy systems of buildings; collection and storage of heat, use of natural light, control of air movement and generation of power; should also be expected. (WBDG Sustainable Committee, 2011) Use of locally produced materials will combine with regional responses to climate and produce regionally

DOI: 10.4018/978-1-5225-9754-4.ch004

based architecture. The determined policies of international resource use and regulations are necessary or sufficient to achieve global sustainability of economic growth (Adeyeye et al., 2007). The fundamental strategy is using less amount of natural resources and constructing less via renewable materials and renewable sources; thereby reducing indoor and outdoor pollution, waste production and embodied energy. Recycling is another approach to waste reduction that has been strongly advocated by many environmentalists. It is a "cradle to cradle" analysis of all building decisions considering not only the extracting phase, but also construction, usage and disposal phase. (Saghafi et al., 2011) The structural complexity made recycling more difficult in many cases like recycling of textiles because of the prevalence of blends of natural and synthetic fibers. (Adeyeye et al., 2007)

The major concepts that guide the selection of materials and components include energy efficiency, sustainability, transferability, cost, aesthetics, availability and mostly human and environmental health. The technological developments create modern constructions and surfaces. (DPTI, 2017) These constructions and surfaces make human life easy in most areas, but they create many problems about environmental and human health. While choosing surface materials, besides the appropriateness of the design, energy, health and economy criteria should be considered carefully. Health and pollution should be of growing concern. (Ragheba et al, 2016) As buildings becomes more complex, liability and responsibility upon the shoulders of the designers and to some extent the constructors of buildings also increase. Risks should be identified at the design stage and appropriate action should be taken to minimize the environmental damage.

BACKGROUND

Life-cycle assessment is defined as "*a system for analyzing environmental aspects of products or processes through their life cycle, from raw material extraction to recycling, which includes manufacturing, use, and end-of-life (EOL) disposal*" (ISO 14040: 2006). While assessing the impacts, first environmentally relevant inputs and outputs are compiled, and then evaluated in order to interpret the results in four interrelated phases; "goal and scope definition phase, inventory analysis phase, impact assessment phase and interpretation phase" as highlighted in Figure 1. (ISO 14040: 2006) (ISO 14044: 2006)

Figure 1. LCA methodology and relationships (ISO 14040: 2006)

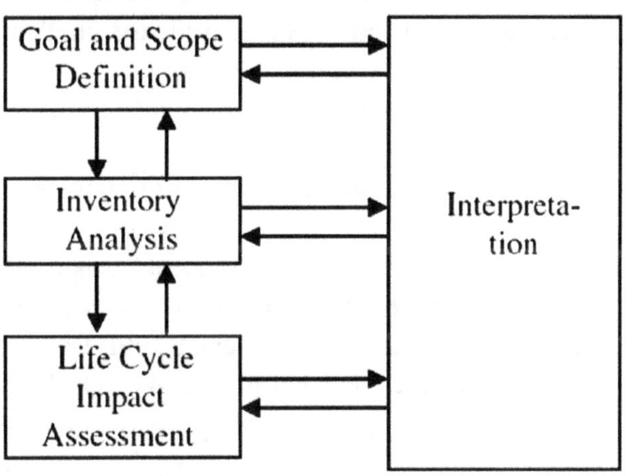

The parameters that are seen in the goal and scope definition phase are the allocation of the functional component and system boundaries. Then, the information of physical material and energy flows throughout its entire life cycle is organized in the second phase. After that, in the third phase, its environmental impacts can be established according to different impact classifications. These classifications are "*climate change, ozone depletion, Eco toxicity, human toxicity, photochemical ozone formation, acidification, eutrophication, resource depletion, and land use*". At the end, the interpretation of results of all phases is derived. (ISO 14044: 2006)

The concept of Reference Service Life of Component (RSLC) concept was defined in ISO 15686-1: 2000 as the "*service life that a building or parts of a building would be expected or predicted to have in a certain set of reference in-use conditions*". While predicting the service life, there are different methodologies to describe the life of materials and components depending on their project-specific and the reference conditions that can be used. These are deterministic methods which contain factor method and graphical method, probabilistic/stochastic methods include Markov chains and engineering methods. (Garrido et al., 2012) (Marteinsson, 2003)

ISO 15686-1: 2011 also determined three kinds of end of life which were "*technical, economical, and functional end of life*". RSLC as the technical service life is ended when the component cannot perform its requirements. On the other hand, if the assembly is exchanged with lesser costs, its economic lifetime is over. And the last one, the functional end of life is occurring when the material does not compensate the claim of people.

Indoor Air Quality (IAQ) discusses the air quality for the sake of occupants' health and comfort within and around buildings. Identifying IAQ problems and their solutions become harder because of hundreds of different contaminants. (US EPA, 2017) Construction materials, furnishings, and equipment can emit individual Volatile Organic Compounds (VOCs) that form new chemicals due to the adequate heat and moisture. These chemicals support the growth of molds and bacteria containing fungal spores, mycotoxins and allergens that can have harmful effects on occupants. On the other hand, it is clear that relatively short exposure time, even to quite weak electromagnetic fields (EMFs) cause physical stress, indicated by the release of stress hormones such as cortisone that leads to headaches, allergies, depression, relationship problems, and even suicide. (NIOSH, 2017)

An Environmental Product Declaration (EDP) is "*a strategy to present the environmental performance of a product or system for decreasing the environmental impacts*". (Sun et al., 2003) The EPD's voluntarily developed due to LCA system, comprehend transparent and comparable information connected to the acquirement of raw materials, energy use, chemical ingredients and any kind of emissions. It is not an evaluation system, but also a verified and registered well-structured numerical data. (Erlandsson and Borg, 2003)

According to ISO 14025:2006, EPD contains:

1. **Life Cycle Impact (LCI):** The energy intensities of the item, the environmental impacts of its life-cycle phases
2. **Material Sources:** Its usage of raw sources and environmental concerns of its manufacturing phase
3. **Environmentally Preferable Ingredients:** The recycling and renewable content of the item, its harmful VOCs and chemicals content and environmental certification
4. **Manufacturing Process:** The land use and labor practices. (ISO 14025:2006)

DETERMINING THE KEY BUILDING MATERIALS

Determining the building materials and preparing guidelines according to characteristics of them play a big role in sustainable and efficient design. To clarify the embodied, including extraction and production, and use impacts of building materials, depending on the habits of the occupants is necessary for energy efficiency measures in various combinations.

Concrete

As structural material concrete's properties of high and calculable strength, foldability and transportability make it a unique product. (Vanderley, 2003) Concrete use can be divided into four separate phases. Mining and production of cement and aggregate, and their delivery to batching plants is determined as phase one. Phase two is an arrangement of forms and reinforcing, phase three is mixing, delivery and placement of fresh concrete, and fourth phase is demolition and disposal or recycling. (John, 1993)

Besides its high durability, exposed concrete must have internal air voids removed (usually by chemical entraining during batching) in northern climates to withstand freeze-thaw surface deterioration, thermal cracking. A brittle material, concrete must be liberally reinforced with steel to withstand seismic forces. Concrete exposed to acids or road salts must be sealed or coated to prevent corrosion of the reinforcing steel. (Naik, 2008)

Cement and Aggregate

Cement is produced by blending materials like limestone and shale and burning the mixture in a kiln at high temperature. Aggregate is obtained from rubble stone mines by machines and then grouped according to size again. (Malhotra, 2004) To reduce the environmental impact the cement industry prefers industrial by-products as sources of raw materials, including aluminum ore refuse, blast-furnace slag, furnace flue dust, cement dust and fly ash. (Naik, 2008) Used tires, used motor fuels and other spent solvents can be utilized as fuel in the kilns that have to be controlled by their smoke emissions to prevent atmospheric pollution (Vanderley, 2003). Calcium chloride is used as an accelerator, gypsum as a retarder, hydroxycarbolic acid as a water-reducer and plasticizer, sulphated melamine formaldehyde (SMF) as a super-plasticizer. Mineral admixtures, called pozzuolana, is a fine ground mineral material and it gets in reaction with calcium hydroxide to form compounds with cement-like properties (Coppola et al., 2004).

Precast Concrete

Depending on the development of the technology and increasing cost and scarcity of good quality timber, wooden concrete forms have been relocated with fiberglass and steel panels which can be used for long years. These are used with steel scaffolding and lightweight aluminum purlins. Although the transportation of heavy precast construction elements requires a remarkable amount of energy, forming, batching and casting at a central location provide a considerable efficiency (Naik, 2008). To crush concrete into small size pieces to utilize in aggregate form is an energy-intensive process and can recycle demolished concrete. Reinforcing steel can be extracted with magnets and reused. Recycled aggregate is not used for new concrete because of the uncertainty of the quality of the demolished material, but mostly used as bedding under asphalt on road construction. Concrete compares well with steel and brick in the energy required for manufacture, transportation and installation. However, throwaway wooden formworks and the removal of demolished concrete create an environmental problem. (Vanderley, 2003)

Autoclaved Cellular Concrete (ACC)

ACC is a lightweight, thermally effective type of concrete made by mixing silica-rich material - fly ash or sand - with lime, cement, aluminum powder and water. The mix is cast in a mold, cut into the desired shape and cured in an autoclave for 10-12 hours. It reduces the construction time by 20%. (Kamal, 2016) The result is a concrete material that weighs as much as 1/5 less than conventional concrete, with insulation values of more than twice that of regular concrete. ACC can be used in reinforced panels as a complete building system providing a flexible, lightweight, thermally resistant system, which helps to decrease nearly 30% of environmental waste and 50% of greenhouse gas emissions as opposed to going with traditional concrete (Kamal, 2016; Downton, 2013).

Stone

This is one of the oldest construction materials. Buildings which are made of stone are highly attractive with thick walls and aged surfaces. Natural stone in the building is highly sustainable due to its low environmental impact, sociological benefits, and low price on a whole-life basis. Despite the conventional quarrying which have many destructive and disastrous environmental problems stone is an enduring material which can be reused for centuries. (SUS01/11, 2011)

Large-scale quarrying causes not only irreparable damage to the natural beauty of many wilderness areas, but also demolish wild life with nature itself. Instead of quarrying new stone from nature old ones should not be preferred to produce noise, dust, etc. Depending on improved stone quarrying and processing technology, utilizing natural stone becomes more affordable. New automated technologies and systems allows the use of thinner natural stone and its installation at less cost and time and safer. (Yates and Bourke, 2005)

Bricks and Tiles

For thousands of years baked-clay bricks and tiles have been used. It is used for multi purposes, but masonry, which is not reinforced, usually makes the required wall thickness unacceptably high and is unsafe in an earthquake zone. On the other hand, its characteristics of modularity, durability and small unit size render it extremely flexible in design, good for filling, and relatively simple in construction. Requiring no formworks is making them an ideal material for small-scale construction.

Pearson (1998) points out that because of their production, transport phases bricks and tiles have a high "embodied energy". Because firing in kilns causes high energy output and smoke emissions. However, its thermal insulation and mass, structural and finish properties are relatively resource-effective. For a long time, the clay brick industry uses recycled waste materials that are nearly 10% of the total amount, from different branches such as *"pulverized fuel ash (PFA), blast furnace clinker, coke breeze and coal slurry, ground glass, sewage sludge, paper ash, bone ash and sawdust"*. (Smith, 2012) By using these materials, some properties of brick like frost resistance or color can be improved whereas some of which are combustible are utilized as a source of fuel, in order to decrease the energy used in kilns. There are some materials like ground glass can be used to decrease the kiln temperatures and processing times, besides emissions to air with 20% savings of energy. (WRAP, 2006)

Bricks and tiles are extremely durable and affordable. Compared to their need of maintenance they last for years. They can often be salvaged and reused as good quality building materials. Even broken bricks can be used as rubble under concrete slabs or landscape finishes. Insulated cavity wall faced with brick is extremely energy-efficient.

Metals

Metals can be divided into two according to iron in them as ferrous (metal made of iron) and non-ferrous (metals other than iron) metals. The main use of ferrous alloys is iron – carbon or cast iron – steel. In structural and civil engineering, the main non-ferrous alloys are aluminium and copper.

Iron and Steel

According to Ayres and Leslie (1996), iron is extracted from natural ores like all metals. These ores mostly have complex chemical compounds. Pig iron, which contains a large amount of carbon, and being industrially useless requires further processing. It is remelted and cast iron, by using sand or moulds. The greatest amount of manufactured metal items contains iron-based materials. Their recyclability capacity is really high, nearly 75% of production of new alloys come from scrap. (Lambert, 2016) Tougher, less brittle and more durable to high-temperature cast irons can be produced by the addition of nickel.

Steel, a kind of alloy of iron and carbon, can be modified by using the elements such as silicon, manganese, tungsten and copper. According their carbon content, their characteristic properties are changed. Low carbon content of steel, below around 0.25%, increases its ability to be cut and shaped (Lambert, 2016). However, this content creates a risk of corrosion because of the effects of moisture and oxygen. Besides using greases and paints as protective coatings, the highly alloyed content can be effective to increase resistance to corrosion. Stainless steel with chromium or weathering steel, with protective oxide layer are examples of them. (McKenzie, 1978)

Domone and Illston (2010) implies that since steel is highly recyclable and has value as scrap, it is accepted as the most common metal that used in construction field. Currently up to 50% or 70% of its energy and pollution can be avoided in steel production. Stainless steel as an alloyed metal is mostly recyclable if it is treated carefully during the process. Recycled steel beam and bar sections the most valuable parts for scrap dealers in the sector to use in non-structural uses.

Aluminum

Ayres and Leslie (1996) verifies that aluminium is an industrially important metal that plays an extremely important part in modern society because its ability to bend, extrude and shape; lightness, resistance to corrosion provide this. It occurs by a wide range of natural compounds and is also easily recycled. There are virtually environmental problems associated with its extraction, use or disposal. The most common aluminium ore is bauxite although the mining of bauxite, alumina is environmentally destructive and the layer of ore is usually rather thin and hardly scarce. (Norgate et al., 2007) The processing of bauxite to alumina generates large quantities of caustic waste; called 'red mud' which has proved hard to dispose of.

LCA comprehensive methodology can determine the environmental problems and benefits, by using raw or waste material in the aluminium industry (Liu and Müller, 2012). On the other hand, aluminium metal smelting is extremely energy intensive. For this reason alternative processes for extracting aluminium, from different sources and by different technical means, should be considered seriously. Aluminium smelting is also responsible for significant problems of atmospheric fluorine pollution. (Olivieri et al., 2006) While largely controllable, it is nevertheless a hazardous, especially for workers and local residents if the technology and its maintenance are not of the highest standard.

The energy saved in processing and carbon dioxide not emitted into the atmosphere during manufacture is not the only benefit of recycling. Other benefits would include reduced demand for caustic soda, reduced demand for fluorine and of course need to dispose of waste 'red mud'. (Leroy, 2009) Through melting up to 85% of the pollution and energy obtained from aluminium can be refrained. (Liu and Müller, 2012)

Wood

The area of the earth's land area covered by forests has declined dramatically in the last two centuries. Parallel with the decline has been a rapid increase in combustion of productive lands worldwide, discharging more and more carbon into the atmosphere. According to Papanek (1995), wood has occupied a unique position in human endeavor. Its very accessibility and ease of use, however, may destroy the capacity of forests. If wood, which is sustainably harvested, is used, and if clear or old growth timber is saved for users, there is an opportunity to develop the healthy life cycles over the years. The major problematic areas of concern for wood use in construction are non-sustainable forestry practices, and the use of toxic resins and glues in the manufacture of some wood products. (Lindner et al., 2010)

Returning more carbon to the atmosphere contributes the rise of the greenhouse effect which is raising the temperature of the earth's atmosphere. The destruction of the habitat of large numbers of plant and animal species, the destruction of the biomass, which cleans the air can be occurred by careless logging, clearing and farming practices. (Rowell, 2012) Besides this, the harvesting of the climax forests for the production of housing is economically and environmentally disastrous.

Wood product manufacturers are using low-density, fast-growing varieties of trees in combination with plastic resins to make uniform, economical structural and finish products. (Lindner et al., 2010) Structural elements from these trees include engineered members like truss joists, I beams laminated lumber with veneer on it (LVL) and parallel strand lumber (PSL). Covering components include, oriented strand board (OSB) and particle boards. Finish materials from scrap, recycled and small-dimension trees include finger-jointed lumber, laminated veneer lumber and several lands of wood fiber reconstituted with plastic resins. (Gustavsson and Sathre, 2006)

The primary cause of wood decay is intermittent contact with water. The best defense against rot is to keep wood dry, well ventilated and raised away from the soil. Traditional detailing has emphasized these aspects of the use of wood, and well-built wood structures have seen many years of service with little or no use of preservatives. (John, 1993) The advent of relatively inexpensive commercially treated wood has encouraged the use of wood in some exterior and damp applications that would previously have been avoided. Pressure treated wood can last about five times as long as untreated wood in a given situation. (Papanek, 1995) The benefits of the use of any preservative, however, including extended life and structural security gained, should be balanced against the risks to manufacturers, users and the environment. There are 3 main protectors in use at the moment, which are pentachlorophenol, inorganic arsenic and creosote which are all carcinogenic and can cause mutation. Because of this, all three materials have been classified as "restricted use only by certified applicators" by the EPA. (Pearson, 1998) (Gustavsson and Sathre, 2006). It is forbidden to use creosote and penta in interior spaces since the major adhesive material in particleboard and plywood is urea formaldehyde, which is classified as a "probable human carcinogen". Off gassing from a given material varies widely with heat, humidity and various combinations of materials; its effect on different individuals varies as well. (Pearson, 1998)

Plastics

Osso et al., (1998) describe plastics as the substance combination of organic materials, any biomass, mostly from fossil-based things like oil, natural gas coal or salt. Examples of other types of plastics include casein from milk, cellulose plastic from wood, resins, gums and rubber latex. (Andrady, 2003) There is a general concern about the usage of some plastic stems because of their toxic wastes during production and disposal. PVCs are widely used in floor tile, window frames, pipes and common office and home products. PVC incineration, however, produces chlorinated hydrocarbons and hydrochloric acid, while land disposal can result in leaching of organochlorine substances into soil and water. (Leadbitter, 2002) As a result of using things once and then discarding, the landfills are filling up with petroleum by-products, whose manufacture phase needs more energy. Simplicity of structure makes a plastic easier to recycle, so that a plastic like PET, with only one polymer, is more recyclable than more complex forms. (Tolinski, 2011)

In the plastic industry, natural gas and crude oil whose extraction and manufacturing are destructive for the environment because of the emissions of greenhouse gases are used. It will be hard for the building industry to avoid the use of petroleum-based plastics until alternate materials are developed. Their use can be limited with care to areas in which the qualities that characterize plastics like rot-resistance, easy workability and lightweight make them the most appropriate choice. (Berge et al., 2009)

Thermal Insulation

Insulation plays a more important role in "envelope-driven" construction, in which the internal and equipment gains play a larger part. (Pearson, 1996) points out the environmental importance of insulation that can be clearly seen by comparing two representative buildings. They are identical except for insulation levels, located in a heating climate, and have high-performance insulating windows (low-e glass, argon-filled). Depending on the type of heat used, the better-insulated building generates six fewer tons of CO2 emissions and consumes nearly four fewer GJs of energy each year.

Savings on other pollutants produced from fossil fuel combustion (sulphur dioxide, nitrous oxides, particulate, etc.) would be proportionate to the CO2 savings. (Sustainable Sources, 2017) It can be clearly seen that, increased insulation levels are environmentally beneficial. However, there are significant environmental problems associated with some types of insulation that also need to be considered. Nearly all isocyanurate, urethane and phenolic foam insulation and some extruded polystyrene products are manufactured with CFC-11, which degrades the ozone layer and contributes to global warming (Kotaji and Loebel, 2010)

The ozone layer in the stratosphere protects the earth from UV radiation. CFCs are stable compounds, which do not break down in the lower atmosphere. It is only when they reach the stratosphere, 15 to 40 kilometers above the earth's surface that high-energy radiation is able to break the CFC molecule apart, releasing the chlorine atoms that destroy ozone. As ozone is lost, the additional UV radiation reaching the earth causes a higher incidence of skin cancer, cataracts, immunologic problems and reduced crop yields. (Anwar et al., 2016)

CFCs are also potent greenhouse gases, both because they absorb long wavelength radiation very effectively and because they last for a longer time in the atmosphere (up to 60 years). Even as less-damaging HCFCs are substituted for CFCs in rigid foam insulation, the foam will still carry an environmental penalty. (Kotaji and Loebel, 2010) In general, these products should be chosen over the CFC-based products they replace, but there are better choices. If a rigid board-stock is needed, choice of rigid fiberglass or expanded polystyrene (EPS) will be better. EPS is foamed with pentane, less environmentally damaging than CFC's or HCFC's, and is often available in higher densities to more closely match extruded polystyrene in its performance. (Engels et al., 2013) Fiberglass and mineral wool have no known problems in the manufacturing process, but their fibres may be carcinogenic. Loose-fill materials can be used where an adequate cavity can be created. Any use should be very carefully sealed from inhabited spaces. Spray technology with acrylic binders can be used to prevent settling. Cellulose is essentially a plant fibre. It can be obtained from recycled wood fibre, recycled newspaper and cardboard. It can be blown dry, mixed with a binder or sprayed wet. (Pearson, 1996)

Moisture Protection: Waterproofing and Dampproofing, Vapor Control, and Sealants

The applications of moisture protection and thermal insulation are often invisible in a finished project, but they are essential elements of every building, and offer a significant area for improving environmental quality. The wide range of materials - metal, plastics, rubber, cellulose, fibreglass, mineral wool, chemicals and foams - used to keep buildings warm and dry make it difficult to develop a comprehensive approach to sustainability. (Berge et al., 2009)

Petroleum has a number of properties that make it almost uniquely suited for rejection of moisture. Virtually unaffected by water, its products also have elastic and self-healing properties that make them ideal for long-term, uninspected installations. It is for all practical purposes a non-renewable resource, with high-embodied energy. Recycled material is available for many applications. (Widyatmoko, 2016)

Building ceilings typically require ventilation between the insulation and the roof deck. There can be a ventilation channel out of recycled plastic that is relatively easy to use. Polyethylene film is widely used as a vapor retarder between inhabited space and floor, roof and wall elements such as Tenoarm and Hygrodiode. (Domone and Illston, 2010) Solvent-release sealants can degrade air quality, so a "neutral curing" material with no solvent release is preferable. Many sealants are petroleum-based; the lines of non-petroleum, non-plastic caulks, putties and sealers should be chosen as available. (Tolinski, 2011)

Finishing Materials

Interior finishing materials depending on their increasing or reducing indoor air pollution capacity are accepted as one of the primary issues that affect the human health. They can be easily worn out and replaced over the lifetime of the building. With this property they are good examples of resource conservation.

Floor Finishing Materials

Terrazzo and ceramics are good examples of floor finish materials with their high durability and extremely low emissions. They can resist abrasion and wear, can be easily cleaned and don't keep odours in them. (Osso et al., 1998) With their long life and minimal maintenance requirement their cost becomes logical when buying. Their settling method, the grout, sealers required for protecting unglazed surfaces are the leading contamination factors. To reduce the high transportation cost local manufacturers should be chosen. In some types up to 70% scrap glass can be recycled. (Li, 2006)

When cement mortars are modified with acrylic additives a safe setting material is obtained. (Malhotra, 2004) Plastic adhesives have solvents in them and this can cause air pollution for indoor. (Andrady, 2003) With their low emissions cellulose-base grouts and cement are very safe. Grout which is modified with epoxy can contain hazardous components. With a porous tile water-dispersed or low-volatility silicone sealers should be chosen to keep it safe. (Pearson, 1996)

Numerous asset effective sorts of wood flooring are accessible, containing rescued, covered, and veneered items. The most vital concern about them are the establishment technique and wrapping up. (Rowell, 2012) Rescued solid wood flooring is generally accessible and requires sanding and resurfacing nearby. New wood flooring materials incorporate an extensive variety of veneered and covered items that have a MDF core with a hardwood surface. (Bianchina, 1997) These are asset productive decisions, however, are less repairable than solid wood. Using a floating system with a steel - track framework makes wood floors simple to remove. Unlike nail-down framework a glue-down framework is preferred because of easiness and decrease extra material usage, however, is needed for parquet flooring. (Li, 2006) Factory prefinished items offers air-quality advantages in light of the fact that no sanding or completing is performed nearby. For completing nearby, the water-scattered urethanes (urethane-acrylic mixes) with low-volatility ingredients should be considered since they are among the most minimal discharge finishes. Solidifying oils, dissolvable varnishes, and corrosive cured varnishes emit delayed discharges of poisons. (Berge et al., 2009)

True linoleum comprises numerous renewable materials has named as cork products. Vinyl, elastic, tile, and cork floors can be maintained easily, and a few sorts are extremely durable. Waste tires also can be used. However, because of the cleaning products and processes, the outgassing can be occurred because of VOCs. (Tolinski, 2011)

There are different carpet types in the market. Polyester and nylon-blended carpets with recycled content can be chosen for resource-efficiency, contamination potential and lower upkeep need, but they should include fewer ingredients in order to separate for recycling. On the other hand, wool-face fibres make the wool carpets more fire resistant and durable. If a carpet tile or roll carpet system is chosen, it is easy to remove and replace them during the renovation. The face fibres are glued with a synthetic latex resin which causes Sick Building Syndrome. So, at this point, needle-punched carpets which are made without latex should be the choice. (Berge et al., 2009)

Wall Finishing Materials

For both outdoor and indoor, paint is a source of toxic waste. By oxidation and evaporating solvents, volatile organic compounds are released from them. Use of alkyds (solvent-based oil paints) should not be preferred instead of waterborne acrylics which are highly durable and create no harmful emissions.

And also the use of paints which contains mercury, lead, chromium, cadmium and should be banned. (Gray, 2017)

Gypsum products are preferred for interior finishing because of their easy application processes, less costs and fire retardance (Osso et al., 1998). By recycling, its commitment to landfill can be reduced if paint and adhesives do not change the content. Gypsum items have less effect on indoor pollution, however the paper covering and adhesives have more. The main problem of gypsum surfaces is their sink effect, first absorbing pollutants and then giving back to the environment whereas which increases its acoustical capacity. (Berge et al., 2009) Durable high-pressure laminate surfaces comprise of laminating paper, colorants and phenolic resin are resource-efficient plastics if their installation is carefully considered in order to minimize the dust and glue emissions from cutting material. (Tolinski, 2011) Ceiling tiles as the most widely recognized ceiling treatment are assessed as far as their resource efficiency capacity and indoor air quality. These tiles are composed of mineral fibre, added clay and gypsum fillers for fire protection and collect dust and absorb odours. (Berge et al., 2009)

ANALYSING BUILDING MATERIALS ACCORDING TO THEIR SUSTAINABILITY GUIDELINES

Energy performance, which is one of obvious priorities, can be supplied by minimizing the use of natural resources, embodied energy, reducing maintenance, transportation and formwork costs, enhancing building durability, reusing and recycling. **Human health performance,** human comfort and productivity are directly related with indoor air quality. The potential indoor air quality benefits are reduced health risk and increase of the productivity of occupants. The most highly volatile organic compound emissions of building materials emit odours, release toxic chemicals, cause indoor air pollution and need of ventilation. **Environmental health performance** can be supplied by prevention of emission of greenhouse gases, elimination of waste and atmospheric, noise and visual pollution. If the environmental pollution is ignored and the maintenance of productivity of natural resources is not taken into account, the sustenance of human life can be endangered.

In order to analyze the characteristics of building materials according to energy performance and environment and human health performances, comparison tables of building materials are prepared. These performances are analyzed in four different stages during their life as manufacturing phase, application phase, usage phase, and demolition phase. The building materials are grouped as follows: construction materials, wall finishing materials, floor finishing materials and others. They are marked to analyze their performances. The marks are put according to their answers to the questions under the headings as positively yes (1) and no (0). During the manufacturing phase, the concepts under energy performance are degradation of natural resources, high embodied energy and transportation cost; the environmental health performances include subheadings as; emission of CFCs and HCFCs, atmospheric pollution because of hazardous gases and dust, destruction of habitats, visual pollution because of on-site construction and noise pollution. The indoor air quality of building materials and their need for ventilation are analyzed under human health performance in this phase. During application phase, the subheadings of energy performance are difficulty of installation, high embodied energy and high costs of formwork and transportation; the subheadings of environmental health and human health performances in this stage are the same with manufacturing phase. In usage phase, the headings as high acoustic performance, high durability, and little maintenance needed can be seen under the energy performance. The environmental

Table 1. Chosen materials in manufacturing phase

CHOSEN MATERIALS			Energy Performance			Environmental Health Performance					Human Health Performance	
			Degradation of Natural Resources	High Embodied Energy	High Transportation Cost	Emission of CFCs and HCFCs	Atmospheric Pollution	Destruction of Habitats	Visual Pollution	Noise Pollution	Indoor air pollution and need of ventilation	Impact Factor
Construction Materials	Rubble Stone				1		1	1	1	1		55,56%
	Baked Brick		1	1	1		1	1	1			66,67%
	Sand & Aggregate				1			1	1	1		44,44%
	Cement		1	1			1					33,33%
	Additives					1	1				1	33,33%
	Pozzuolana		1		1		1	1	1		1	66,67%
	Precast Concrete Panels		1	1	1		1			1		55,56%
Insulation	Rigid Foam Insulation		1	1		1	1				1	55,56%
	Mineral Fibre Insulation			1							1	22,22%
	Cellulose			1		1	1				1	44,44%
Wall Finishing Materials	Paint	Oil-Based	1	1							1	33,33%
		Water-Based		1								11,11%
	Ceramic Tile	With Adhesive	1		1		1				1	44,44%
		With Mortar	1		1				1			33,33%
	Engineered Composite Wood Panel		1		1			1				33,33%
	Gypsum Boards		1					1	1		1	44,44%
	Wallpaper with Adhesive		1	1				1				33,33%
Floor Finishing Materials	Carpet	Glued	1			1	1				1	44,44%
		Punched	1									11,11%
	Resilient Flooring	Glued	1	1		1					1	44,44%
		Punched	1	1		1						33,33%
	Wood Flooring	Varnished	1		1			1			1	44,44%
		Not varnished	1		1			1				33,33%
	Ceramics, Terrazzo	With Adhesive	1		1		1				1	44,44%
		With Mortar	1		1				1			33,33%
Others	Aluminium		1	1			1		1			44,44%
	Steel		1	1			1		1			44,44%
	PVC		1	1		1					1	44,44%

(By author)

Table 2. Chosen materials in application phase

CHOSEN MATERIALS			Energy Performance				Environmental Health Performance					Human Health Performance	
			Difficulty of Installation	High Cost of Formwork	High Embodied Energy	High Transportation Cost	Emission of CFCs and HCFCs	Atmospheric Pollution	Destruction of Habitats	Visual Pollution	Noise Pollution	Indoor air pollution and need of ventilation	Impact Factor
Construction Materials	Rubble Stone		1			1			1	1			40,00%
	Baked Brick		1			1			1	1			40,00%
	Concrete			1	1	1		1	1	1	1		70,00%
	Concrete with Additives			1	1	1		1	1	1	1	1	80,00%
	Pozzuolana			1		1			1	1		1	50,00%
	Precast Concrete Panels			1	1	1					1		40,00%
	Plaster				1					1			20,00%
	Mortar				1					1			20,00%
Insulation	Rigid Foam Insulation		1		1		1					1	40,00%
	Mineral Fibre Insulation		1					1				1	30,00%
	Cellulose		1		1			1					30,00%
Wall Finishing Materials	Paint	Oil-Based			1			1		1		1	40,00%
		Water-Based			1			1		1			30,00%
	Ceramic Tile	With Adhesive	1		1	1		1		1		1	60,00%
		With Mortar	1		1	1				1			40,00%
	Engineered Composite Wood Panel		1		1	1						1	40,00%
	Gypsum Boards			1	1								20,00%
	Wallpaper with Adhesive		1					1				1	30,00%
Floor Finishing Materials	Carpet	Glued	1					1				1	30,00%
		Punched									1		10,00%
	Resilient Flooring	Glued	1		1			1				1	40,00%
		Punched	1		1						1		30,00%
	Wood Flooring	Varnished	1			1		1			1	1	50,00%
		Not varnished	1			1					1		30,00%
	Ceramics, Terrazzo	With Adhesive	1		1	1		1		1	1	1	70,00%
		With Mortar	1		1	1				1	1		50,00%
Others	Aluminium				1	1					1		30,00%
	Steel				1	1					1		30,00%
	PVC				1	1					1		30,00%

(By author)

Table 3. Chosen materials in usage phase

CHOSEN MATERIALS			Energy Performance			Environmental Health Performance		Human Health Performance						
								Bad indoor air quality						
			Low acoustic performance	Low durability	Maintenance needed	Habitat unfriendly	More pollutant emissions	Less VOC emissions	More VOC emissions	Adsorption of odours	Need of ventilation	Difficulty of wall and floor breathing	Toxicity of using cleansers	Impact Factor
Construction Materials	Rubble Stone							1			1			18,18%
	Baked Brick													0,00%
	Concrete		1			1	1	1			1		1	54,55%
	Sand & Aggregate		1				1							18,18%
	Cement		1			1	1	1				1		45,45%
	Additives			1	1	1	1	1		1		1		63,64%
	Pozzuolana			1	1		1							27,27%
	Precast Concrete Panels		1	1	1	1	1	1				1	1	72,73%
Insulation	Rigid Foam Insulation		1	1	1	1	1	1			1	1		72,73%
	Mineral Fibre Insulation			1	1		1	1			1			45,45%
	Cellulose			1	1		1							27,27%
Wall Finishing Materials	Paint	Oil-Based	1			1	1		1	1	1	1	1	72,73%
		Water-Based	1	1	1			1		1	1			54,55%
	Ceramic Tile	With Adhesive			1	1	1		1	1	1	1	1	72,73%
		With Mortar			1		1	1		1	1	1		54,55%
	Engineered Composite Wood Panel			1	1		1	1		1	1	1	1	72,73%
	Gypsum Boards		1	1	1		1	1		1	1	1	1	81,82%
	Wallpaper with Adhesive		1	1	1		1		1	1	1	1	1	81,82%
Floor Finishing Materials	Carpet	Glued	1				1	1		1	1	1	1	63,64%
		Punched	1					1			1		1	36,36%
	Resilient Flooring	Glued	1	1	1		1		1	1	1	1	1	81,82%
		Punched	1	1	1				1	1	1	1	1	72,73%
	Wood Flooring	Varnished			1		1		1	1	1	1	1	63,64%
		Not varnished		1	1			1		1			1	45,45%
	Ceramics, Terrazzo	With Adhesive			1		1		1	1	1	1	1	63,64%
		With Mortar			1		1	1		1	1	1	1	63,64%
Others	Aluminium		1	1	1	1								36,36%
	Steel		1		1	1								27,27%
	PVC				1	1	1	1		1	1		1	63,64%

(By author)

Table 4. Chosen materials in demolition phase

CHOSEN MATERIALS			Energy Performance			Environmental Health Performance			
			Less Reusing Capacity	Less Recycling Capacity	Having no Recycled and Reused Ingredients	Increasing Waste	Increasing Natural Resources Use	Increasing Air Pollution During Recycling	Impact Factor
Construction Materials	Rubble Stone				1	1	1		27,27%
Construction Materials	Baked Brick			1	1	1			27,27%
Construction Materials	Concrete		1	1		1	1		36,36%
Construction Materials	Sand & Aggregate					1	1		18,18%
Construction Materials	Cement		1	1	1		1	1	45,45%
Construction Materials	Additives		1	1			1	1	36,36%
Construction Materials	Pozzuolana				1	1	1		27,27%
Construction Materials	Precast Concrete Panels		1	1			1		27,27%
Insulation	Rigid Foam Insulation		1	1		1	1	1	45,45%
Insulation	Mineral Fibre Insulation		1	1	1	1	1		45,45%
Insulation	Cellulose			1	1	1	1		36,36%
Wall Finishing Materials	Paint	Oil-Based	1	1				1	27,27%
Wall Finishing Materials	Paint	Water-Based	1	1					18,18%
Wall Finishing Materials	Ceramic Tile	With Adhesive	1	1				1	27,27%
Wall Finishing Materials	Ceramic Tile	With Mortar	1	1				1	27,27%
Wall Finishing Materials	Engineered Composite Wood Panel			1			1	1	27,27%
Wall Finishing Materials	Gypsum Boards			1	1			1	27,27%
Wall Finishing Materials	Wallpaper with Adhesive		1			1	1	1	36,36%
Floor Finishing Materials	Carpet	Glued	1	1	1		1	1	45,45%
Floor Finishing Materials	Carpet	Punched		1	1		1		27,27%
Floor Finishing Materials	Resilient Flooring	Glued	1	1	1		1	1	45,45%
Floor Finishing Materials	Resilient Flooring	Punched		1	1		1	1	36,36%
Floor Finishing Materials	Wood Flooring	Varnished		1	1		1		27,27%
Floor Finishing Materials	Wood Flooring	Not varnished		1	1		1		27,27%
Floor Finishing Materials	Ceramics, Terrazzo	With Adhesive		1	1		1		27,27%
Floor Finishing Materials	Ceramics, Terrazzo	With Mortar		1	1		1		27,27%
Others	Aluminium					1		1	18,18%
Others	Steel					1		1	18,18%
Others	PVC		1		1	1			27,27%

(By author)

health performance in this phase is analyzed according to their pollutant emission rates and hazards to habitat. Whereas the concepts under human health performance are good indoor air quality, need of ventilation, difficulty of wall and floor breathing and toxicity of using cleansers. Lastly, in demolition phase, the subheadings under energy performance are high reusing and recycling capacity and using recycled and reused ingredients; under environmental health performances, they are reducing waste and natural resources use and increasing air pollution.

After getting the answers to the questions under the necessary subheadings, the material's impact factor is calculated in percentage according to these results in four different stages during their life.

CONCLUSION

The growing green development ethic is based on the principles of resource efficiency, health and productivity. This chapter aims to search guidelines of selecting green and energy efficient building materials. Constructions as an effectively a building organism that interacts with the inhabitants, and with the environment. The problems of construction materials related to energy and human and environment health are discussed in the tables. It can be said that, life cycle assessment, the embodied energy of materials and their chemical properties are important concepts that must be considered in order to get rid of health problems and related illnesses such as sick building syndrome, legionnaires' disease. Both natural and industrial materials should be analyzed carefully for health and energy viewpoints like in the tables. The advantages and disadvantages must be examined and appropriate and efficient ones must be chosen. The research results of the study can be beneficial for designers and contractors and guide them when selecting appropriate building materials among natural or industrial ones. The cost performance should not be the only significant criterion. The energy and health performances of building materials are as important as the cost performance.

ACKNOWLEDGMENT

This research received no specific grant from any funding agency in the public, commercial, or not-for-profit sectors.

REFERENCES

Adeyeye, K., Osmani, M., & Brown, C. (2007). Energy conservation and building design: The environmental legislation push and pull factors. *Structural Survey*, *25*(5), 375–390. doi:10.1108/02630800710838428

American Society for Testing and Materials. (1996). *Standard Practice for Developing Accelerated Tests to Aid Prediction of the Service Life of Building Components and Materials*. West Conshohocken, PA: Author.

Andrady, A. L. (2003). *Plastics and the Environment.* Hoboken, NJ: John Wiley & Sons.

Anwar, F., Chaudhry, F. N., Nazeer, S., & et al, . (2016). Causes of Ozone Layer Depletion and Its Effects on Human. *Atmospheric and Climate Science*, *6*(01), 129–134. doi:10.4236/acs.2016.61011

Ayres, R. U., & Leslie, W. (1996). *Industrial Ecology: Towards Closing the Materials Cycle*. Brookfield, VT: Edward Elgar Publishing.

Berge, B., Butters, C., & Henley, F. (2009a). Fossil Oils. In B. Berge, C. Butters, & F. Henley (Eds.), *The Ecology of Building Materials* (2nd ed., pp. 139–156). Philadelphia, PA: Taylor & Francis.

Berge, B., Butters, C., & Henley, F. (2009b). Surface Materials. In B. Berge, C. Butters, & F. Henley (Eds.), *The Ecology of Building Materials* (2nd ed., pp. 299–355). Philadelphia, PA: Taylor & Francis.

Berge, B., Butters, C., & Henley, F. (2009c). Paint, varnish, stain and wax. In B. Berge, C. Butters, & F. Henley (Eds.), *The Ecology of Building Materials* (2nd ed., pp. 381–407). Philadelphia, PA: Taylor & Francis.

Bianchina, P. (1997). *Builder's Guide to New Materials and Techniques*. New York, NY: McGraw Hill

Coppola, L., Cerulli, T., & Salvioni, D. (2004). Sustainable development and durability of self-compacting concretes. *8th CANMET/ACI International Conference on Fly Ash, Silica Fume, Slag and Natural Pozzolans in Concrete,* 1, 29-50.

Domone, P., & Illston, J. (2010). *Construction Materials: Their Nature and Behaviour* (4th ed.). London, UK: Spon Press.

Downton, P. (2013). Autoclaved aerated concrete. Retrieved from http://www.yourhome.gov.au/materials/autoclaved-aerated-concrete

DPTI, Government of South Australia. (2017). *Environmentally Sustainable Building Materials – Selection.* Retrieved from https://www.dpti.sa.gov.au/__data/assets/pdf_file/0009/293688/Environmentally_Sustainable_Building_Materials.pdf

Edwards, B. (1998). *Sustainable Architecture: European Directives and Building Design*. Oxford, UK: Architectural Press.

Engels, H., Pirkl, H. G., & Albers, R. (2013). Polyurethanes: Versatile Materials and Sustainable Problem Solvers for Todays Challenges. *Angewandte Chemie International Edition*, *52*(36), 9422–9441. doi:10.1002/anie.201302766 PMID:23893938

Erlandsson, M., & Borg, M. (2003). Generic LCA-methodology applicable for building, construction and operation services-today practice and development needs. *Building and Environment*, *38*(7), 919–938. doi:10.1016/S0360-1323(03)00031-3

Erlandsson, M., & Levin, P. (2005). Environmental Assessment of Rebuilding and Possible Performance Improvements Effect on a National Scale. *Building and Environment*, *39*, 1453–1465.

European Commission. (2003). EPD: The International EPD System. Retrieved from http://www.environdec.com/en/What-is-an-EPD

Garrido, M., Paulo, P., & Branco, F. (2012). Service life prediction of façade paint coating in old buildings. *Construction & Building Materials*, *29*, 394–402. doi:10.1016/j.conbuildmat.2011.10.057

Gray, J. (2017). Environmentally Friendly, Non-Toxic Paint. Retrieved from http://www.sustainablebuild.co.uk/nontoxicpaint.html

Gustavsson, L., & Sathre, R. (2006). Variability in energy and carbon dioxide balances of wood and concrete building materials. *Building and Environment*, *41*(7), 940–951.

International Organization for Standardization. ISO 14025: 2006. (2006). Environmental labels and declarations -- Type III environmental declarations - Principles and procedures. Brussels, Belgium: CEN (European Committee for Standardisation).

International Organization for Standardization. ISO 14040: 2006. (2006). Environmental Management - life cycle assessment - principles and framework. Brussels, Belgiuim: CEN (European Committee for Standardisation).

International Organization for Standardization. ISO 14044: 2006. (2006). Environmental Management - life cycle assessment – requirements and guidelines. Brussels, Belgium: CEN (European Committee for Standardisation).

International Organization for Standardization, ISO 15686-1. (2011). *Buildings and Constructed Assets. Service Life Planning. Part 1: General Principles and Framework.* Switzerland.

Kamal, M. A. (2016). *Autoclaved Aerated Concrete (AAC): A Sustainable Building Material.* Retrieved from: https://www.masterbuilder.co.in/autoclaved-aerated-concrete-aaca-sustainable building-material-2/

Kotaji, S., & Loebel, O. (2010). Sustainability of Polyurethane Thermal Insulation – Performance Assessment at Building and Building Component Level. In *Central European Sustainable Building Conference.* Prague, Czech Republic. Sustainable Building Society.

Lambert, P. (2016). Sustainability of metals and alloys in construction. In J. Khatib (Ed.), *Sustainability of Construction Materials* (2nd ed.; pp. 105–127). Sawston, UK: Woodhead Publishing. doi:10.1016/B978-0-08-100370-1.00006-8

Leadbitter, J. (2002). PVC and sustainability. *Progress in Polymer Science*, *27*(10), 2197–2226.

Leroy, C. (2009). Provision of LCI data in the European aluminium industry methods and examples. *The International Journal of Life Cycle Assessment*, *14*, 10–44.

Li, Z. (2006). A new life cycle impact assessment approach for buildings. *Building and Environment, 41*, 1414–1422.

Lindner, M., Suominen, M., & Palosuo, T. (2010). ToSIA—A tool for sustainability impact assessment of forest-wood-chains. *Ecological Modelling, 221*(18), 2197–2205. doi:10.1016/j.ecolmodel.2009.08.006

Liu, G., & Müller, D. B. (2012). Addressing sustainability in the aluminium industry: A critical review of life cycle assessments. *Journal of Cleaner Production, 35*, 108–117.

Malhotra, V. M. (2004). Role of supplementary cementing materials and superplasticizers in reducing greenhouse gas emissions. In *ICFRC International Conference on Fibre Composites, High-Performance Concrete, and Smart Materials* (pp. 489-499). Indian Institute of Technology.

Marteinsson, B. (2003). Durability and the factor method of ISO 15686-1. *Building Research and Information, 31*(6), 416–426.

McKenzie, M. (1978). The corrosion performance of weathering steel in highways bridges: TRRL Laboratory report 857. Wokingham, UK: Transport Research Laboratory.

Naik, T. R. (2008). Sustainability of Concrete Construction. *Practice Periodical on Structural Design and Construction, 13*(2), 98–103. doi:10.1061/(ASCE)1084-0680(2008)13:2(98)

Napier T. / Whole Building Design Guide. (2016). Construction Waste Management. Retrieved from https://www.wbdg.org/resources/construction-waste-management

Norgate, T. E., Jahanshahi, S., & Rankin, W. J. (2007). Assessing the environmental impact of metal production processes. *Journal of Cleaner Production, 15*(8-9), 838–848. doi:10.1016/j.jclepro.2006.06.018

Nunen, N., Hendriks, A., & Erkelens, P. (2004). Service Life as Main Aspect in Environmental Assessment. In *CIB Congress, editor. Building for the Future: The 16th CIB World Building Congress* (vol. 1, pp. 10). Toronto, Canada: Academic Press.

Olivieri, G., Romani, A., & Neri, P. (2006). Environmental and economic analysis of aluminium recycling through life cycle assessment. *International Journal of Sustainable Development and World Ecology, 13*(4), 269–276.

Osso, A., Walsh, T., & Gottfried, D. (1998). *Sustainable Building Technical Manual*. New York: Public Technology Inc.

Papanek, V. (1995). *The Green Imperative Ecology and Ethics in Design and Architecture*. Singapore: Thames and Hudson.

Pearson, D. (1996). *The Natural House Catalog: Where to Get Everything You Need to Create an Environmentally Friendly Home*. New York, NY: Simon & Schuster.

Pearson, D. (1998). *The New Natural House Book: Creating a Healthy, Harmonious, and Ecologically Sound Home*. New York, NY: Simon & Schuster.

Quadrennial Technology Review, D. O. E. (2015). *Increasing Efficiency of Building Systems and Technologies*. Retrieved from https://www.researchgate.net/file.PostFileLoader.html?id=595770fb3d7f4b73007a48f9&assetKey=AS%3A511249461858304%401498902778200

Ragheba, A., El-Shimyb, H., & Raghebb, G. (2016). Green Architecture: A Concept of Sustainability. *Procedia: Social and Behavioral Sciences, 216*, 778–787. doi:10.1016/j.sbspro.2015.12.075

Rowell, R. M. (2012). Handbook of Wood Chemistry and Wood Composites (2nd ed.). New York, NY: CRC Press. doi:10.1201/b12487

Saghafi, M. D., Hosseini, T., & Teshnizi, Z. S. (2011). Recycling value of building materials in building assessment systems. *Energy and Building, 43*(11), 3181–3188. doi:10.1016/j.enbuild.2011.08.016

Smith, A. (2012). Materials from Alternative, Recycled and Secondary Sources (MARSS) 2005–2010: 6 Year Review. Retrieved from http://www.lucideon.com/newsroom/news-releases/ceram-publishes-a-6-year-review-of-the-substitute-raw-materials-used-in-uk

St. John, A. (1993). *The Sourcebook for Sustainable Design: a Guide to Environmentally Responsible Building Materials and Processes*. Boston, MA: The Boston Society of Architects.

SUS01/11. Stone Federation Great Britain, London, UK. (2011). *Natural Stone: The Oldest Sustainable Material.* Retrieved from http://www.stonefed.org.uk/uploads/Natural%20Stone.%20The%20oldest%20natural%20material.pdf

Sun, J., Han, B., Ekwaro-Osire, S., & Zhang, H. (2003). Design for environment: Methodologies, tools and implementation. *Journal of Integrated Design & Process Science, 7*(1), 59–75.

Sustainable Sources. (2017). Insulation: Sustainable Sources: 20 Years of Online Green Building Information. Retrieved from http://insulation.sustainablesources.com

Thormark, C. (2007, September). Energy and resources, material choice and recycling potential in low energy buildings. In *CIB SB07 Sustainable Construction, Materials and Practices, Lisbon, Portugal.* (Vol. 1, pp. 759–766). IOS Press.

Tolinski, M. (2011). *Plastics and Sustainability: Towards a Peaceful Coexistence between Bio-Based and Fossil Fuel-Based Plastics*. New York, NY: John Wiley & Sons.

United States Environmental Protection Agency. Indoor Air Quality (IAQ). (2017). Retrieved from: https://www.epa.gov/indoor-air-quality-iaq/introduction-indoor-air-quality

Vanderley, M. J. / UNEP Industry and Environment. (2003). *Sustainable Building and construction: On the sustainability of concrete.* Retrieved from http://www.cbcs.org.br/userfiles/bancoDeConhecimento/on%20the%20sustainability%20of%20concrete%20UNEP%20Sustainable%20concrete%20005-098.pdf

WBDG Sustainable Committee. (2011). *Sustainable. Whole building design guide.* Retrieved from: http://www.wbdg.org/design/sustainable.php

Widyatmoko, I. (2016). Sustainability of Bituminous Materials. In J. Khatib (Ed.), *Sustainability of Construction Materials* (2nd ed., pp. 343–370). Sawston, UK: Woodhead Publishing.

Workplace Safety and Health Topics. (2017). Retrieved from https://www.cdc.gov/niosh/pdfs/sec_2.pdf

WRAP. (2006). Use of Recycled Glass. Retrieved from http://www.wrap.org.uk

Yates, T., & Bourke, K. (2005). Sustainability and whole life costing for stone. *Arcamedia*, 49-52.

ADDITIONAL READING

Neveen, H., & Greenwood, D. (2009). Energy conservation regulations: Impacts on design and procurement of low energy buildings. *Building and Environment*, *44*(5), 929–936. doi:10.1016/j.buildenv.2008.06.010

Chapter 5
A Comprehensive Overview of IoT-Based Green Buildings: Issues, Challenges, and Opportunities

Rajalakshmi Krishnamurthi
Jaypee Institute of Information Technology, India

Dhanalekshmi G.
Jaypee Institute of Information Technology, India

ABSTRACT

Green building defines the way buildings are designed, constructed, operated, and maintained in such a way that the negative impacts are reduced and eliminated. The green building ensures to impart the sustainability of a building environment for quality living of citizens. It is to be noted that, not only the structure of the building, but its entire life cycle—including designing, constructing, operating, maintaining, renovating, and demolishing—must ensure responsibility towards environmental and natural resources. However, there are numerous factors that influence the effect of green building. Further factors such as pollution control, air quality monitoring, adaptability to evolving environment, and materials used also need to be handled by the green building. Hence, this chapter focuses on exploring the various issues, challenges and opportunities of green building concepts. Further, this chapter addresses how IoT-based green building will assist in achieving the goal through other emerging technology such as cloud computing.

INTRODUCTION TO GREEN BUILDING

Buildings have a substantial impact on the environment in all the stages. Starting from their construction, occupancy, revamp, repurposing, and devastation, buildings uses energy, water, and raw materials, produce waste, and discharge potentially toxic atmospheric releases. Thus propagates the positive impacts on the natural surroundings and weather conditions to enhance the quality of life of the occupants of the building. The building industry is a significant energy-consuming sector in the industry. It consumes energy not only during the construction, but the whole of its life as well — the high consumption of energy

DOI: 10.4018/978-1-5225-9754-4.ch005

for everyday life like lighting, air-conditioning challenges the architects, civil engineers and constructors plan environment-friendly and resource-efficient constructions. These challenges have triggered the creation of green building aimed at mitigating the impact of buildings on the natural environment through sustainable design.

Green building is a real-world and mindful environment approach to building design. It consumes less water, generates less waste, conserves the right amount of energy, uses natural resources and provides a comfortable and healthier living environment for human beings. In the conventional way of constructing a building pay a very little attention towards the impact that has on the environment as they use a large amount of non-renewable and toxic materials for the construction. The green building considers the environment at every stage of its construction starting from design, implementation, operation and maintenance. The serious environmental problems, the constant demand for the improvement of the air quality encourages the rapid development of green buildings. Firstly, the efficient uses of non renewable resources like electricity, water and gas. To overcome this, Green Building promotes usage of alternate renewable energy source likes solar energy.

Several organizations, standards towards Green Building exist across the globe, namely World Green Building Council (WGBC), Indian Green Building Council (IGBC), Leadership in Energy and Environmental Design (LEED), US Green Building Council, British Building Research Establish Environment Assessment Method (BREEAM), Excellence in Design for Greater Efficiency (EDGE), Green Building Council of Australia (GBCA), Green Building Index (GBI) of Malaysian Government, ISO/TS 21931 2006, etc.. The major objectives of these organizations and standards are to facilitate sustainability, durability, economy and comfort through green building construction and maintenance.

In this aspect, Internet of Things (IoT) has proven to be an excellent opportunity to maintain and control buildings. According to ABI research report, the count of such IoT based opportunistic systems for Green Buildings is expected to grow over 64 million by 2021. The Green building along with IoT things will soon be evolved as connected everything that is belong to the buildings. And such connection will be supported by intercommunication of IoT things through Internet technology. The IoT based Green Building will ensure the enhancement of Quality of Occupant Life (QOL) in a green building environment. Some of the prominent areas where IoT can be implemented are cognitive modelling-based building maintenance and construction, air quality monitoring within building, energy usage monitoring, fire safety, security and surveillance, lighting system within given building, measuring and verification of inaccessible areas of the building using IoT systems. Hence, this chapter focuses on exploring the various issues, challenges and opportunities that are existing in present Green Building concepts. Further, this chapter address how the IoT based Green building will assist in achieving various goals of the Green Building.

The key contribution of this chapter are to:

- Introduce the concept of Green Building, its objectives, issues and challenges associated
- Need for IoT to achieve and overcome issues of Green Building.
- Discuss various IoT frameworks that are existing in literature for Green building.
- Address various Green Building Organizations Standards across international and national level.
- Real Time data collection through IoT systems from Green Building.
- Cognitive IoT approaches for maintenance of Green Building.
- IoT system based automated optimization, recommendation and replacement mechanisms for supply chain management of assets and equipments within a Green Building.

GREEN BUILDING CONCEPT

The primary objective of green building is to improve the quality of air and water for people living inside the buildings. According to the Environmental Protection Agency (EPA) report, the indoor air pollution is more compared to outdoors. It is mainly due to the raw materials and furnishing used in the construction of the building. Taking human health factors into consideration the green building uses the recyclable or biodegradable construction materials in its construction. Green building use land and energy efficiently as renewable energy for most of the energy requirements. It conserves water and improves the air quality in indoor and outdoor and use the recycled and renewable materials in its construction.

Definition of Green Building

Although the definition of a green building is continually growing, The Office of the Federal Environmental Executive (Howard, J. L. (2003)) defines the term as follows: It is the practice of (1) using the energy, water, and materials efficiently, and (2) reducing negative impacts on occupants health and the environment by better architectural design, construction, operation, maintenance, and removal—the complete building lifecycle. Likewise, the Environmental Protection Agency defines the green building as the practice of creating environmentally responsive and resource-efficient structures in the entire life-cycle starting from the design, construction, maintenance, reconstruction and renovation.

Kibert, C. J. (2016) defines the green building as "healthy facilities designed and built on a resource-efficient manner, using ecologically based principles". Robichaud, L. B., & Anantatmula, V. S. (2010) defines that there are four pillars of green building as (i) minimization of impacts on the environment, (ii) enhancing the health conditions of the occupants, (iii) the return on the investment to developers and local community, and (iv) life cycle considered during the planning and development process.

Chatterjee, A.K. (2009) defines the "green building" as a method to make buildings and infrastructure by minimizing the use of resources, lessen harmful effects on the ecosystem, and create better environments for occupants. It must display a high level of environmental, economic, and engineering performance. These include energy efficiency and conservation, improved indoor air quality, resource and material efficiency, and occupant's health and productivity. Common elements in all the above definitions are the concern about efficient energy usage, improving the positive impacts on the environment and sustainability, health issues and impacts on the community.

Objectives of Green Building

The objective of green building is to re-integrate humanity in nature. In a natural cyclical system, the waste product of one stage can be input to the other stage. The goal of green building concept is to develop buildings which use the natural resources to the minimal starting at the time of construction to operate. The built environment has an enormous effect on the natural environment, human health, and the economy. The strategies in green building try to maximize both economic and environmental performance. Even though the Green construction methods can be integrated into buildings at any stage, from design and construction to renovation and deconstruction, the most important benefits can be attained if the design and construction integrated at the earlier stages of the building project.

The parameters responsible for achieving the green building concepts are:

- **Efficient Use of Energy**: The buildings on the globe use around 40% of the energy (Huovila, P., 2007). The major part of this energy is generated using non-renewable sources such as coal, gas and oil. The non-renewable resources become more expensive as they become more scarce and difficult to extract. Conversion of these fossil fuels into energy in most cases produces greenhouse gases (GHGs) which contribute to global warming. By employing renewable energy sources such as solar heaters, solar photovoltaic panels which convert sunlight into electrical energy, wind energy, bio energy and any other type of clean energy contribute to the preservation of the environment as clean as possible.
- **Water Reproduction:** Conservation of water is an integral part of every sustainable construction. The various methods like water harvesting, recycling and conservation reduce water wastage are now part of the green building concept. Water drained from bathrooms, washing machines and kitchen can be collected and reused for irrigation purposes. Choosing sanitary fixtures in the apartments which minimize the water consumption will also reduce the water wastage.
- **Indoor Air Quality:** One of the primary objectives of green building is to improve the indoor air quality to reduce the negative impact on the occupants of the building. A conventional home built without any green building considerations will have emissions that may affect the health and well-being of the individuals. Also, the poorly ventilated buildings and the facilities like do not sufficiently filter the recirculated air often harmful to the occupants. The various steps were taken by the green building to reduce the toxic effect and improve the indoor air quality. A good ventilation and filtration system to allow fresh air to flow inside the house and filter the hazardous particles in the circulated air, installing an exhaust system for radon gas, using a zero or low volatile organic compounds (VOC) for interior paint solvent-free finishes.
- **Waste Management:** Green Building concept emphasizes improving the design of the product, re-using and recycling materials. It results in a tremendous waste reduction and also helps to reduce the environmental impact of the building. Green building generally tries to minimize the amount of construction and demolition (C&D) waster they generate by recycling or reusing. For example, the inactive demolition material can be used as base material for parking and roadways. A plan should be generated to use the large number of demolition materials such as wood,concrete, and other types of drywall.
- **Construction Materials:** The green building concepts use the materials that are generally composed of renewable rather than non-renewable resources and are environmentally responsible. It is because their impacts are considered over the life of the product. Also, the materials used in the construction of green building cause less maintenance and replacement costs over the life of the building. It should save energy and improve resident health and output. The materials are chosen by evaluating the characteristics such as reused and recycled content, zero or low off-gassing of harmful air emissions, zero or low toxicity, sustainable and rapidly renewable harvested materials, high recyclability, durability, longevity, and local production. Locally sourced recycled materials can be a good choice. Recycled materials could be reclaimed metal or wood. Reclaimed wood can be used for furniture, wall panelling, on ceilings, flooring and much more. The recycled engineering materials minimize environmental impacts.

ORGANIZATIONS STANDARDS FOR GREEN BUILDING

In order to incorporate the various objectives of Green Building, several organizations has contributed towards formalized rating parameters and certification mechanisms. Further, these organizations ensure a building and associated maintenance systems are green aware. The standards and certification are complied with irrespective of the building. The worldwide, there exists both international organizations as well as specific to each nation. In fact, these organization exhibits different and unique perfection for green building guidelines.

The most established and widely used green building rating system is Leadership in Energy and Environmental Design (LEED) System (U.S. Green Building Council (USGBC), 2003). It was launched in 2000 and is presently managed by the Green Building Certification Institute (GBCI), a third-party organization created by USGBC to administer certification and accreditation programs related to green building. It is a point-based rating system. LEED allocates points in a wide range of sustainable criteria; including water efficiency, energy performance, and indoor environmental quality. The GBCI designates the building as Certified, Silver, Gold, or Platinum (the highest achievement level) according to the points achieved by the building.

The Green Building Initiative (GBI) (Bowyer, J. L. (2007)). is another organization which is committed to sustainable construction for residential and commercial buildings. It developed a green certification system known as Green Globes in 2004. It is also based upon meeting certain sustainable objectives like LEED; an applicant can achieve up to four "globes" depending on points awarded during the assessment process.

United Kingdom started the Building Research Establishment (BREEAM) in 1990. It is one of the first international rating and certification standards for sustainable buildings. BREEAM has covered the sustainability requirements for many different building types has expanded its reach throughout Europe (to include Germany, the Netherlands, Norway, Spain, and Sweden). The certification standards for sustainability are similar to LEED standards, with levels ranging from zero to six.

CASBEE (Comprehensive Assessment System for Building Environmental Efficiency), which was developed in Japan in 2003 (Murakami, S. et al. 2003). It collaborates with nongovernmental entities (such as industrial organizations and academia). It is based on BREEAM concepts, but tailored to meet the needs of Japan and Japanese buildings. It assesses the predesign, new construction, existing building, and renovation. Comprehensive assessments are graded at five levels: Class C (poor), Class B–, Class B+, Class A, and Class S (excellent). 35+

The International Code Council released a model code titled International Green Construction Code (IgCC) in 2012. The IgCC works in concurrence with other model codes (such as the International Energy Conservation Code) to promote minimum green standards for all buildings. Specifically, the IgCC needs to reduce the carbon footprint through reductions in energy usage and water usage and also pursues to ensure the health and safety of building occupants through a detailed commissioning process. Also, the IgCC includes land use regulations that address issues such as greenfield, preservations.

Green Star Rating System designed by The Energy And Resources Institute (TERI), INDIA, it identifies projects that have proved a promised sustainability in the design, construction, GRIHA certification system consists of 34 criteria of the rating under four categories. The categories are (i) site selection and planning, (ii) building planning and construction, (iii) building operation and maintenance, and (iv) innovation. Credits are awarded within each category. Based on the acquired points different levels of certification (one star to five stars) are awarded. The minimum required points for certification is

50. Points between 50 to 60 will get one star, between 61 to 70 points will get a two star, between 71 to 80 will get a three star, between 81 to 90 will get a four star and points 91 to 100 points will get the maximum rating, i.e. five stars.

APPLICATIONS OF GREEN BUILDING

The underlying concept of Green building refers to environmental protection applied to the design, construction and use of the building. Its ultimate aim is to prepare green technologies to achieve green building energy conservation, and sustainable development.

The main focus of Green building is (1) Energy saving (2) To protect the environment, by reducing environmental pollution and reduce the amount of carbon dioxide emissions; and, (3) healthy living space to the people. Green building from the design, construction, use, maintenance to the demolition of each link has a variety of energy saving and environmental requirements. In the design phase it is necessary to focus on the use of environmental factors, but also to minimize the adverse effects of the construction process on the environment. In the operational phase, provide people with low consumption, comfort, health and space. Reduce the degree of damage to the environment caused by demolition.

- **Energy Consumption:** The green technology aims to achieve higher energy efficiency. In the new construction, the green concept plays a role in every phase of the development. Every aspect like site, design, materials used, the system used to run and maintain are chosen to be sustainable and energy-efficient as much possible. For the existing constructions, the green concept can be incorporated by integrating the control systems that are already in place in the facility. Developing automation systems with advanced software features to control the electrical, mechanical and security features can incorporate the green concepts to save energy. For example, typically 30% -40% of a commercial building are unoccupied at any given time. Adopting green technology like automatically shutting -off lights and adjusting HVAC, cooling, heating and ventilation options can achieve 30% of savings in the energy expenses by removing unnecessary energy use in this manner.
- **Indoor Air Quality Monitoring:** Use of green building materials ensures the health and safety of the occupants. The products meet the following standards, including: (i) Building materials that are naturally sourced like wood, bamboo, cork, etc., or no-toxic ingredients are considered as green. (ii) Materials that emit zero or low levels of Volatile Organic Compounds (VOCs). (iii) The materials that are moisture resist that dry quickly should be used for green construction.
- **Smart Lighting**: Generally, the commercial buildings, the lighting consumption accounts more than 30% of the total power consumption. Artificial daylight lighting could be reduced by introducing the sunlight into the interior of a building which reduces the power consumption and environmental pollution.

CHALLENGES OF GREEN BUILDING

Despite green building practices are gaining increasing acceptance in the construction industry for the health and environment-friendly buildings, the acceptance of Green Building technologies is not as deceptive as it would have been.

- **Limited Awareness:** Most of the users are ignorant of the concept of green building and its long-term benefits. Also, those who know little about these concepts feel that it is expensive to achieve.
- **Technological Challenge:** Some of the technical risks involved in the Green building concepts vegetative roofs and indoor air quality (IAQ). In vegetative roofs, it must be carefully designed and due to a constantly wet condition may become riskier than conventional roofs. It may result that reusing or recycling the existing building components or recycled components, may not be approvingly integrated with the new materials or may not provide optimum performance. The field-test of new green construction material has not been performed over time is the risk factor associated with the technical issue. It is mandated that the new materials and the risks associated with it need to be tested compared to conventional building construction. Another technical issue may be due to increased ventilation to meet indoor air quality (IAQ) that may affect the humidity level during the hot or humid weather. The proper care should be taken in the design to dehumidify the building when exceeding the minimum outside air requirements for heating, to refrigerate and air-conditioning as per the recommendations.
- **Financial Challenge:** The most important factor in the implementation process of the green construction involves cost efficiency (Meryman et al., 2004). In the process of decisions related to the implementation of green construction, cost efficiency is considered to be the most important factor (Meryman et al., 2004). One main issue in the implementation of green construction is that it involves an extra cost (Ofori, et al., 2004). The cost of water and energy saving equipment's and high performance of insulation protects costs high. According to (Tagaza, et al., 2004) the capital costs are usually 1 - 25% higher in green projects in comparison with conventional projects. The green materials are s are normally 3 - 4% higher if compared with conventional construction materials (Zhang, et al., 2011). In addition, the complexity in design and modeling also causes an extra cost. The project manager should manage the high cost of the green construction as the budget is pre-allocated to them and the manager has to control the project under the budget.
- **Policy and compliances:** In real life scenario, the building consists of many parts such as office, business center, campus, clubs, departments, cafeteria, entertainment arena, shopping complex, homes, etc. These components of building inhibit different requirements in terms of building management perspective. Thus, different policies, access control mechanism, security measures, fault tolerance are required to be framed with respect to the building components. (Jianli Pan et al., 2015) discussed the tree like structure for adopting energy saving policy at the various building components. For example, in case of the housing complex, each family may have different requirements and preferences to be taken account in the IoT system for each individual family. Hence, there exists a hierarchy among the policies and adjustments at various levels of this hierarchy. This hierarchy basically insists policy controllers at different levels enforcing to adopt policies for energy saving.

ARCHITECTURE OF IOT FOR GREEN BUILDING

The IoT based Green Building architecture consists of four primary layers, namely capillary network layer, Access Network layer, middleware layer and service layer (Gu et al., 2017, Bunning et al., 2017; Azar et al., 2017) and is depicted in figure 1 below.

Capillary Network Layer

It is important to understand the basic distinguish between IoT and Wireless Sensor Networks (WSN). The WSN is a homogeneous network of sensors and are meant for single objective. These homogeneous sensors are embedded in the physical environment to perform certain defined functions. In IoT networks, the sensors are having heterogeneous features and autonomous functionalities. In IoT based Green building architecture, the capillary network is the lowest layer. The capillary network layer consists of IoT sensors, actuators, computing devices and WSNs. However, the capillary network components are limited in resources such as computing power, battery operated, memory, limited storage and networking capability. The IoT sensors perform different task by fetching different readings from environment at different time and space scales. The main benefits of a heterogeneous network of IoT sensors provide extensive options of services that are domain specific and context aware IoT applications. In fact, the IoT sensors and actuators are constantly evolving according to the wide variety of application services to meet human needs. Thus, the capillary network layers provide a voluminous amount of data gathered through IoT sensors. The sensor data collected from the environment are then sent to the computing nodes through the access network layer.

Access Network Layer

The collection of sensor data in capillary network layer must be performed both asynchronous and synchronous operational modes depending on the diverse application services. Hence, the main challenge at the access network layer is to have a flexible middleware configuration in order to manage the end-to-end transmission for a wide range of applications at different modes of operations. Further, the access network layer should be flexible for different domain controls and also optimized towards load balancing.

Middleware Layer

The middleware layer incorporates a building management system that integrates with sensors and actuators to provide a unified platform for various building services. The middleware layer includes operating systems, run time environments, libraries and domain specific software. The main objective of middleware is to bridge the different software services and perform the intended operations in a coherent way, to get the services done as per requirements of IoT applications. Basically, there exists a unified flow of data and information from various dimensions of the building. Middleware provides well defined insight about the building through sensors; even for those parts of the building that human cannot be accessed. For this purpose, the continuous monitoring of the building is performed. Then, the monitored data is collected, stored, computed and analyzed. In the IoT based framework, the IoT protocol MQTT, CoAP, REST APIs, XMPP is interfaced at the middleware according the requirement specification of Green building.

Service Layer

The objective of the service layer is to provide data handling and knowledge generation as per the different requirements of Green building application users using IoT smart devices. The service layer incorporates programming applications which perform live data streaming of real world parameters

Figure 1. Basic architecture of iot for green building

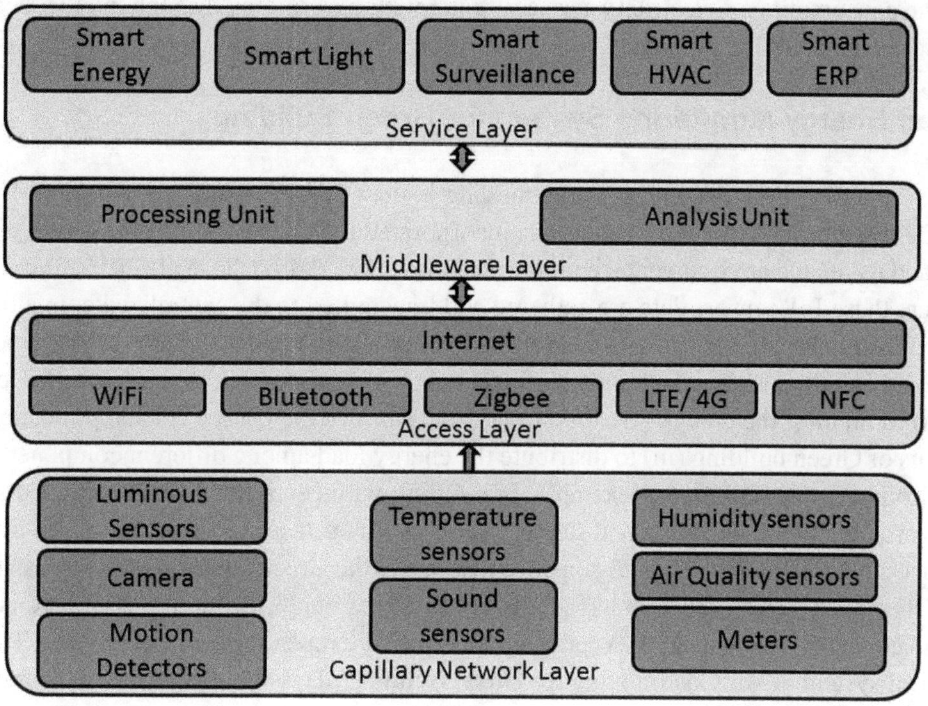

through IoT sensors. Next, the query engine present at the middleware layer calls respective application program interfaces to process the data and send it to the remote storage databases. The services such data access, data control, data computation and data analysis are performed. The service layer incorporated with powerful advanced data visualization on to a single centralized platform at the convenience of the user web application or mobile application.

ROLE OF IOT FOR GREEN BUILDING

In recent years, the IoT application for smart building attracts researchers and industrial developers to study and explore the adaptability of this concept into everyday scenarios (Kerdan et al., 2016). This section provides overview of incorporating IoT into the building management, challenges and opportunities of such integration. The green building using IoT technology aims to explore the design challenges for energy conservation, such that there is a significant reduction in the energy consumption of various components of buildings. Thus, the main objectives for integrating IoT with building management systems is to provide a green concept in terms of monitoring and reducing the consumption of resources such as electricity, gas and water (Barbon et al., 2016). Further, to increase the efficiency of the building by monitoring and analysing various operations such as lighting system, building occupancy, air quality, noise monitoring, temperature and humidity monitoring, heating system, air-conditioning and ventilation systems of buildings. To handle various dynamic requirements of building during the maintenance process with the help of IoT technology and predictive analysis and tools. To monitor and track movement of various activities and occupancies those are happing within and around the buildings. Further, to optimize the consumption of electricity, gas, and water within a building through intelligence statistical analysis mechanism and IoT systems.

IoT Based Energy Monitoring System for Green Building

One of the common problems under Green building is the energy saving. In order to perform energy saving, different types of IoT sensors nodes are incorporated to monitor the current level of energy consumption and its related environmental parameters (Rosa et al., 2017, Cho et al., 2016, Lawerence et al., 2017). Then, these IoT sensors data are gathered and transmitted to the centralized computing centres. This computing centre further consists of control systems. These control systems are incorporated with predictive and cognitive techniques using intelligent algorithms. The objective of these IoT control systems are (i) to monitor the energy consumption (ii) to optimize the energy consumption among different elements of Green building (iii) to distribute the energy load among different components of Green building (Rodden et al., 2003). For example, in building with centralized air-conditioned system, the objective is to effective monitoring and saving of energy. In such scenario of centralized systems, it is obvious that the AC is switched ON in empty rooms, while the Green building concept insist that AC is switched ON only in those rooms that has occupants. In this example, the monitoring of AC is associated with a head count of occupants in the room and the external temperature recordings also. Thus, the IoT based control system of Air Conditioning for Green Building plays significant role in energy savings.

IoT Based Lighting System for Green Building

(Bates et al., 2012; Gupta et al., 2010; Singhvi et al., 2005) discussed about smart lighting systems of indoor scenarios using illumination-based sensing. The highlights of this work are that both daylight and the human occupancy parameters are considered together. The objective of this works is to have optimized control over the illumination of the room along with minimizing the overall energy consumption. The article has discussed both centralized and decentralized approaches for IoT based lighting system. (Strengers et al., 2014) discussed about the reinforced Q-learning techniques for performance improvement of lighting control systems. The paper also presented how user comfort can be improved along with reduction in energy consumption. Further, the work proposed integrated mechanism of reducing heating as well as cooling of HVAC systems along with lighting systems. (Churchill et al., 2013) has proposed occupancy, based control of lighting systems in office spaces. The authors discuss how the lighting system can be developed, tested and post occupancy evaluation mechanism based on the human head count in the room. The basic principle behind occupancy based lighting system is that, whenever there is a motion within the coverage area, the lights will be switched ON. Further, depending on the proximity of human, the lights will be dimmed gradually or switched off if no human detected after a certain threshold of a period. However, the studies conducted by National light product information program (NLPIP) there exist a large number of invalid cases of motion sensing based light control systems. Further, these mechanisms of lighting systems fail to detect the human motion beyond certain coverage range contradicting to what the product manufacturers provides in the market.

IoT Based Surveillance System for Green Building

The objective of Surveillance system is to observe and learn the human activity in a way energy consumption can be controlled (Wayes Tushar et at., 2018). Hence, based on IoT sensors and devices, the surveillance can be performed by using machine learning techniques, by learning about people in the building and hence effectively utilization of energy based on these factors (Paul et al., 2008; Liu et al., 2009). In literature, the human surveillance system is carried out using two main techniques, namely Open CV based and transfer learning based. For both these techniques, the primary mechanism is to perform image processing up on the images captured through the video camera. Usually, fish eyed camera lens are used for capturing the video under the coverage area. The images are captured at the rate of 60- 30 frames per second. These captured images are then processed further.

The OpenCV provides a wide range of image processing libraries which can be used for applications such as detecting human from random images. OpenCV uses colors based detection methods. Along with this, the background substitution technique to filter out the noises and unwanted objects such as doors, wall, furniture, etc. been identified. However, the efficiency of this standalone OpenCV technique is very poor. Hence, it requires enhanced techniques such deep learning and convolution neural networks (CNN). Although these techniques have made significant contributions in improving performance of the image processing, it involves complex programming. Also, it becomes inefficient at handling large scale image processing. Further, these techniques are constrained by the threshold level of illumination over images captured.

In the transfer learning method, to perform human detection based on image processing, two parameters are incorporated namely feature maps and aspect ratio of the images (Jiang et al., 2009). Further, SSD Multibox is prebuilt software platform provides efficient methods that work on the principle of feature maps and aspect ratio. SSD Multibox can detect a wide range of objects based on six sets of feature maps extracted from an input image. A match is detected based on Jaccard threshold technique. SSD models involve transfer learning and fine-tuning mechanisms. The transfer learning works on the principle of feedback based on pre-trained weight assignment for extraction of various features from images. Then, using these pre-trained weights the values are used for retraining through remaining iterations of image processing.

However, (Wayes Tushar et at., 2018) demonstrated that SSD Multibox and OpenCV as such failed to detect human from video images under low illuminating conditions. Whereas SSD mailbox enhanced with transfer learning has outperformed in both bright and low illuminated scenarios of human identification.

CHALLENGES OF IOT FOR GREEN BUILDING

The primary challenges of IoT form Green Building are quality of data, computing models, fault tolerance, middleware components and these are depicted in figure 2 below.

- **Quality of Data:** The IoT for Green Building has a major challenge as the quality of data that are sensed through the IoT sensors and devices (Wan et al., 2011; Cho et al., 2016; Lawrence et al., 2017). The data obtained through IoT sensors are influenced by temporal and spatial details, that dedicates how the entire IoT system should behave. The temporal factor of data determines the lifetime. Further, indicates when the data needs updating, deletion, validation, sampling rate and how long to be stored. While the spatial factor of data includes location of data sensed, space in storage system, accuracy, and lifetime. The lack of these factors on the IoT sensed data will lead to misinterpreted or misused due to the lack in the quality of data.

- **Computing Models:** There are two types of computing models, namely centralized or decentralized. These two models have their own advantages and disadvantages (Flouquet, 1992; Capehart et al., 2014). Hence, the decision of selecting an appropriate model depends on the objective of the IoT system and its domain of application. Centralized mode is independent of user involvement. The update of software and hardware deployment is done at one centralized server. However, the demerits is that less prone to fault tolerant. Further, the failure of the centralized server will lead huge loss of data and IoT system process. The scalability is also limited in centralized model. On the other hand, decentralized model facilitates the distributed services across the IoT devices. The Internet is not required as such, because the system is already distributed extensively through access network, like wireless or mobile communication technology. However, the implementation of the decentralized model involves complex programming and computing techniques to be incorporated in embedded systems, IoT sensors and Devices and also at computing centres. Further, the performance metrics such as delay, throughput, and algorithm efficiency are challenging to be achieved.
- **Fault Tolerance:** (Lazarova et al., 2016) discussed IoT systems for green Building may be prone to faults due to circumstances or voluntary activities. The IoT systems are heterogeneous and also autonomous, thus implementation of fully effective fault tolerance mechanism is tedious. Because of the IoT system fault, the regular lifestyle of users may be affected or endangered. Hence, monitoring of fault is essential challenging issue in IoT system of Green Building.
- **Middleware:** The middleware of the IoT green building must be efficiently designed to handle and incorporate privacy and security of data, customer access, storage devices and data centres (Motegi et al., 2003; Papantionious et al., 2014). Moreover, the access controls of various computing devices, control systems and IoT sensors need to be handled by middle wares.

Figure 2. Primary challenges of IoT based green building

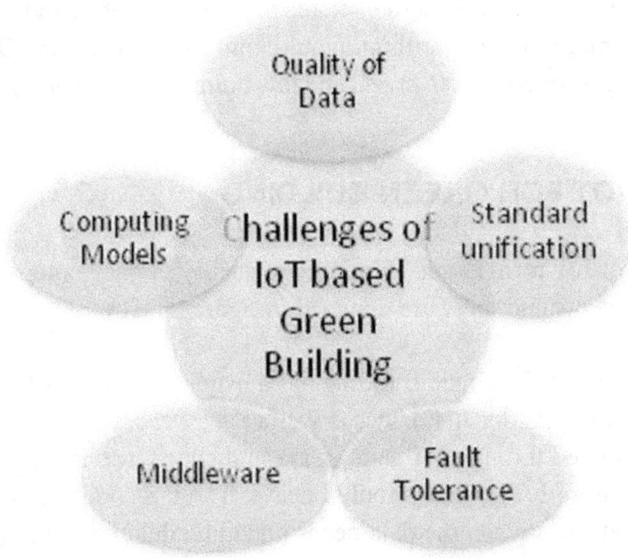

- **Standard Unification:** In the growing trend of IoT based Green building, there exist several components and role players in terms of communication protocols, edge devices, networking capabilities, and medium accessing standards. Hence, the unified framework for accommodating these proprietary and diversified standards and protocols within one umbrella is quite challenging. Particularly, the interoperability among different techniques is highly tedious. Each of these standards is heterogeneous and has unique requirements and specifications. Thus, unification incurs huge amount of economic cost and time consuming.

CLOUD COMPUTING FOR IOT BASED GREEN BUILDING

The concept of Green building ascertains the sustainability in the business of architecture. This goal can be achieved through a combination of several emerging technologies like IoT, Could Computing, Big Data and Cyber physical systems (Curry et al., 2013). Further, the integration of these technologies ensures that the technical standards and codes for energy efficiency and other factor green building are incorporated. Basically, the IoT system has the capability to integrate various stakeholders of the Green building such as occupants, environment, physical component and sensors. However, different types of applications, automation, monitoring and control modules are involved in the Green Building management systems. Further, the complexity is high in the construction of the high raised building. Also, these buildings have a huge impact on the socio economical factor; it has several open problems yet to be handled. In addition, the factors such as limited manpower and constrained resources are also considered in IoT based systems in Green building. On the other handle, the technical aspects like the communication of IoT devices, data gathering through heterogeneous IoT systems and storing of these data for analysis are quite challenging. Figure 3 depicts the sample Integrated Framework of IoT and Cloud based for Green Building

As a solution to this, in recent trends of Green building the construction involves autonomous Building Management System (BMS) that involves integrating IoT system along with remote cloud computing platform. The building managing system basically involves server and client-based cloud architecture. The functional components are contained by the remote cloud servers. These functional components perform operations such as monitoring of building and environment surrounding it. In addition, management of data, controlling of IoT systems, data processing over gathered data, and providing services as per the IoT system requirements. The user of the building management system can operate and perform monitoring and controlling of various building components through simple and user-friendly client modules.

According to (Ming Wang et al., 2017), the cloud based IoT system for Green Building involves four basic layers, namely device, networking, cloud platforms and service layer. The device layer incorporates hardware and software enabled sensors, computing devices and actuators. These devices has support of intelligence mechanism to sense the building and its surrounding environment, control various equipment of building, measure energy consumption, occupant related information. These sensed data and its information are relayed over the network to the controlling nodes. The environmental measurement involves data, such as carbon monoxide, benzene, illumination level, temperature, occupant head count, soil particles in the air, and humidity of air. The control related parameters are like monitoring and controlling electrical equipments and appliances in building, electrical meters. The electrical measurement involves measuring voltage, current and power usages. All these devices and measurements gathered by the device layer constitute huge amount of data to be stored and processed. Hence, all the integrated

Figure 3. IoT and cloud based integrated framework for green building

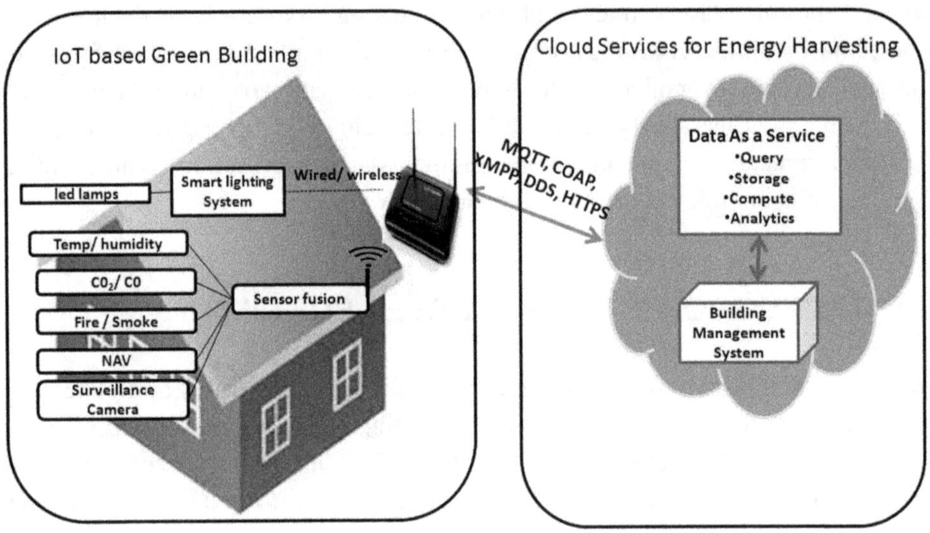

components are networked to the remote cloud server for the Green building management system. The cloud platform provides services for green building in the form of application program interfaces (API), access control and policy mechanism, registration and authentication processing, and optimization of different functional modules. Finally, at the cloud service level, various functionalities as monitoring, controlling, knowledge gathering, data analysis, data visualization, alerting and information dissemination across IoT devices and users are provided as services. These services are customized as per the application requirements by simple procedures on the application browser of the user interface on the client side.

CONCLUSION

In this chapter, we explored the issues, challenges, and opportunities in incorporating IoT towards the objective of Green Building. To begin with, we presented the collection of definitions of Green Building as followed by different organizations and standard bodies. Then presented the objectives, applications, and organizations of Green Building. The chapter addressed various Green Building Organizations, Standards across international and national level. Next, we focused on the emerging technologies and its role towards green building. Further, the chapter presented proliferation of IoT as technology to facility green building concepts. However, there exits various challenges while integrating IoT with Green building such as the quality of data, fault tolerance of the green building management system, computing models and unified middleware solutions. Further, the chapter presented how the amalgamation of other enabling technology like cloud computing with IoT facilitates the Green building life cycle, namely automation, optimation, recommedation, maintenance and replacement of building elements.

REFERENCES

Azar, E., & Ansari, H. A. (2017). Framework to investigate energy conservation motivation and actions of building occupants: The case of a green campus in Abu Dhabi, UAE. *Applied Energy*, *190*, 563–573. doi:10.1016/j.apenergy.2016.12.128

Barbon, G., Margolis, M., Palumbo, F., & et al, . (2016). Taking Arduino to the Internet of Things: The ASIP programming model. *Computer Communications*, *89-90*, 128–140. doi:10.1016/j.comcom.2016.03.016

Bates, O. (2012). Accounting for EnergyReliant Services within Everyday Life at Home. *Proc. Int'l Conf. Pervasive Computing*, 107–124. 10.1007/978-3-642-31205-2_8

Bowyer, J. L. (2007). Green building programs-Are they really green? *Forest Products Journal*, *57*(9), 6.

Bünning, F., Sangi, R., & Müller, D. (2017). A Modelica library for the agent-based control of building energy systems. *Applied Energy*, *193*, 52–59. doi:10.1016/j.apenergy.2017.01.053

Capehart, B. L., & Brambley, M. R. (2014). *Automated Diagnostics and Analytics for Buildings*. Lilburn, GA: Fairmont Press.

Chatterjee, A. K. (2009). Sustainable construction and green buildings on the foundation of building ecology. *Indian Concrete Journal*, *83*(5), 27–30.

Cho, J., Kim, J., Lee, S., & Koo, J. (2016). A bi-directional systematic design approach to energy optimization for energy- efficient buildings. *Energy and Building*, *120*, 135–144. doi:10.1016/j.enbuild.2016.03.066

Churchill, E. F. (2012). From Data Divination to Data-Aware Design. *Interaction*, *19*(5), 10–13. doi:10.1145/2334184.2334188

Curry, E., O'Donnell, J., Corry, E, & . (2013). Linking building data in the cloud: Integrating cross-domain building data using linked data. *Advanced Engineering Informatics*, *27*(2), 206–219. doi:10.1016/j.aei.2012.10.003

EPA. (n.d.). *Green Building*. Available at http://www.epa.gov/greenbuilding/pubs/about.htm

Flouquet, F. (1992). Local weather correlations and bias in building parameter estimates from energy-signature models. *Energy and Building*, *19*(2), 113–123. doi:10.1016/0378-7788(92)90005-2

Gu, W., Wang, J., & Lu, S. (2017). Optimal operation for integrated energy system considering thermal inertia of district heating network and buildings. *Applied Energy, 199,* 234-246.

Gupta, S., Reynolds, M. S., & Patel, S. N. (2010). Electrisense: Single-Point Sensing Using EMI for Electrical Event Detection and Classification in the Home. *Proc. 12th ACM Int'l Conf. Ubiquitous Computing*, 139–148. 10.1145/1864349.1864375

Howard, J. L. (2003). *The federal commitment to green building: experiences and expectations*. Washington, DC: Federal Executive, Office Of The Federal Environmental Executive.

Huovila, P. (2007). *Buildings and climate change: status, challenges, and opportunities*. UNEP/Earthprint. Available at http://smap.ew.eea.europa.eu/media_server/files/ R/S/UNEP_Buildings_and_climate_change.pdf

Jiang, X., Ly, M. V., & Taneja, J. (2009). Experiences with a high-fidelity wireless building energy auditing network. *Proc. 7th ACM Conf. Embedded Netw. Sensor Syst. (SenSys'09),* 13–126. 10.1145/1644038.1644050

Kerdan, I. G., Raslan, R., & Ruyssevelt, P. (2016). An energy-based multiobjective optimisation model for energy retrofit strategies in nondomestic buildings. *Energy, 117*, 506–522. doi:10.1016/j.energy.2016.06.041

Kibert, C. J. (2008). *Sustainable construction: green building design and delivery*. Hoboken, NJ: John Wiley and Sons.

Lawrence, T. M., Watson, R. T., Boudreau, M., & Mohammadpour, J. (2017). Data flow requirements for integrating smart Buildings and a smart grid through model predictive Control. *Procedia Engineering, 180*, 1402–1412. doi:10.1016/j.proeng.2017.04.303

Lazarova-Molnar, S., Shaker, H. R., Mohamed, N., & Jørgensen, B. N. (2016). Fault Detection and Diagnosis for Smart Buildings: Trends and Challenges. *Proc. the 3rd MEC Int'l Conference on Big Data and Smart City (ICBDSC 2016).*

Liu, L., Kuo, S. M., & Zhou, M. (2009). Virtual sensing techniques and their applications. *IEEE International Conference on Networking, Sensing and Control*, 31-36.

Meryman, H., & Silman, R. (2004). Sustainable engineering - using specifications to make it happen. *Structural Engineering International, 14*(3), 216-219.

Motegi, N., Piette, M. A., Kinney, S., & Herter, K. (2003). Web-based energy information systems for energy management and demand response in commercial buildings. University of California: Lawrence Berkeley National Laboratory. doi:10.2172/823939

Murakami, S., (2003). Study on Comprehensive Assessment System for Building Environmental Efficiency Part 1-9. *Summaries of Technical Papers of Annual Meeting Architectural Institute of Japan*, 993-1006.

Ofori, G., & Kien, H. L. (2004). Translating Singapore architects' environmental awareness into decision making. *Building Research and Information, 32*(1), 27–37. doi:10.1080/09613210210132928

Pan, J. (2013). Toward an energy-proportional building prospect: Evaluation and analysis of the energy consumption in a green building testbed. *Proc. IEEE EnergyTech*, 1–6. 10.1109/EnergyTech.2013.6645358

Papantoniou, S., Kolokotsa, D., & Kalaitzakis, K. (2015). *Building optimization and control algorithms implemented in existing BEMS using a web-based energy management and control system. Energy and Buildings, 98,* 45-55.

Paul, S., Jain, R., Pan, J., & Bowman, M. (2008). A vision of the next generation internet: A policy-oriented perspective. *Proc. Brit. Comput. Soc. (BCS) Int. Conf. Vis. Comput. Sci.*, 1–14.

Robichaud, L. B., & Anantatmula, V. S. (2010). Greening project management practices for sustainable construction. *Journal of Management Engineering, 27*(1), 48–57. doi:10.1061/(ASCE)ME.1943-5479.0000030

Rodden, T., & Benford, S. (2003). The Evolution of Buildings and Implications for the Design of Ubiquitous Domestic Environments. *Proc. SIGCHI Conf. Human Factors in Computing Systems*, 9–16. 10.1145/642611.642615

Rosa, T. R., Betim, F. S., & Ferreira, R. Q. (2017). Development and application of a lab made apparatus using open-source "arduino" hardware for the electrochemical pretreatment of boron-doped diamond electrodes. *Electrochimica Acta, 231*, 185–189. doi:10.1016/j.electacta.2017.01.180

Singhvi, V., Krause, A., & Guestrin, C. (2005). Intelligent light control using sensor networks. *Proc. 3rd Int. Conf. Embedded Netw. Sensor Syst.* 10.1145/1098918.1098942

Strengers, Y. (2014). Smart Energy in Everyday Life: Are You Designing for Resource Man? *Interaction, 21*(4), 24–31. doi:10.1145/2621931

Tagaza, E., & Wilson, J. L., (2004). *Green buildings: drivers and barriers - lessons learned from five Melbourne developments.* Report Prepared for Building Commission by University of Melbourne and Business Outlook and Evaluation.

Tushar, W., Wijerathne, N., & Li, W.-T. (2018). Internet of Things for Green Building Management: Disruptive Innovations Through Low-Cost Sensor Technology and Artificial Intelligence. *IEEE Signal Processing Magazine, 35*(5), 100–110. doi:10.1109/MSP.2018.2842096

U.S. Green Building Council. (2003). *Green Building Rating System for New Construction & Major Renovations.* (LEED-NCTM) Version 2.1. Available at http://www.usgbc.org/Docs/LEEDdocs/LEED RS v2-1.pdf

Wan, K., Li, D., & Lam, J. (2011). Assessment of climate change impact on building energy use and mitigation measures in subtropical climates. *Energy, 36*(3), 1404–1414. doi:10.1016/j.energy.2011.01.033

Wang, M., Qiu, S., Dong, H., & Wang, Y. (2017). Design an IoT-based building management cloud platform for green buildings. *2017 Chinese Automation Congress (CAC)*, 5663-5667. 10.1109/CAC.2017.8243793

Zhang, X. L., Shen, L. Y., & Wu, Y. Z. (2011). Green strategy for gaining competitive advantage in housing development: A China study. *Journal of Cleaner Production, 19*(1), 157–167. doi:10.1016/j.jclepro.2010.08.005

Chapter 6
Green Building Efficiency and Sustainability Indicators

Mir Sayed Shah Danish
https://orcid.org/0000-0002-5196-2280
University of the Ryukyus, Japan

Tomonobu Shah Senjyu
https://orcid.org/0000-0003-4494-6773
University of the Ryukyus, Japan

ABSTRACT

A rapid change in technology trend and lifestyle around the globe has induced a drastic increase in energy production, delivery, distribution, and consumption. That forced building planners, designers, scholars, researchers, and practitioners to come up with a sustainable solution within constrained economic and environmental dimensions. With a proper definition and usage of efficiency and sustainability dimensions in terms of green building design and construction, global challenges (global warming, climate change, poverty, global health and education, etc.) can be mitigated, leading to long-run sustainability. This chapter presents indicators to define, manage, measure, and enhance efficiency and sustainability phenomena for proposing a green building. A primary objective of this study is to identify influencing factors and set forth viable indicators and framework in terms of energy-efficient green building from different standpoints hiring innovative tangible and non-tangible tools and technique.

INTRODUCTION

Energy consumption is an indicator of the economic growth of a nation. According to International Energy Outlook (IEO), world energy consumption will increase by 53% between 2008 and 2035 (Jeffs, 2017). Energy efficiency in building refers to save energy for producing the same service or useful output (Patterson, 1996). According to (Liu and Mi, 2017), buildings consume 40% of the total energy and emit 30% of the greenhouse gases (GHGs) worldwide. Therefore, since many decades' energy efficiency

DOI: 10.4018/978-1-5225-9754-4.ch006

and management in the building remains an interesting topic for researchers around the globe. Recently, energy consumption in buildings estimated one-third of total primary energy resources (Danish, Senjyu, Ibrahimi, Ahmadi, & Howlader, 2019). Therefore, energy efficiency and saving become a matter of focus on well-managed energy consumption and supply.

Measuring building energy efficiency is a problematic endeavor due to lack of definitive-quantitative measurement (McNeil & Rue du Can, 2016). While, a series of energy efficiency indicators facilitate to assess a building energy efficiency. As a matter of fact, efficiency and sustainability indicators are not solely data or simple information that can be defined conveniently (International Atomic Energy Agency, 2005). Efficiency and sustainability indicators in the context of a green building are beyond basic statistics to deal with an in-depth analysis of multidimensional aspects. Hence, these indicators establish a relationship between a parameter's variables to regulate important aspects related to green building efficiency and sustainability. These dimensions are associated with green building behavior in term of energy production, delivery, distribution, and consumption; considering efficiencies (technical, technological, ecological, and etc.) and sustainability pillars (economic, social, institutional, technical, and environmental).

As a matter of fact, by proper application of the proposed efficiency and sustainability indicators within management disciplinary, which backed by theories and practices, it can offer the practitioners, scholars, researchers, scientist, and students with an exhaustive investigation from every aspect of a situation to adapt in real-life challenges. This chapter offers the opportunity to understand a systematic approach of efficiency and sustainability indicators and their proper application. As well as, to diminish energy losses by any means, and ensure efficiency and reliability, reduce pollutant emissions, and buoyant socio-economic development. Besides, this chapter explores efficiency and sustainability dimensions in detail along with the importance of state-of-the-art building innovations.

The background section explores previous studies on deployment of renewable energy in term of sustainable building, followed by energy-efficient building indicators and examples. This section as a glance at the literature looks for assessing world-wide practices and success stories of the green building in order to figure out an optimum solution to be fit problems in this chapter. Next section discusses the significance of efficiency and sustainability indicators. This section deals with four main subsections: scoping objectives, elucidating boundaries, develop indicators, indicators localization, which are mainly associated with economic, environmental, technical, regulatory, social, and efficiencies constraints. This section sums up with an appropriate roadmap for development of efficiency and sustainability indicators by ensuring affordability, accessibility, marketability, efficiency, durability, conform, and many other aspects. The Efficiency and Sustainability indicators formulation section offers a roadmap for a green building planning, design and implementation based on measurable and manageable indicators. This section aims to provide a viable framework for development and considering applicable indicators for a green building. The following section describes the green building framework backed by different approaches and sceneries: consensual-based approach, efficiency oriented-based approach, and sustainability oriented-based approach. This section aims to propose an emerging solution for an energy-efficient building based on an exhaustive investigation of different approaches. Then, energy-efficient green building within real-life scenarios and practical case studies are discussed. As well as, this section explores the criteria for deployment of renewable energy in an energy-efficient building. Building Sustainability

Requirements section covers five pillars of sustainably with related energy indicators in detail. Besides, in this section, criteria for determining sustainability indicators are discussed as well. After, greening through decarbonization, climate change, and global warming is concerned. This section sums up with the available solution to overcome these global challenge trough decarbonization and motivating green building followed by sustainability and efficiency appraisal justification for an energy efficient building. This section reviews criteria (appropriately, preciseness, practicability, affordability, simplicity, reliability, sensitivity, obviously, testability, and acceptability) that can evaluate building sustainability and efficiency. Afterward, the lifecycle analysis of green building construction from initiation to complete construction is discussed. This section mainly focuses on the project management of a green building in view of technical and technological efficiencies improvement. Finally, an overall recommendation is proposed for renewable energy deployment from literature and the authors' past experience and studies. These recommendations are calcified into five classes, which are discussed in detail. At last, the overall summary, results, and discussion continued with the outcome of the chapter is concluded.

BACKGROUND

There are various approaches to assess a building performance from sustainability and high-quality standpoints. These approaches are mainly relied on the weighting and scoring methodologies based on different codes (Altomonte & Schiavon, 2013; Fagin &Wimmers, 2000). According to Danish, Senjyu, Ibrahimi, Ahmadi, & Howlader (2019), to establish the minimum level of energy efficiency for new buildings, and for renovations to existing buildings must demonstrate an 18%, and 14% improvement, respectively. This chapter can be counted a good reference for practitioners and researchers as a concise set of efficiency and suitability indicators, lifecycle, process analysis, and policy development guide with real-world green building application.

(Huo, Yu, and Wu, 2018) pointed out Kendall's coefficient of concordance (Kendall's W) approach for a green building site planning and design based on land use, site assessment, open space, passive building design, green-vehicle parking, landscaping and irrigation, stormwater management, neighborhood daylight access, cultural heritage, microclimate around buildings, reduced parking footprint, ecological value and protection, and environmental management plan factors.

(Sharma, 2018) reported the main challenges faced green building in the developing countries-India, Malaysia, and Saudi Ariba. These barriers are highlighted as decision-making (on time, and budget and risks), external drivers (corporate, property, project and individual levels), economic, political, behavioral, and cultural, lack of awareness from green building.

(He, Kvan, Liu, and Li, 2018) studied and compared green building rating system since the 1990s. The authors tried to benchmark findings based on Green star, the Leadership in Energy and Environmental Design (LEED), and the Assessment Standard for Green Buildings (ASGB) bodies of standards. Under a specific standard, special indicators are evaluated for example: Green stare (management, indoor environment quality, energy, transport, water, materials, land use and ecology, emissions), LEED (location and transportation, sustainable sites, water efficiency, energy and atmosphere, material and resources, indoor environmental quality), for ASGB (land saving and outdoor environment, energy saving and energy utilization, water saving and water resource utilization, material saving and material resource utilization, indoor environment quality).

(Jensen, Maslesa, Berg, and Thuesen, 2018) investigated sustainable buildings within main three options (heating and cooling demand reduction, energy efficient equipment, low energy technologies, and renewable energy supply), followed by benefits of merits and challenges of the proposed options. Different assessment and planning tools were supposed for a sustainable building within constrained environmental and ecosystem aspects. In which, the Building Research Establishment Environmental Assessment Methods (BREEAM) was the first to be launched in 1990 (Lambrechts, Gelderman, Semeijn, & Verhoeven, 2019).

THE SIGNIFICANCE OF AN INDICATOR

Indicators provide a clear picture of an intricate phenomenon in term of quantitative and qualitative measures (Popovic, Barbosa-Póvoa, Kraslawski, & Carvalho, 2018; Tripathi, 2016). Indicators used to measure the level or state of a phenomenon in term of tangible and nontangible, positive or negative result or outcome. Since decades, indicators are frequently applied in form of qualitative and quantitative measures in a broad term of applications. In a great deal, these indicators seek to evaluate the existing state of a case or situation in comparison of a predefined benchmark. Whereas, each indicator highlights a specific target, a set of these indicators can have an exhaustive consequence (International Atomic Energy Agency, 2005).

Scoping Objectives

Scoping the objectives of a building project in view of client expectations within constrained local and global criteria. These constrain can be local and national government law and regulations' limitations, efficiencies (technical, technological, ecological, etc.) requirements, and overall sustainability exigency.

Elucidating Boundaries

Well-defining the boundaries of a building in term of efficiency and sustainability, and their applicable terms and pillars. The next step is selection of optimum-viable parameters to be ensured for building design and implementation.

Develop Indicators

Correlating boundaries in term of measurable and manageable indicators inconsistent with overall objectives is an important endeavor. At this point, core indicators clarify, and suggest ways how to adapt, measure, and report objectives' outcome using these indicators (Adaptation Fund Board, 2011). In the meantime, selection of proper tools and techniques to realize the indicators for monitoring, control, and quality assurance are known pertinent.

Indicators Localization

Referring to the statistics about energy poverty, it seems this concept is perceived somehow different from its original purpose. Eventually, energy poverty defines lack of access to energy for any routine consumptions at the household and individual levels. In the present sufficient energy supply, constraints that can be caused to prevent using available energy by any means including cost affordability, energy-efficient appliances accessibility, east in primary energy resources accessibility, institutional and cultural barriers, and etc. are known energy poverty factors (Rademaekers et al., 2016). Energy power has significant social, ecological, health, and economic consequences. Therefore, addressing energy poverty indicators for green-sustainable building design is important. These indicators localize and define case by case for specific climate, economic, culture and so on. Among all indicators, supply and demand infrastructures, policy and regulatory effectiveness, and demographical suitability factors are influential.

INDICATORS FORMULATION

This section formulates a logical and systematic framework to sequence indicators in form of an emerging solution for green building planning and design. Among various approaches, few are hired here considering suitability and pertinency of these approaches to sustainability and efficiency. An indicator formulation itself, more importantly setting indicators framework are a complex and transdisciplinary endeavor. This chapter applied breakdown methodology and exposed indicator formulation based on indicator origination and foundation. The proposed framework is developed based on the benchmarking technique using efficiency and sustainability parameters. The benchmarking approach regulates and compares available parameters based on a predefined standard baseline (Duverge, Rajagopalan, Fuller, & Woo, 2018). This technique is successfully evaluated for building energy efficiency in the past (Chung, Hui, & Lam, 2006; Shabunko, Lim, Brahim, & Mathew, 2014). The benchmarking range is not given since it can be varied case to case, and place to place. Therefore, a consensual-based approach is targeted to deal with this subject.

As per recent studies in Table 1 (Danish, Senjyu, et al., 2019; Waas et al., 2014), efficiency and suitability indicators are at the fourth layers of data analysis. It means, development of indicators needs consequence steps of analysis and a comprehensive infestation.

Table 1. The indicators hierarchy

Phases of Information	Hierarchy
Phase 1	Information
Phase 2	Primary data
Phase 3	Secondary data
Phase 4	Indicator
Phase 5	Index

Efficiency analysis and developing indicators required historical data in order to figure out the performance of a building. These data and information can be listed as follows (Danish, Senjyu, et al., 2019):

- Billing month
- Reading data
- Days in billing cycle
- Kilowatt-hours (or kilo-volt-ampere-hours, if billed on this basis)
- Billing kW demand (or kVA demand, if billed on this basis)
- Actual kW demand (or kVA demand, if billed on this basis)
- Kilo-vats (actual and billed)
- Kilo-var-hours (actual and billed)
- Power factor (average or peak, as billed)
- Load factor (average use compared to peak use)
- Power bill (broken down into the above categories along with fuel cost)
- Occupancy level
- Heating or cooling degree days
- An electricity usage history, including appropriate remarks (such as vacation periods).

Studies report (Danish, 2018; Frame, Tembo, Dolan, Strachan, & Ault, 2011) that structure and size of a managed energy system are driven by the demand for energy services that can be determined by driving forces, including:

- Economic structure, economic activity, income levels and distribution, access to capital, relative prices, and market conditions.
- Demographics such as population, age distribution, labor force participation rate, family sizes, and degree of urbanization.
- Geography, including climatic conditions and distances between major metropolitan centers.
- Technology base, the age of existing infrastructure, the level of innovation, access to research and development, technical skills, and technology diffusion.
- Natural resource endowment and access to indigenous energy resources.
- Lifestyles, settlement patterns, mobility, individual and social preferences, and cultural mores.
- Policy factors that influence economic trends, energy, the environment, standards and codes, subsidies, and social welfare.
- Laws, institutions, and regulations.

GREEN BUILDING FRAMEWORK

Green building is defined as a catalyst in the adaption of green planning, design, construction of a building across the globe in an innovative manner towards efficient use of materials and process (Sharma, 2018). The framework layers conceptualize and prioritize a systematic roadmap for efficient deployment of the framework. Layers implement in order, as well as parallel for particulars circummundane. Defining targets in coordination of green building criteria and proposed indicators confirm future success. Defined targets for a green building describe particular values to be accomplished within a timeframe using proper indicators (Adaptation Fund Board, 2011; Zajac, 2016). Defining targets need a baseline that often relies on standards, or current situation. Setting proper targets and objectives for building planning, design, and construction foster overall process, prevent unrealistically high expectation, and maintain attainable resources (Zwikael, Chih, & Meredith, 2018).

Consensual-Based Approach

This approach targets to bring a consensus among influential factors for an entire process from a green building initiation to the completion. Therefore, this trend acts as a mother approach that follows all upcoming phases and approaches within the lifecycle of a green building.

Considering the life cycle of a green building, balancing between energy expenditure and expected lifecycle output in consensus of inhabitant income and lifestyle is essential. So, expenditure-based and expectation/outcome-based analysis can contribute building planners, owners, contractors, and policy-makers. In the expenditure-based building design, the cost of energy supply and consumption adjusts in accordance with the income of building residence in an affordable cost-effective manner (Rademaekers et al., 2016). At this point, expectations levelized to the outcomes and overall lifecycle cost along with many other factors.

Efficiency Oriented-Based Approach

Building automation as the main part of energy saving and efficiency improvement effort is an integrated control of a building's facilities such as lighting, vitalization, air conditioning, security systems, and building management systems (Aparicio-Ruiz, Barbadilla-Martín, Salmerón-Lissén, &Guadix-Martín, 2018). There are many ways to deal with building automation, in which some important points are listed as follows:

- Building real-time monitoring and control
- Building performance optimization
- Energy conservation management
- Development and utilization of renewables
- Improvement of efficiency in production, delivery, distribution, and consumption
- Energy storage and load management

In addition, routine facilities of a building play a significant role in energy efficiency that refers to consumption efficiency. These facilities are lighting, heating and Colling comfort (thermal), Energy consumption awareness (using efficient appliances), and so on. The cost of energy is correlative to the importance of energy conservation in a building in term of socio-economic and ecological efficiencies.

Sustainability Oriented-Based Approach

According to Danish, Senjyu, et al. (2019), for an efficacious invention of a sustained plane in every field, setting measurable sustainability indicators is indispensable. In the context of a green building, the below factors are essential to meet indicators in order to ensure sustainability. Transition to green building by the deployment of renewable energy resources will be time-consuming due to some barriers. For instance, these barriers are namely reported in the authors previous publications (Danish, Sabory, et al., 2017; Klein & Coffey, 2016; Kundur, 2004).

ENERGY-EFFICIENT GREEN BUILDING

With today's available opportunity for improving energy efficiency in a building, sustainability culture must propound in design and planning. In this regards, green building can contribute to the trends of climate mitigation with referring to the past success lessons-learned.

BUILDING SUSTAINABILITY REQUIREMENTS

Previous studies (Danish et al., 2016; Danish, Senjyu, et al., 2019; Hossain, 2019; Svensson et al., 2018) defined sustainability in term of use of energy resources in a way to be sufficient for now, and do not compromise the ability of future generations to meet their needs. According to Danish et al. (2019) and Danish, Senjyu, et al. (2017), the most effective sub-dimensions of suitability are accessibility, affordability, disparity, safety, use efficiency, supply and production efficiency, cost-effectiveness, and environmental impacts on air, water, and soil quality. In which, this sustainability concept is broken down into five pillars as follows (Danish, Sabory, et al., 2017):

- Technical sustainability
- Economic sustainability
- Institutional sustainability
- Environmental sustainability
- Social sustainability.

For a competent green building design and construction, deployment of all above-mentioned pillars associated with efficiency approaches (technical, technological, ecological, and etc.) are known exigence (Danish et al., n.d.; Waas et al., 2014). To achieve efficiency and sustainability objectives within a green building, setting manageable and measurable indicators are an inevitable task to be followed. According to Danish et al. (2019) and Waas et al. (2014), criteria for determining sustainability indicators can be listed as follows:

- Representativeness and interpretation capability
- Simple and easy to interpret
- Scientifically valid
- Able to show trends over time
- Give an early warning, and influence about irreversible trends where possible
- Sensitive to change in the environment, society or economy it is meant to indicate
- Based on readily available and adequately documented data
- Capable of being updated at regular intervals
- Have a target level or guideline against which to compare.

The concept developed in this chapter is targeted to contribute the 2030 Sustainable Development Goals (SDGs) as well which acts as basis for residential lifestyle change and improvement living facilities through green buildings. For efficacious invention of a sustained plane for a sustainable building, setting proposed indicators is indispensable. Thus, these indicators are developed with confirming the 2030 Sustainable Development Goals (SDGs), which is adapted from (International Atomic Energy Agency, 2005) as shown in Table 2.

Energy security is concerned for a sustainable supply, which it defined as the accessibility of energy supply at an affordable price in an environmentally friendly atmosphere (Danish, Sabory, et al., 2017; Selvakkumaran & Limmeechokchai, 2011). According to Danish et al. (2016), energy accessibility is a factor of energy sustainability that measures the acceptability and accessibility of energy supply by all in the society.

GREENING THROUGH DECARBONIZATION

A sudden increase in the concentration of carbon dioxide in the atmosphere has been observed in the last twenty years which forced energy production utilities to consider environmentally friendly factors (accessibility, affordability, disparity, safety, use efficiency, supply and production efficiency, and cost-effectiveness) (Danish, 2018).

Green building is one of a significant contributor for green energy deployment to achieve global warming targets that mainly backed by renewable energy technologies deployment. Renewable energy resources can be utilized in form of different technologies for various type of applications as shown in Table 3(Danish, 2018).

These resources and technologies produce environmentally friendly energy with a trivial environmental impact; such as solar, wind geothermal, biogas, biomass, and hydroelectric source. On the other hand, these technologies are overloading our atmosphere with global warming emissions and steadily drive up the planet's temperature that causes unexpected environmental and climate changes (Danish, Sabory et al., 2017). Therefore, among many options of decarbonization, renewable energy deployment at a significant size of supply are preferred. Renewable energy resources for building energy supply is available anywhere of the earth. But, selection of optimal resources with high efficiency and cost-effectiveness remain a challenge of decision-making.

Table 2. Sustainability five pillars with relevant indicators

No.	Sustainability Pillar	Categories	Indicator
1	Technical	Production, delivery, distribution and consumption efficiencies	Enhancing an energy system overall performance
		Energy security	Ensuring generation, transmission, and distribution security in accordance to the energy intensity
		Energy intensity	
		System reliability	Attaining an energy system reliable operation and stable performances
		Automation	Making a building comfort in term of automation, and save energy losses
2	Institutional	Strategies and policies	Assessing the present policies through comparing with a baseline; in order to explore further improvement opportunities
		Regulations and procedures	Interpreting policies in form of viable regulations and procedure
		Standardizing	Aligning policies with standards and codes
		Proper management	Applying worldwide accepted management methodologies and practices
3	Environmental	Climate change	Mitigate climate change trough available approaches within global agreements
		Global warming	
		Air quality	Prioritize saving the world by improving air, water and soil quality
		Water quality	
		Soil quality	
		Waste generation and management	Ensure to keep run the triple R concept (Reduce, Reuse, and Recycle) active, and manage waste to the best possible option.
4	Social	Accessibility	Access to commercial clean energy
		Affordability	Access to energy with an affordable price that meet the income and expenditure balance
		Disparities	Balancing energy use as per income group
		Safety	Reliable-save access to energy source within a safe consumption method
5	Economic	Cost-effectiveness	Assessment of energy cost in adapt with residence income
		Economic efficiency	Meet demands with less cost and sufficient supply
		Transform efficiency	
		Supply efficiency	
		Consumption efficiency	
		Subsidy	Supportive transition program to clean energy as well as swapping to energy-efficient appliances
		Supportive program	

Table 3. Renewable energy sources and technologies

No	Energy Sources	Types of Technology	Type of Energy
1	Solar energy	Photovoltaic	DC electricity
		Collectors	Heating
		Concentrators	AC electricity
2	Wind energy		AC electricity
3	Hydropower		AC electricity
4	Ocean energy	Waves	AC electricity
		Tides	AC electricity
5	Geothermal	Low-temperature heat	Heating
		High-temperature heat	AC electricity
6	Biomass	Synthetic fuels	
		Low-temperature heat	Heating
		High-temperature heat	AC electricity
7	Organic waste	Synthetic fuels	
		High-temperature heat	AC electricity

SUSTAINABILITY AND EFFICIENCY JUSTIFICATION

By using the benchmarking technique and proper indicators can measure and justify a building greenery degree within sustainability and efficiency criteria. For this purpose, selecting and genuine usage of indicators, plus outcome assessment is crucial (Borup et al., 2013). Meanwhile, defining an interlink between indicator selection and an indicator outcome ensure a high accuracy level as shown in Table 4(Adaptation Fund Board, 2011). To ensure an indicator sufficiency, and in-depth analysis is recommended as shown in Tables 4 and 5.

According to a green building scope (greenery measures, building size, location, utilization, decarbonization method, heating and cooling facilities, automation and smartness, isolation, architecture and etc.) parameters defined in light of efficiency and suitability criteria (Atkinson, Crawford, & Ward, 2006). Then a specific indicator with defined threshold in adaption with proposed parameter is selected. For achieving the objective, often indicators backed by supplementary tools and techniques of multidisciplinary are sum up with a final result (Danish, Zaheb, et al., 2019). Finally, obtained results and outcomes differentiate with targeted scope and desired expectations.

GREEN BUILDING CONSTRUCTION LIFECYCLE

The next step toward achieving greenery goals of an energy-efficient building is construction lifecycle. Construction of a green building broken down into six phases as listed below (Danish, Zaheb, et al., 2019):

Table 4. Energy efficiency and sustainability indicators justification

No.	Indicator		Question to Address
	Selection Criteria	Adaptability and Measurability Limits	
1	Appropriately enough		Is this indicator lead to expected outcome?
2	Preciseness		Is this indicator clearly acceptable for the proposed application? Or, is it accepted and used worldwide for the same purpose?
3	Practicability		Is this indicator logically and scientifically fit the problem?
4	Affordability		Are information and data accessible at permitted cost and budget?
5	Simplicity		Is it straightforward to measure and calculate?
6	Reliability		Is it reliable at any condition within a series of time?
7	Sensitivity		Is it enough stable or sensitive with system parameters? How much is sensitivity tolerance?
8	Obviously		Is it generate a clear outcome for making a decision?
9	Testability		Is the indicator demonstrate with testability behavior options through different approach or methodology?
10	Acceptability		It is acceptable for stakeholders as a mean of measurement?

Table 5. Energy efficiency and sustainability indicators proper application guide

Parameter	Indicator	Objective	Adaption Assessment	Measurement Threshold	Supplementary Tool and Technique	Outcome	Remark
Name of parameter	Related indicator	Objective of the proposed indicator	Result of adaption assessment	Threshold for measurement and evaluation criteria	Used tools and techniques to obtain the results	Tangible and nontangible outcomes by hiring the indicators	

Develop a Business Model

- Define business need and real-world demand.
- Officially approve and confirm start date.
- Agree on high-level milestones, acceptance criteria, objectives and requirements.

Plan the Project

- Identify essential project plans to be developed.
- Describe boundaries for project directing, managing, controlling, and closing.
- Officially put forward with implementation constraints and high-level risks.

Interpret the Scopes and Make Viable

- Recognize crucial and optional requirements.
- Plan requirements and define scopes.
- Make scopes manageable into small chunks.
- Define procedure for validating and controlling the scopes.

Conceive and Manage Constraints

- Define, sequence, and estimate deliverables and resources (human, technical, technological, etc.) within constraints (time, cost, quality).
- Define criteria to validate and control constrains.
- Minimize risk and its impact of the constraints.

Shape the Project Smoothly Endeavors

- Well-defined stakeholders, and their influence, impact and communication mean.
- Investigate resources availability, affordability, accessibility, procurement and expected and unexpected risk and alternative solutions.

Successfully Handover the Project

- Close the project
- Archive success, failure and lesson-learned.
- Compare the outcomes of the project under the program.
- Align future actions and precautions accordingly.

All these phases are backed by tools and techniques, which bring smartness and accuracy for complex and interrelated activities. These tools and techniques are not limited to the following list, the most worldwide accepted parameters are as follows(Danish, Zaheb, et al., 2019):

- Alternative Generation
- Analogous Estimation Technique
- Analytical Techniques
- Benchmarking
- Bottom-up Estimation Technique
- Change Control
- Context Diagram
- Contingent Response
- Cost Aggregation Technique
- Critical Path Method
- Decomposition

- Dependency Determination Technique
- Diagramming Technique
- Direct and Indirect Observation
- Document Analysis
- Expert Judgment
- Facilitated Workshops
- Facilitation Technique
- Focus Group
- Group decision-making Techniques
- Historical Relationship Method
- Inspection
- Interview
- Leads and Lags Technique
- Make-or-Buy Analysis
- Meetings
- Modeling Technique
- Network Analysis
- Networking
- Parametric Estimation Technique
- Precedence Diagramming Method
- Probability and Impact Matrix
- Prototype
- Quantitative Risk Analysis and Modeling Techniques
- Questionnaires and Surveys
- Reserve Analysis Technique
- Resource Optimization Technique
- Schedule Compression Techniques
- Statistical Sampling
- SWOT Analysis
- Three-point Estimation Technique
- To-complete Performance Index
- Variance Analysis Technique

These tools and techniques are used for a particular objective, while they can also be used in an integrated mechanism as well.

CONCLUSION

The main objective of this chapter was to introduce the most crucial indicators of energy efficiency and sustainability for green building planning, design, and construction. With hiring a special measure for development of these indices within five pillars of suitability and several aspects of efficiencies, this chapter tries to draw an exhaustive roadmap in compliance with resiliency, sustainability, and efficiency criteria throughout life-cycle of a green building. In fact, this chapter has integrated a hierarchy sequence that covers all aspects of green building energy efficiency, conversion, and saving as a complete solution. Meanwhile, this chapter deals with a comprehensive study of sustainability, global warming, climate change, decarbonization, and renewable energy deployment in form of an emerging solution. This chapter argued cases based on different scenarios to prove its validity and effectiveness. Therefore, this chapter can be counted a good reference for practitioners and researchers as a concise set of energy efficiency, conservation, saving, automation, and management guide throughout the lifecycle, of green building with real-world application.

ACKNOWLEDGMENT

Any This study is a part of the research efforts, which is financially and technically supported by the Center for Strategic Research Project, University of the Ryukyus (Senbaru 1, Nishihara, Okinawa, 903-0213, Japan) under the research budget of JFY 2018 (98394640).

REFERENCES

Adaptation Fund Board. (2011). Project level results framework and baseline guideline document (No. AFB/B.14/Inf.6). Author.

Altomonte, S., & Schiavon, S. (2013). Occupant satisfaction in LEED and non-LEED certified buildings. *Building and Environment*, *68*, 66–76. doi:10.1016/j.buildenv.2013.06.008

Aparicio-Ruiz, P., Barbadilla-Martín, E., Salmerón-Lissén, J. M., & Guadix-Martín, J. (2018). Building automation system with adaptive comfort in mixed mode buildings. *Sustainable Cities and Society*, *43*, 77–85. doi:10.1016/j.scs.2018.07.028

Atkinson, R., Crawford, L., & Ward, S. (2006). Fundamental uncertainties in projects and the scope of project management. *International Journal of Project Management*, *24*(8), 687–698. doi:10.1016/j.ijproman.2006.09.011

Borup, M., Klitkou, A., & Andersen, M. M. (2013). *Indicators of energy innovation systems and their dynamics. A review of current practice and research in the field (Radar report)*. EIS.

Brenna, M., Falvo, M. C., & Foiadelli, F. (2012). *Sustainable Energy Microsystem (SEM): preliminary energy analysis. In 2012 IEEE PES Innovative Smart Grid Technologies* (pp. 1–6). Washington, DC: IEEE; doi:10.1109/ISGT.2012.6175735

Chung, W., Hui, Y. V., & Lam, Y. M. (2006). Benchmarking the energy efficiency of commercial buildings. *Applied Energy*, *83*(1), 1–14. doi:10.1016/j.apenergy.2004.11.003

Danish, M. S. S. (2018). *A Managed Energy Framework for Least Developed Countries: Resilience to Energy Sustainability (PhD)*. Okinawa, Japan: University of the Ryukyus.

Danish, M. S. S., Sabory, N. R., & Ershad, A. M. (2016). The Least Developed Countries Needs for Changing the Passive Trend of Renewable Energy Exploitation to a Proactive Trend. *International Journal of Energy and Power Engineering*, *5*(6), 215–221. doi:10.11648/j.ijepe.20160506.17

Danish, M. S. S., Sabory, N. R., & Ershad, A. M. (2017). Sustainable Architecture and Urban Planning trough Exploitation of Renewable Energy. *International Journal of Sustainable and Green Energy*, *6*(3), 1–7. doi:10.11648/j.ijrse.s.2017060301.11

Danish, M. S. S., Senjyu, T., & Ibrahimi, A. M. (2019). A managed framework for energy-efficient building. *Journal of Building Engineering*, *21*, 120–128. doi:10.1016/j.jobe.2018.10.013

Danish, M. S. S., Senjyu, T., & Sabory, N. R. (2017). Afghanistan's aspirations for energy independence: Water resources and hydropower energy. *Renewable Energy*, *113*, 1276–1287. doi:10.1016/j.renene.2017.06.090

Danish, M. S. S., Zaheb, H., & Sabory, N. R. (2019). The Road Ahead for Municipal Solid Waste Management in the 21st Century: A Novel-standardized Simulated Paradigm. In *The 3rd International Conference on Energy and Environmental Science 2019*. Seoul, South Korea: IOP Conference Series: Earth and Environmental Science. (in press)

Duverge, J. J., Rajagopalan, P., Fuller, R., & Woo, J. (2018). Energy and water benchmarks for aquatic centres in Victoria, Australia. *Energy and Building, 177*, 246–256. doi:10.1016/j.enbuild.2018.07.043

Fagin, R., & Wimmers, E. L. (2000). A formula for incorporating weights into scoring rules. *Theoretical Computer Science, 239*(2), 309–338. doi:10.1016/S0304-3975(99)00224-8

Frame, D., Tembo, K., & Dolan, M. J. (2011). *A community-based approach for sustainable off-grid PV systems in developing countries. In 2011 IEEE Power and Energy Society General Meeting* (pp. 1–7). Detroit, MI: IEEE. doi:10.1109/PES.2011.6039593

Frini, A. (2014). A multi-criteria multi-period approach for energy project selection in sustainable development context. In *2014 First International Conference on Green Energy ICGE 2014* (pp. 65–71). Sfax, Tunisia: IEEE. 10.1109/ICGE.2014.6835399

He, Y., Kvan, T., Liu, M., & Li, B. (2018). How green building rating systems affect designing green. *Building and Environment, 133*, 19–31. doi:10.1016/j.buildenv.2018.02.007

Hossain, M. F. (2019). Sustainable technology for energy and environmental benign building design. *Journal of Building Engineering, 22*, 130–139. doi:10.1016/j.jobe.2018.12.001

Huo, X., Yu, A. T. W., & Wu, Z. (2018). An empirical study of the variables affecting site planning and design in green buildings. *Journal of Cleaner Production, 175*, 314–323. doi:10.1016/j.jclepro.2017.12.091

International Atomic Energy Agency. (2005). Energy Indicators for Sustainable Development: Guidelines and Methodologies. Vienna, Austria: IAEA.

Jeffs, E. (2017). Green Energy: Sustainable Electricity Supply with Low Environmental Impact (1st ed.). Boca Raton, FL: CRC Press.

Jensen, P. A., Maslesa, E., Berg, J. B., & Thuesen, C. (2018). 10 questions concerning sustainable building renovation. *Building and Environment, 143*, 130–137. doi:10.1016/j.buildenv.2018.06.051

Klein, S. J. W., & Coffey, S. (2016). Building a sustainable energy future, one community at a time. *Renewable & Sustainable Energy Reviews, 60*, 867–880. doi:10.1016/j.rser.2016.01.129

Kundur, P. (2004). Sustainable electric power systems in the 21st century: requirements, challenges and the role of new technologies. In *IEEE Power Engineering Society General Meeting, 2004* (vol. 2, pp. 2298–2299). Denver, CO: IEEE. doi:10.1109/PES.2004.1373295

Lambrechts, W., Gelderman, C. J., Semeijn, J., & Verhoeven, E. (2019). The role of individual sustainability competences in eco-design building projects. *Journal of Cleaner Production, 208*, 1631–1641. doi:10.1016/j.jclepro.2018.10.084

Liu, M., & Mi, B. (2017). Life cycle cost analysis of energy-efficient buildings subjected to earthquakes. *Energy and Building, 154*, 581–589. doi:10.1016/j.enbuild.2017.08.056

McNeil, M. A., & Rue du Can, S. (2016). Energy Efficiency Indicators and Impact Metrics (No. LBNL-1005695). Lawrence Berkeley National Laboratory.

Patterson, M. G. (1996). What is energy efficiency? Concepts, indicators and methodological issues. *Energy Policy*, *25*(5), 377–390. doi:10.1016/0301-4215(96)00017-1

Popovic, T., Barbosa-Póvoa, A., Kraslawski, A., & Carvalho, A. (2018). Quantitative indicators for social sustainability assessment of supply chains. *Journal of Cleaner Production*, *180*, 748–768. doi:10.1016/j.jclepro.2018.01.142

Rademaekers, K., Yearwood, J., & Ferreira, A. (2016). Selecting Indicators to Measure Energy Poverty. Rotterdam, The Netherlands: Trinomics.

Selvakkumaran, S., & Limmeechokchai, B. (2011). Assessment of energy security and low carbon society scenarios in Thailand and Sri Lanka. In *2011 International Conference & Utility Exhibition on Power and Energy Systems: Issues and Prospects for Asia (ICUE)* (pp. 1–6). Pattaya, Thailand: IEEE. 10.1109/ICUEPES.2011.6497738

Shabunko, V., Lim, C. M., Brahim, S., & Mathew, S. (2014). Developing building benchmarking for Brunei Darussalam. *Energy and Building*, *85*, 79–85. doi:10.1016/j.enbuild.2014.08.047

Sharma, M. (2018). Development of a 'Green building sustainability model' for Green buildings in India. *Journal of Cleaner Production*, *190*, 538–551. doi:10.1016/j.jclepro.2018.04.154

Svensson, G., Ferro, C., & Høgevold, N. (2018). Framing the triple bottom line approach: Direct and mediation effects between economic, social and environmental elements. *Journal of Cleaner Production*, *197*, 972–991. doi:10.1016/j.jclepro.2018.06.226

Tripathi, V. (2016). A literature review of quantitative indicators to measure the quality of labor and delivery care. *International Journal of Gynaecology and Obstetrics: the Official Organ of the International Federation of Gynaecology and Obstetrics*, *132*(2), 139–145. doi:10.1016/j.ijgo.2015.07.014 PMID:26686027

Waas, T., Hugé, J., & Block, T. (2014). Sustainability Assessment and Indicators: Tools in a Decision-Making Strategy for Sustainable Development. *Sustainability*, *6*(9), 5512–5534. doi:10.3390u6095512

Zajac, P. (2016). Indicators and Measures of Energy Efficiency the Warehouse. In P. Zajac (Ed.), *The Energy Consumption in Refrigerated Warehouses* (pp. 23–28). Cham, Switzerland: Springer International Publishing. doi:10.1007/978-3-319-40898-9_2

Zwikael, O., Chih, Y.-Y., & Meredith, J. R. (2018). Project benefit management: Setting effective target benefits. *International Journal of Project Management*, *36*(4), 650–658. doi:10.1016/j.ijproman.2018.01.002

Chapter 7

Green and Smart Buildings:
A Key to Sustainable Global Solutions

Rudra Rameshwar
Thapar Institute of Engineering and Technology, India

Arun Solanki
Gautam Buddha University, India

Anand Nayyar
https://orcid.org/0000-0002-9821-6146
Duy Tan University, Vietnam

Bandana Mahapatra
https://orcid.org/0000-0002-1388-5413
SOA, India

ABSTRACT

Buildings across the world consume a significant amount of energy which is equivalent to one third of total primary energy resources available. This has led to lots of challenges with regard to supplies of energy, energy resources quick depletion, increase in building service demands, improvised comfort lifestyle along with time increase spend in builds; this all has increased the energy consumption. Even the global sustainability is also pushing the implementation of green buildings in the real world. Researchers and scientists have been working on this issue for a very long time, but still the issue is prevalent. The aim of this chapter is to present comprehensive and significant research conducted to date with regard to green buildings. The chapter provides in-depth analysis of design technologies (i.e., passive and active technologies) that lay a strong foundation for green building. The chapter also highlights the smart automation technologies which help in energy conservation along with various performance metrics.

DOI: 10.4018/978-1-5225-9754-4.ch007

CONCEPT OF SMART BUILDING

The temperature of the earth is increasing continuously and this effect is known as Global warming, which is caused by the emission of greenhouse gases (Pachauri & Reisinger, 2008). The buildings are also the one of the major provider to worldwide carbon emissions, accounting for about 40% of the world's total carbon footprint (Janda, 2009). Majorly, in developed nations, commercial buildings alone represent close to 20 percent, about half of the total.

Commercial buildings are also costly. After salaries, buildings are one of the biggest operational expenses for organizations. Energy plays a significant part in this. (Michaels, 2019)

A more efficient building portfolio can improve the value of real estate assets, help the bottom line, cut emissions, and bolster the corporate image. So, there is acute need of building that can work intelligently i.e. smartly. A Smart Building minimizes energy and water consumption; minimizes waste and maximizes recycling; provides healthy living conditions and promotes environmental performance.

Smart Buildings are termed as "Digital Extensions" to all sorts of engineering and architectural activities. Smart Building is termed as structure facilitating automated processes to automatically control all sorts of building's operations like: Security, Lightning, Air conditioning, Heating, Ventilation, etc. A Smart building makes use of sensors, actuators, and microchips to manage everything. The utilization of sophisticated hardware facilities improvement in asset reliability, performance and in turn reduces energy utilization. The following points highlight the benefits of smart buildings:

1. **Predictive Maintenance:** Sensors can detect all sorts of technical performance of the building and can automatically activate the maintenance procedures in case of any sort of malfunction.
2. **Energy Saving:** Data sent by the sensors can be carefully analyzed, doing prompt actions for temperature and lighting settings.
3. **Effective Monitoring:** With smart buildings, all the equipment's can be effectively monitored and can be replaced on time, making maintenance highly cheap and effective.
4. **Optimized Site Cleaning:** All sorts of presence sensors can optimize cleaning operations. The data sensed by sensors can alert facility manager to undertake all cleaning operations whenever or wherever required.
5. **Redesigned Space:** Sensors enable easy identification of the overused and underused area of the building and can recommend some adjustments by making use of modern techniques like Artificial Intelligence and Machine Learning.

PROGRESSION FROM INTELLIGENT TO SMART BUILDINGS

Evolving definitions of Smart Buildings have been developed since the 1980s. Various researchers defined Intelligent Building as being:

"A building which totally controls its own environment". This seems to imply that it is the technical control of heating and air conditioning, lighting, security, fire protection, telecommunication and data services, lifts and other similar building operations that are important – a control typically given to a management computer system. Such a definition for a conventionally Intelligent Building does not suggest user interaction at all.

Figure 1. Primitive building to smart building

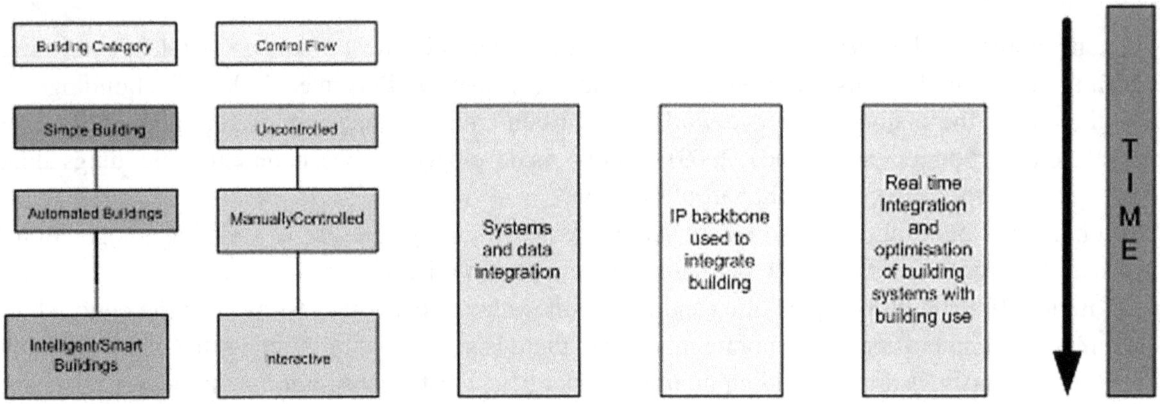

The early definitions of Intelligent Buildings are what would be expected, since at that time architects and building engineers were progressing from what can now be seen to be automated buildings, as demonstrated in Figure 1.

(Everett, 2008) defined the Intelligent Building as: *"A dynamic and responsive architecture that provides every occupant with productive, cost effective and environmentally approved conditions through continuous interaction among its four basic elements: places (fabric; structure; facilities); processes (automation; control; systems) people (services; users) and management (maintenance; performance) and the interrelation between them."*

(Clements-Croome, D. (2011), developed the following definition which he later repeated:

An Intelligent Building is one that is responsive to the requirements of occupants, organizations and society. It is sustainable in terms of energy and water consumptions besides being lowly polluting in terms of emissions and waste: healthy in terms of well-being for the people living and working within it; and functional according to the user needs.

(Katz and Skopek, 2009) show that Intelligent Buildings contain aspects of automation and similarly intelligence is an important aspect of Smart Buildings

Intelligence, enterprise, materials & design and Control are the four essential features which makes a building smart. All of these are integrated over adaptability and with the help of the driver the building become smart. The Adaptability comprises the short, mid and long-term factors. The ability to adapt in response to information gathered from building use is essential to a Smart Building's successful operation.

Enterprise in the context of non-domestic buildings consists of a combination of hardware and/or software used to overcome fragmented, non-compatible, non-proprietary legacy systems (Robey et al., 2002), in order to allow the building operation to be optimized towards the building function.

Control in the context of manual or automatic. Buildings relying upon human control assume that the occupants will use the building in the way it was designed for; automated buildings tend to be designed to the theoretical climatic conditions, occupancy and use. Both types are subject to changes during construction and commissioning that differ from the design intent. Therefore, both categories are susceptible to decreases in performance during change of occupancy, use or climatic conditions. The materials and

Figure 2. Features of smart buildings

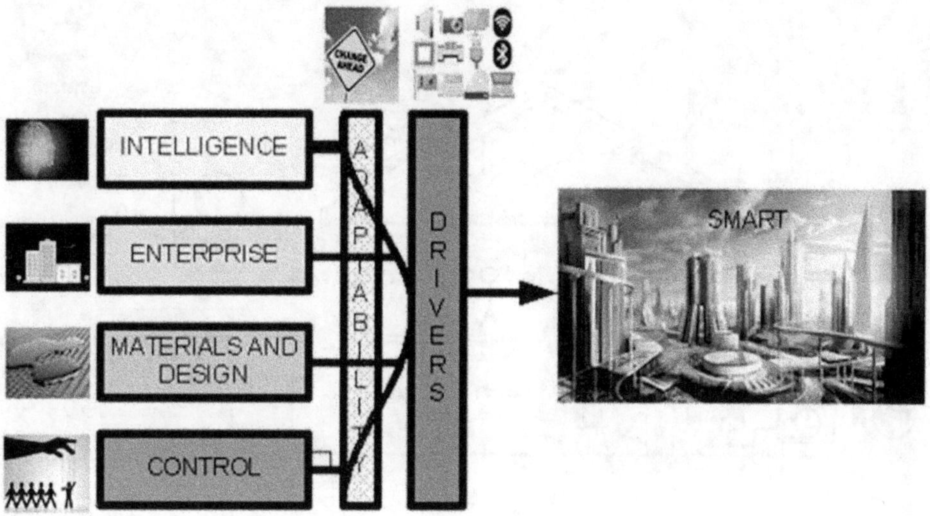

construction feature within the Smart Building definition represents the built form. The construction of a Smart Building needs to reflect and house the smart functions within it.

ESSENTIAL COMPONENT OF SMART BUILDINGS

The hardware, software and network are the essential components of the Smart Buildings. These three components make a building smart.

- **Hardware:** First of all, "smart" buildings need an ability to recognize what's happening with an environment (inside and outside a building), needs something like human senses.
- **Software:** Sensors and meters provide only raw information. A "smart" building needs to extract useful information, learn from this information, make decisions and even predict the future state of the environment and people activities. It is done by special software which is an artificial intelligence of a building.
- **Network:** To allow the building to act as a whole- a communication network is required. It connects all devices between each other and with the artificial intelligence component. It is the nervous system of a building.

GREEN SMART BUILDINGS

Along with the growth of research interest in areas of sustainable green building, a novel energy ecosystem concept is emergent that merges the smart green building and smart grid concept together giving an overall optimized energy flow (Adalberth et al., 2001).

The Smart building and Smart grid technology inter-wined together offer huge benefits in terms of better functioning buildings, which are all the more comfortable as well as safe place to work and live

Figure 3. Elements of green smart buildings

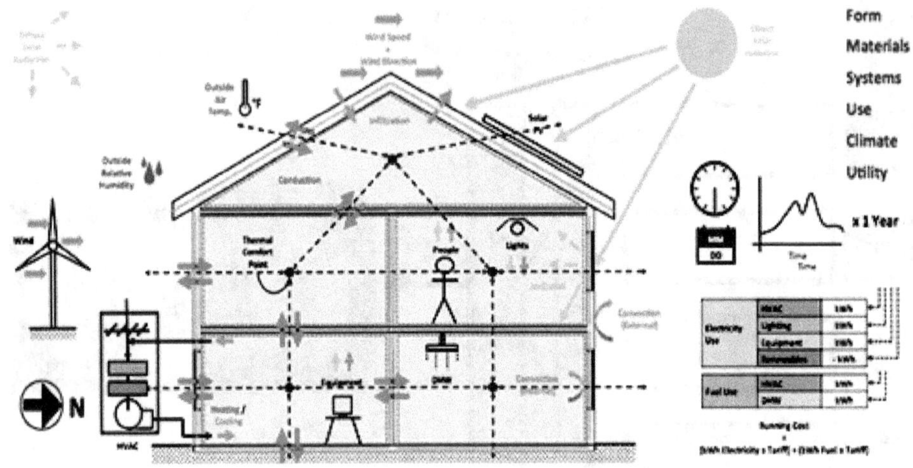

in. The concept of automating the green buildings offers promising opportunities for designing as well as proposing technologies like energy efficient, mass scale renewable generation, automated demand response systems. Apart from them concepts like, low cost, off-peak times, a peak power generation is often proposed by the highly polluting power plants (Annual Energy Review, 2004).

Systems incorporating automated demand response in design make usually the building block contributing into built up of energy efficient programs. They carry the minute detailing of all the information that overall makes the smart energy decision system as shown in Figure 3.

The Smart Green Energy Building Design is said to be completely automated eradicating any human efforts required for its operation introducing a new energy ecosystem.

A smart building may be defined as "*A building that connects its various subsystems together via information technology operating independently while sharing information giving the optimised performance as a result*". The Smart Building Concept is built upon building blocks that spread out much beyond the four walls of the buildings. They are well connected as well as responsible to smart power grid, interacting with building operators, and occupants in order to empower them with advanced visibility and actionable information.

The whole of smart building is based on the building management system (BMS) or Building Automation System (BAS) that constitute the basis but vital aspect of the green building construction.

The BAS or BMS consist of the following essential elements:

- **Data Acquisition:** Sensor, meters, etc.
- **Data Base Management:** Analytics, visualization, controls, etc.

ORIGIN AND HISTORICAL HIGHLIGHTS

The concept of sustainable development has its origin in 1960s & 70s during the challenges faced due to the energy crisis and high level of pollution generating a sort of alarm among the human Rachel Carsoris book "Silent spring during 1962, has first discussed and described the concept and need of sustainable development with regards to green building. Thereafter there has been several factors like environmental, economic, social, benefits that have had positive influence over the conceptualization of green building (Carson, 1962).

The green building movement initiated in US was mainly due to the need and desire of adopting energy efficient and environmentally friendly construction practices. The current sustainability initiatives has integrated as well as synergized both the old as well as new concept giving a retro fitted combination of old and new technology. The approach has well integrated the building development lifecycle with the green practice employee creating a synergized modern practices.

The technical uplifting contributing to our day to day life style have proved to be more economically sound. They revolve around new software protocols that has inculcated the desperate control system with wireless technology that completely wiped out the cost and need of rewiring. "Jack McGowan, president of energy control Inc. and chairperson of Grid Wise Architecture Council(GWAC) have stated that, Smart Buildings are Green Buildings who wanted to design and develop protocols in order to connect the previous pieces of Smart Grid Technology.

GWAC, US Department of energy effort mainly aims at protocol development that could connect the various pieces of smart grid technology. According to New Energy Independence and Security Act (ESCI), 2007 McGowan stated the goals for growth of non-zero energy buildings. These buildings may be labelled as energy generators, users, which transmit energy to the grid equal to the amount drawn out of it. The background idea of the building is, it can give as well as take energy. It is believed that, with the growth of Intelligent Building, which can be credited as a self-manager managing both energy and revenue flows, net-zero building concept would also grow to existence (Independence, 2007).

GREEN BUILDING PROSPECTS

The Green Building is basically, creation of effective building element and adopting appropriate process that are environmentally responsible as well as resource efficient. These methods are used at every phase of building, life cycle arranging, all over from design to deconstruction.

The green building concept is all about increasing the efficiency with which natural resources are being used such as energy, water, material etc, reducing its adverse impact over the human health as well as the environment throughout the lifecycle of building. The green building concept is all about increasing the efficiency with which the natural resources such as energy, water, material, etc. are used in the building process and reducing its adverse impact over the human health as well as the environment throughout.

The Green building concept is extended far beyond the mere building structure. The various concepts covered in Green smart building is shown in Figure 4.

Its concept involves site planning, community and land use planning as well. The growth and development of the residential communities incur huge impact over the natural environment. The building process like manufacturing, designing construction as well as any basic operation of the building consumes a lot of natural resource (Ahmad et al., 2019).

Figure 4. Concepts covered by green smart building

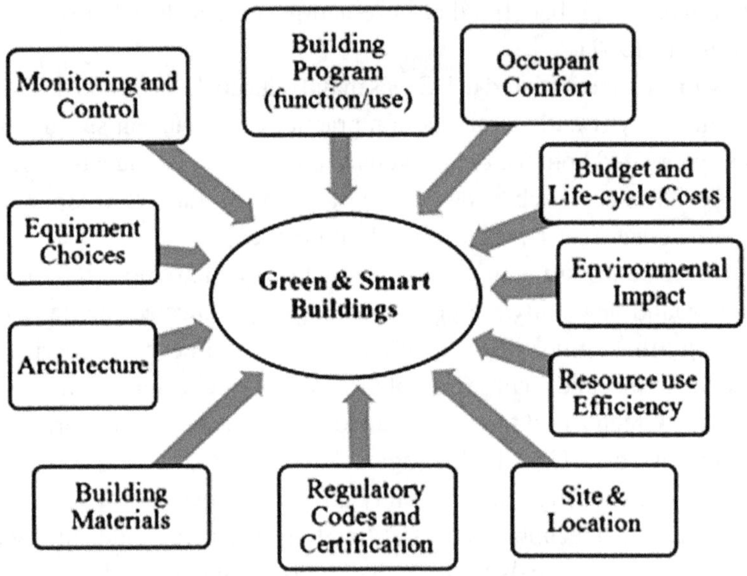

DESIGN TECHNOLOGIES BEHIND GREEN SMART BUILDINGS

The Concept of integrated design or whole building design has each building element first optimized; thereafter the impact as well as the interrelationship among various elements or systems constituting the building and sites are re-evaluated, integrated as well as optimized as a part of the whole building solution. *Example:* Interconnection between building sites, site features, location, path of sun, and construction elements like windows, and shading devices have a significant impact over the quality as well as effectiveness of natural day lighting. The elements also cause an impact over quality and effectiveness over natural day lighting.

The elements also cause an impact over the quality and effectiveness over natural day lighting. The elements also cause an impact over the quality and effectiveness of natural day lighting. The elements do effect, direct solar loads, overall energy performance of the life of a building (Hossain, 2018). Previously ignoring these issues in design process resulted in a partially optimized design giving a rather inefficient building as shown in Figure 5.

An integrated design process makes it compulsory for all of the design professionals into cooperative performance for attaining common goal. Apart from the standards specified every designers and professionals holding their personal goals to attain, where many goals may go against each other. But by clear communication and collaboration early while conceiving the project may help the designers to confirm with common goals avoiding high cost implications using delays.

A successful integrated design offers us the following outcome (Zhao, 2019):

- Establishing a multidiscipline team that can work together towards listing common goals to achieve.
- Early agreement regarding project design properties.
- Setting of project performance goals.

Figure 5. Areas of expertise to be represneted in the integrated design charrette

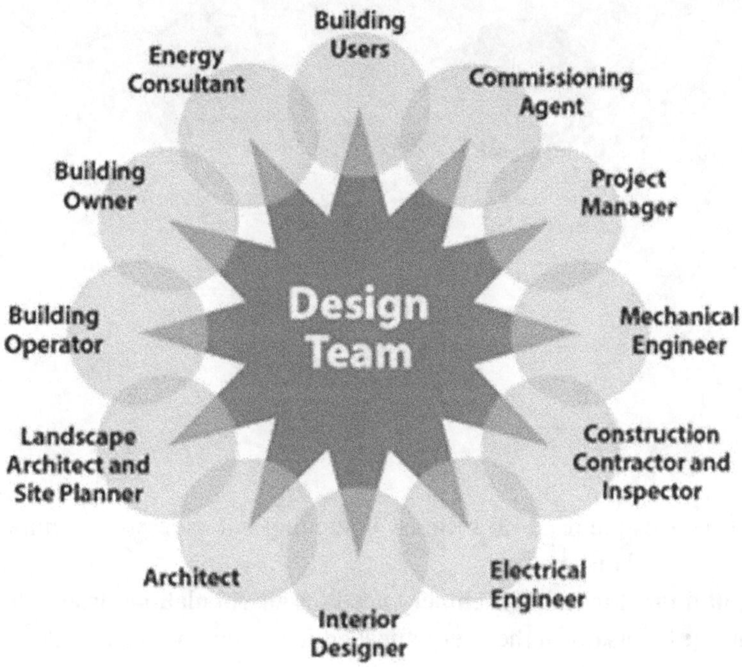

- Generating a quantifiable metrics in order to measure the final energy and environmental outcome against performance goals.
- It outlines the goals to achieve at the performance metrics
- It gives the potential impact over the various design strategies
- It introduces the integrated design process which result in reduction of project cost, schedules as well as obtains the best energy including the environmental performance
- It recognizes the available project strategies with its associated cost, time required and the required expertise in order to eradicate expensive requirements incurring later in the design and construction processes.
- It identifies partners, available grants, as well as potential collaborations, which offers expertise, funding credibility, support etc.

Passive Design Technologies

The terminology passive design denotes to various strategies, technological aspects, as well as solutions referring to design which efficiently utilizes the advantage of various environmental conditions that result in raise of energy and lowers the cost without compromising the core building facility and provisions (Chen et al., 2015) as shown in Figure 6.

Key elements of passive design for green building: The environmental conditions that provide various advantages or disadvantages while designing a building are:

Figure 6. Layout of green building design

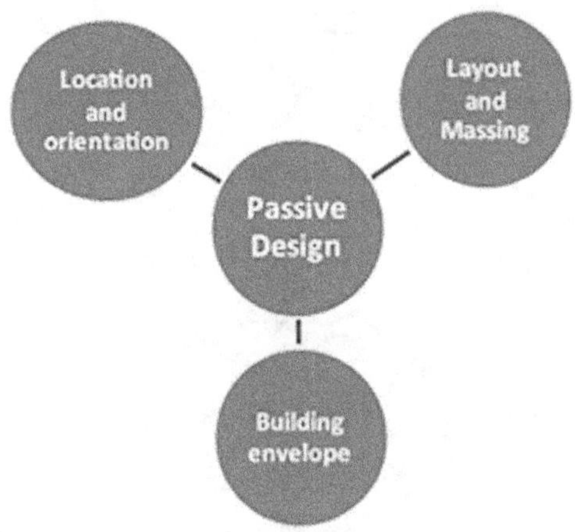

- **Day Light:** It reduces the need of artificial lightening but excessive or improper exposure may lead to glare and visual discomfort.
- **Natural Ventilation: It** reduces machinery need for air circulations around the room
- **Natural Cooling:** It cuts down the energy needed for air conditions to reduce the heating effect.
- **Natural Heating:** It uses the solar energy to keep the house warm.
- **Shading:** It may reduce the heat being directly exposed during hot summer.

The major factors of the passive designs are:

- Building of location as well as orientation of site
- Building layout and massing
- Building Envelope

The passive design elements are generally the massive layers that support to attain the high-level performance.

- **Passive Design Technologies:** The term passive design denotes the specific way in order to construct buildings using natural resources available around.

Example: Subtle movement of air, heat, solar gains, to maintain a health internal environment. The passive solutions are helpful either to reduce or eradicate the use of mechanical systems and the energy demand of about 80% along with the CO_2 emissions. Building these houses involves major planning, including the monitoring of basic principle like orientations, overhangs, shadings, insulation, double or triple glazing, thermal mass, etc. the passive house designing and construction may be defined as 'building where thermal comfort may be generated by post heating or post cooling, where the fresh air and mass flow is needed in order to maintain a good air quality.

Apart from European countries, these standards of house building techniques cannot be applied directly over other parts of the world. These standards must be incorporated considering the geographical areas as well as the climatic conditions should undergo proper study to attain the best passive solution. Integrating

Figure 7. Ray diagram of summer and winter midday shadow

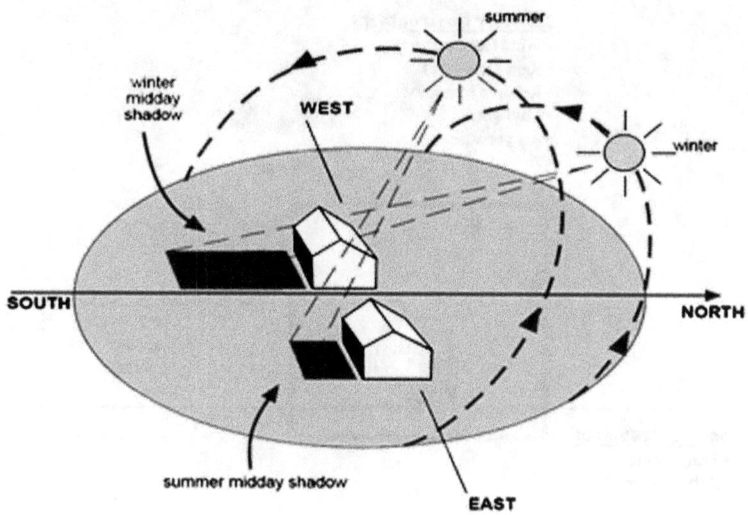

the passive design solution at a very nascent stage brings along almost minor or no cost to the building construction. These solutions slowly contribute to reduction of CO_2 emission from the building as well as the use of mechanical systems for both heating as well as cooling down of the house (Kibat, 2016).

Few of the key elements of passive modes of designing a building are:

- Building location and orientation of the site
- Basic plan and layout of the building
- Window design
- Thermal mass
- Shading
- Ventilation.

These elements work in coordination with others in order to achieve comfortable temperatures and good indoor air quality (Ghaffarian Hoseini, 2013) as shown in Figure 7.

- **Building Location and Design:** Few factors while building on site, like orientation site environment may have heavy impact over its performance.
- The amount of natural sunlight
- The gush of wind flowing around and needed for cross ventilation
- Views aiding the look
- Options of varied communication mode
- Impact of neighboring buildings or construction

The range of natural sunlight and source of heat obtained by the building determines the major direction of the building. Based on topography and weather, the design may differ from building to building. The natural light breeze reduces the need to obtain them using artificial method. Though it is always desirable to avoid solar heat gain to cut down on energy requirement for air conditioning it is always

Figure 8. Exterior and indoor environment layout

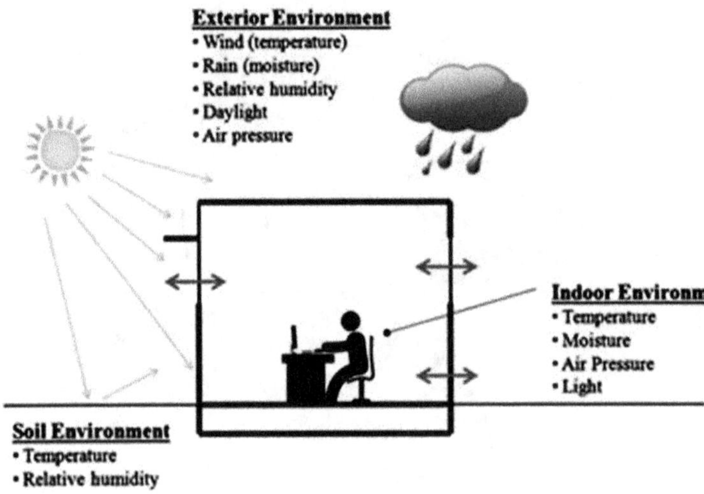

harness sun or solar energy generation. It is equally necessary to take into account the surrounding site topology as well as elevation that would affect the wind conditions. *Example:* a building at high elevated floor enjoys consistent natural breeze where as in the leeward side of hill would give stagnant wind irrespective of orientation, passive design location and orientation layout and massing building envelop.

Building Envelope and Connected Technologies

The building envelop can be defined as the skin of the building which works as physical cooperation between the interior and exterior built environment.

The major elements of building envelop are (Xu & Dessel, 2008):

- **Roof**: The uppermost part of the building
- **Wall**: The vertical constructed structure that connects roof and the floor
- **Floor**: The grounding surface of the building used for walking purpose
- **Fenestration:** Openings in the building structure like windows, skylights, doors.

Factors that decide the choice of envelop are climate, culture, aesthetic, and the available materials. They occur a major impact over the building performance. It specifies the amount of sunlight/heat and air ingress over the building as shown in Figure 8. The material selected for building envelop heavily determines its carbon foot prints as well as environmental friendliness. In order to build envelop it is necessary to understand the properties of the material like insulin, thermal resistance, heat gain coefficient, air infiltration, and visual light transmittance.

Passive Heating Technologies

The concept of passive heating considers solar energy for designing the comfort of the resident instead of utilizing the machineries as a substitute. The sunlight may heat a space through heat absorbed by

Figure 9. Thermal mass wall layout with glazing effect

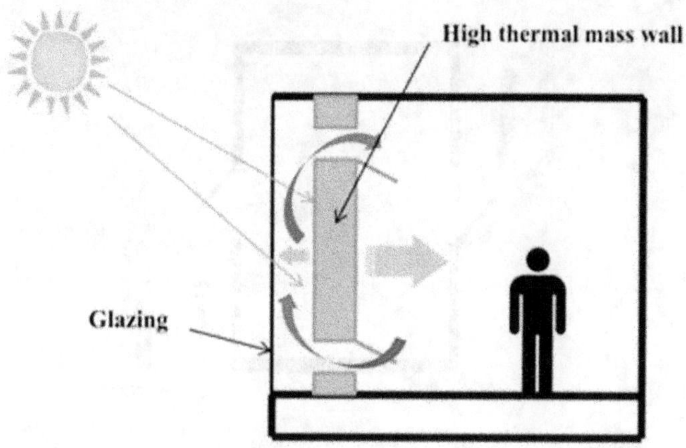

walls and roofs of the building, or fenestration, and heat the interior surfaces. Certain solar lights are long in wavelength infra-red (IR) radiations, i.e. heat. Apart from that light of any wavelength consumed by internal surfaces turns into heat in those materials as shown in Figure 9. The heat thereafter warms the air that carries it by process of connection and radiation. This process is generally termed as 'Solar Heat Gain' (Sadineni, 2011).

Passive Cooling Technologies

The passive cooling technology is all about providing a cool and comfortable atmospheric, temperature at an indoor of a building through conventional use of natural energy resources, see Figure 10. The main modes of passive cooling are natural cross ventilation, using ambient air. The various ways methods of inducing a natural airflow and natural elements are (Chan et al., 2010):

- Natural ventilation and cooling that uses the atmospheric air, movement and pressure variations, to passively cool and ventilate a building without needing electronic fans or AC.
- Allowing gush of air inside so as to replace the stale air that can result in heat accumulation or collection of harmful gasses etc.
- Providing wind speed, comforting the human residents making it suitable for activities carried out in spaces.
- Obtaining good temperature and air quality comfortable enough for targeted space.

Passive Lightning Technologies

The method of passive lighting is all about utilizing the design for proper illumination of the indoor spaces rather depending on artificial lights to do the same.

This method reduces hugely the energy usage in a building where people perform their day to day activities. The natural light apart from the lighting a space has other significance, psychological and physiological contributions see Figure 11. The natural day light may be significantly differentiated from

Figure 10. Wind layout and phenomenon

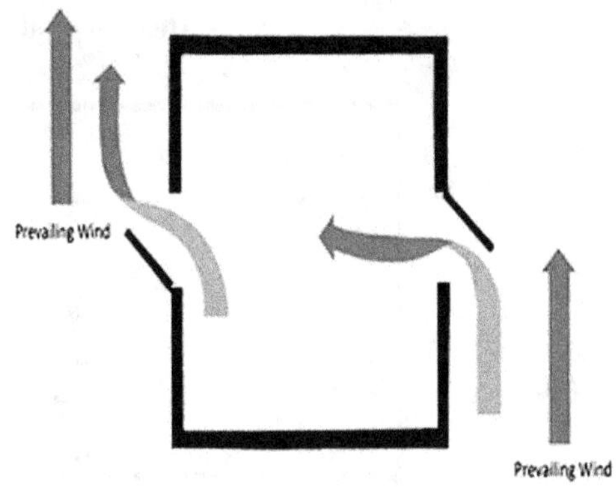

direct sun light where the latter is a hyper intensification of the pervious resulting in heating up spaces (Jadhav, 2016).

Active Design Technologies

The active design technology is all about providing us with the benefits of passive design through the usage of artificial electronic devices. Here modern electronic machines are needed which consumes high amount of energy. The concept of green building is therefore based upon aspects of energy efficiency that is ratio of output produced to the input consumed.

The active design technology mainly revolves around the following area:

- Heating, cooling & ventilating
- Lighting
- Building service equipment's like pumps, lifts, escalators etc.
- Plugs, loads or receptacle loads

Generally, within a building the comfort of residents is obtained in a stringent manner through usage of machines referred to as Heating, Ventilation and Air Condition (HVAC) equipment.

Figure 11. Indoor space illumination without artificial lighting

The aspects of human comfort is quite subjective to deal with, dependent on topology and climatic changes, e.g. in tropical and humid climate de-humidification of air is needed where as in warm climate cooling is mostly needed may change to warming during winters.

The various factors needed to be considered while deciding over the thermal comforts of the building residents are:

- Temperature of air surrounding the occupant
- Relative humidity or moisture content in air
- Air velocity or rate of air movement
- Radiant temperature surrounding occupants
- Metabolism of energy released from human body
- Clothing insulation provided by clothes the person wear

Though the personal requirement vary from person to person and is hard to consider they cause an impact over the residents overall perception of the thermal comfort.

HVAC

Incurring a passive house principles and technologies provides some additional investment cost to offer comfort like high-performance envelope insulation, triple glazing windows, air-tight construction, heat recovery ventilators, stringent construction details etc. However, the asymptotic increase in investment in cost may be balanced by avoiding costs of investing in sophisticated heating, ventilation and air conditioning (HVAC) systems and their added cost of operation. Besides putting money on HVAC systems, passive houses invest in better building envelopes, which may improvise the building's longevity. A passive house is generally said as being cost effective when "the combined capitalised costs does not outgrow the cost of an average new home" (Ehlers & Beaudet, 2006).

Lightning

Visual comfort is another important aspect of buildings where artificial or active lighting is required so as to provide an optimal visual comfort throughout the whole day. The green building aims at providing an optimal visual need at minimum energy consumption.

The following parameters may be considered while measuring the brightness of the light:

- **Luminous Flux:** it is measured in 'Lumens'. This is the quantification of light that comes from the various directions.
- **Luminous Intensity:** This is the measure of amount of life coming from various sources which travels in certain direction. It is measured as 'Candelas'
- **IL Luminance:** The amount of light falling over the surface measured in 'lux' (Lumen/m2) or 'foot candles' (Lumen/ FL^2).
- **Luminance:** The amount of light reflected by a surface. It is measured in candles per square meter (cd/m^2) or Nits (in imperial units)

Plug and Process Resource Management

The plug and process load in a building consume a major share of the building overall energy use. These are the basic energy loads apart from the HLAV appliances and they are not used for providing comfort to the residents. These loads are purely the contribution of electronic machines like refrigerators, microwaves, computers, printers, projectors, audio equipment's, television, chargers, etc. Understanding, managing and optimizing these loads have taken over as a major area of research interest by the scientist. These equipment's solve the dual purpose of directly saving energy as well as cooling energy due to the heat generated by use of these appliances.

The major issue in managing these loads are the known usage patterns dependent completely over them whim fancy of the resident using them.

Environmental Impact of Conventional Building Materials: Why Sustainable Green Buildings?

Basically, in today's global environment, there is an important dialogue across globe about environmental impact of conventional building materials; mostly materials of construction are based on instant building solutions in order to achieve multi objectives. Major applications by cement in this framework of conventional building materials, are resulted in the shortage but achievable in long run on concepts of modern construction materials system. This impact is severe, negatively impacting atmosphere day by day with the increase of conventional building construction materials as follows:

- Maximum Greenhouse Emission
- Excess Water Usage
- Excess Energy Used and Consumed by Conventional Plant
- CO_2 Emissions of Construction Materials (*Mined Manufacturing Process*)
- Maximum Buildings Demolition and Construction Effects

Thus, there is strong need for green buildings in India and other developing countries because nations are in highlighted position of population blast, increasing every day, need to focus on Sustainable Business Strategies (SBS) and its development in India.

CONCLUSION

Progressively moresustainability aspects are going and popular among and across technocrats, policy makers and industry practitioners, who are continuously striving for excellence towards zero-tolerance energy construction initiatives with respect to net zero energy consumption followed by strong monitoring of energy consumption patterns through sustainable development goals in order to understand and measure greenhouse gas emissions across disciplines of heavy engineering works, and world should think about minimization of their regular practices in our day to day life achieving sustainability in true sense. Companies involved in construction of building, should quickly shift to Green Smart Building initiatives without compromising needs for sustainable development.

According to United Nations – Sustainable Development Goals (SDGs), green smart building initiatives in major goals linking fibre –based materials and its processing units with state of art facilities, technologies to new end-use concepts in modern world. Skill development is another aspect, which can lead to sustainable development fulfilling dreams of Green Smart Building in the field of Construction integrating related major areas of engineering, heavy works, machinery, scientists, project developers, technocrats, researchers, policy makers, architects, business development managers etc.

REFERENCES

Adalberth, K., Almgren, A., & Petersen, E. H. (2001). Life cycle assessment of four multi-family buildings. *Intl J Low Energy Sustain Build, 2*, 1–21.

Ahmad, T., Aibinu, A. A., & Stephan, A. (2019). Managing green building development–A review of current state of research and future directions. *Building and Environment, 155*, 83–104. doi:10.1016/j.buildenv.2019.03.034

Carson, R. (2009). *Silent spring*. Boston, MA: Houghton Mifflin.

Chan, H. Y., Riffat, S. B., & Zhu, J. (2010). Review of passive solar heating and cooling technologies. *Renewable & Sustainable Energy Reviews, 14*(2), 781–789. doi:10.1016/j.rser.2009.10.030

Chen, X., Yang, H., & Lu, L. (2015). A comprehensive review on passive design approaches in green building rating tools. *Renewable & Sustainable Energy Reviews, 50*, 1425–1436. doi:10.1016/j.rser.2015.06.003

Clements-Croome, D. (2011). Sustainable intelligent buildings for people: A review. *Intelligent Buildings International, 3*(2), 67–86. doi:10.1080/17508975.2013.786874

Ehlers, G. A., & Beaudet, J. (2006). *U.S. Patent No. 7,130,719*. Washington, DC: U.S. Patent and Trademark Office.

EIA. (2005). *Annual energy review 2004*. U.S. Department of Energy, Energy Information Administration. Retrieved from www.eia.doe.gov/emeu/aer/contents.html

Everett, R. (2008). The 'building colleges for the future' program. Delivering a green and intelligent building agenda. *New Review of Information Networking, 14*(1), 3–20. doi:10.1080/13614570902953549

GhaffarianHoseini, A. H., Dahlan, N. D., & Berardi, U. (2013). Sustainable energy performances of green buildings: A review of current theories, implementations and challenges. *Renewable & Sustainable Energy Reviews, 25*, 1–17. doi:10.1016/j.rser.2013.01.010

Hossain, M. F. (2018). Green science: Advanced building design technology to mitigate energy and environment. *Renewable & Sustainable Energy Reviews, 81*, 3051–3060. doi:10.1016/j.rser.2017.08.064

Independence, E. (2007). Security act of 2007. *Public Law, 110*(140), 19.

Jadhav, N. Y. (2016). *Green and Smart Buildings: Advanced Technology Options*. Springer. doi:10.1007/978-981-10-1002-6

Janda, K. B. (2009). Worldwide status of energy standards for buildings: a 2009 update. *Proceedings of the ECEEE Summer Study*, 1-6.

Katz, D., & Skopek, J. (2009). The CABA building intelligence quotient programme. *Intelligent Buildings International, 1*(4), 277–295. doi:10.3763/inbi.2009.SI05

Kibert, C. J. (2016). *Sustainable construction: green building design and delivery*. Hoboken, NJ: John Wiley & Sons.

Michaels, J. (2012). *Commercial Buildings Energy Consumption Survey: Energy Usage Summary.* Academic Press.

Pachauri, R. K., & Reisinger, A. (2008). [Synthesis report. Contribution of Working Groups I, II and III to the fourth assessment report.]. *Climatic Change*, 2007.

Robey, D., Ross, J. W., & Boudreau, M. C. (2002). Learning to implement enterprise systems: An exploratory study of the dialectics of change. *Journal of Management Information Systems*, *19*(1), 17–46. doi:10.1080/07421222.2002.11045713

Sadineni, S. B., Madala, S., & Boehm, R. F. (2011). Passive building energy savings: A review of building envelope components. *Renewable & Sustainable Energy Reviews*, *15*(8), 3617–3631. doi:10.1016/j.rser.2011.07.014

Xu, X., & Van Dessel, S. (2008). Evaluation of a prototype active building envelope window-system. *Energy and Building*, *40*(2), 168–174. doi:10.1016/j.enbuild.2007.02.027

Zhao, X., Zuo, J., Wu, G., & Huang, C. (2019). A bibliometric review of green building research 2000–2016. *Architectural Science Review*, *62*(1), 74–88. doi:10.1080/00038628.2018.1485548

Chapter 8
Cloud-Based IoT Architecture in Green Buildings

Abhinav Singh
Netaji Subhas University of Technology, India

Utsha Sinha
Netaji Subhas University of Technology, India

Deepak Kumar Sharma
https://orcid.org/0000-0001-6117-3464
Netaji Subhas University of Technology, India

ABSTRACT

The development and expansion of mankind in addition to the advancement of technology have a substantial impact on the environment. The construction, design, and operation of buildings account for a large consumption of natural resources. Due to the exploitation of natural resources on a large scale through these buildings, it has become necessary to have a better-designed building for the efficient use of resources. The concept of "green building" solves the aforementioned issues apart from promoting eco-friendly activities. IoT makes the idea of having buildings that are energy sufficient possible through networked sensors that not only help in managing the assets better but also reducing harmful impacts on human health and the environment. This chapter talks about the concept of the green building and the smart automation achieved through IoT as well as cloud architecture for the green building also referred to as green cloud. While it explains the basic cloud architecture in green building, it also proposes future challenges for the aforementioned subject.

INTRODUCTION

With the growth of demand for internet data and high-speed network in the current world scenario, there is a need for large-scale data storage providers which is being met by the high-processing data center's which in turn consists of a large number of servers along with organized and structured facilities. Various multi-national organizations such as Amazon, eBay, Google and Yahoo run these enormous data

DOI: 10.4018/978-1-5225-9754-4.ch008

center's throughout the world. Numerous business enterprises traditionally invested an immense amount of resources and capital in making such massive data center's for data storage (Kumar & Buyya, 2012). With the advancement in technology and the formation of the network in IoT, there is a desperate need to store all the data in an efficient manner and process it optimally. Cloud computing is an ideal architecture for data hosting, computing and storage infrastructure which eventually makes it cost-effective. The highly scalable, dynamic field of ICTs can be converted into green cloud computing by working to improve energy efficiency and reducing carbon emissions and electronic waste.

The history of cloud computing traces back its roots from the 1960s (Mohamed, 2018). However, the aforementioned term became popular worldwide in recent years due to its ability to overcome the drawbacks of data center's such as high-maintenance cost, proximity challenges, huge infrastructure expenditures, etc. Even though various business corporations are migrating from data center services to cloud computing for its ease of use but there remains a considerable amount of companies that still prefers the traditional data center services. The advancement in the technologies has in various cases led both producers as well as the users to neglect the environmental issues (Radu, 2017). For instance, there are about 44 million servers incorporated in various data center's that consume about 0.5 percent of all the electricity (Forrest, Kaplan, & Kindler, 2008). It is interesting to note that the data center's or server farms alone stood for 14% of the global CO2 emission (Uddin, Abdul Rahman, & Memon, 2011). Considering the aforementioned harmful effects of technology, it is, therefore, the need of the hour to use green technologies. Green technology is defined as the technology which can perform the same functions as any other technology, but is energy efficient while executing its task. While taking into the account, the increase in energy consumption it can be inferred that Information Communication Technology or ICT is largely accountable for it. This expansion of the mobile phone network and ICT services lead to an increase in energy consumption. Since the internet has become a large part of the humankind, cloud computing has surfaced a useful tool to keep up with the rising demand for data storage. Since cloud computing emerged around the world, it becomes a necessity to come up with green computing (Prashant, 2017). The idea of the green building goes along with the need of the hour that is low power consumption and being environment-friendly.

This chapter focuses on cloud computing architecture via IoT in green buildings. The remaining chapter is structured as- Section 1 explains the three main concepts of the chapter, i.e. green building, IoT devices, and green cloud computing. In section 2, the smart automation in green building and methods to minimize energy consumptions are discussed. Section 3 & 4 covers the IoT architecture in green buildings and the basics of green cloud computing. Finally, the last section i.e. section 5 explains the green cloud architecture in green building.

BACKGROUND

It is a well-known fact that data center's consumes a large amount of electricity. Also, most of the electricity is produced by fossil fuels and therefore causes carbon emissions, which is harmful to the environment. Thus, many companies are moving toward building a "green" data center like Apple (Bostic, 2013) and McGraw-Hill (Miller, 2014) constructed 20MW and 14MW solar arrangement for their organizations, respectively. There some cloud providers that use solar or wind farms to power their data center's ("Wind-Powered Data Centers", n.d.). Since the production of electricity depends upon fossil fuels, which causes carbon emissions the idea of the green building is getting popular all over the world.

The green building, however, is limited to a few renewable energy methods like solar power over fossil fuels. Apart from being eco-friendly, the green building might result in an extra added cost of production. Not all professionals value energy efficient methods and other factors such as components of green building, primarily temperature control, ventilation control, abundant daylighting, and lighting control, which becomes another challenge. Greener solutions are the demand of the day. However, since they're still adapting to it, it becomes a little difficult for the construction companies to quickly gain green building expertise (Nichols, 2016). The system is advancing from Europe and Asia to the US which permits the number of carbon emissions that also enables an organization to acquire the rights of carbon emission of another organization, enhancing the scope of building green. For instance, the UK government mandated the same system involving companies consuming more than 6 GWh per year (Department for Business, Energy & Industrial Strategy and Environment Agency, 2015); i.e. a company having a data center as small as 700-kW must take part. While some researchers focused on assimilating renewable energy resources into data center's, there are several studies that were carried out on load distribution between that are geographically distributed in order to gain from green energy produced on-site (Le, Bianchini, Nguyen, Bilgir, & Martonosi, 2010). There were two main findings from these works which are: (1) in order to increase the availability and reduce the response time, redundant services are made available on various geographically distributed and (2) heterogeneous green energy generation due to differing local weather conditions (Berral, Goiri, Nguyen, Gavalda, Torres, & Biu, 2014). Since most of the works consider an important factor i.e. energy efficiency, they do not focus on carbon emissions. This chapter highlights the cloud architecture in detail while focusing on carbon emission.

Green Building

A building that creates a positive impact on our climate or natural environment, and which eliminates or reduces the negative impacts in its architecture, construction, or operation, is known as a 'green' building. Green buildings help in better quality of living and preserve natural resources. The green building, or sustainable building, aims to utilize resources efficiently, maintaining the ecological balance, without harming the society, and is also economically feasible through optimal design, construction and maintenance. The following factors contribute to making a building 'green' ("What is green building", 2019). The factors contributing to making a building 'Green' are highlighted in figure 1.

- Efficient utilization of energy, water and other resources.
- Renewable resources usage.
- Determination of pollution and waste reduction and facilitation of recycling.
- High environmental air quality inside the premises.
- Sustainable and non-toxic material utilization.
- Environmental consideration in design, construction and operation.
- Adaptable design to a changing environment.

Green Computing

'Green Computing' is defined as the study and practice to design IT equipment efficiently and effectively which has minimal or no environmental impact (Saha, 2018). The classification of green computing is

Figure 1. Factors affecting green building

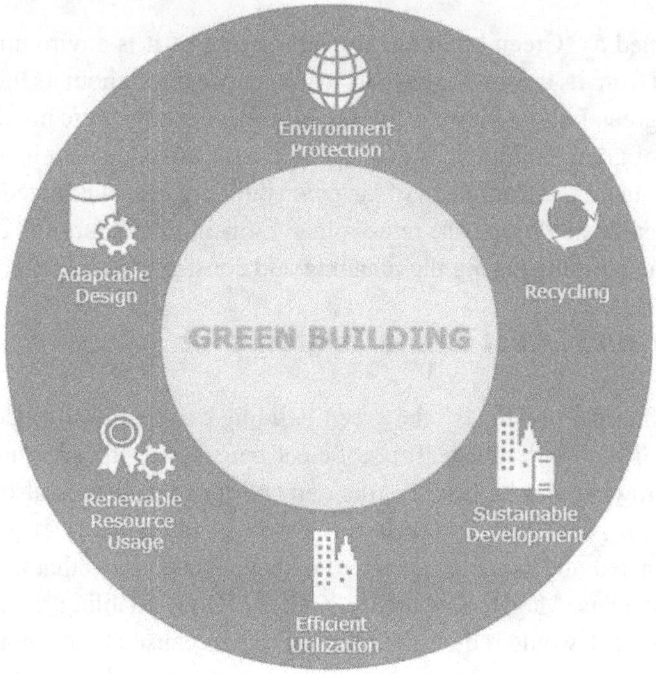

Figure 2. Division of green computing

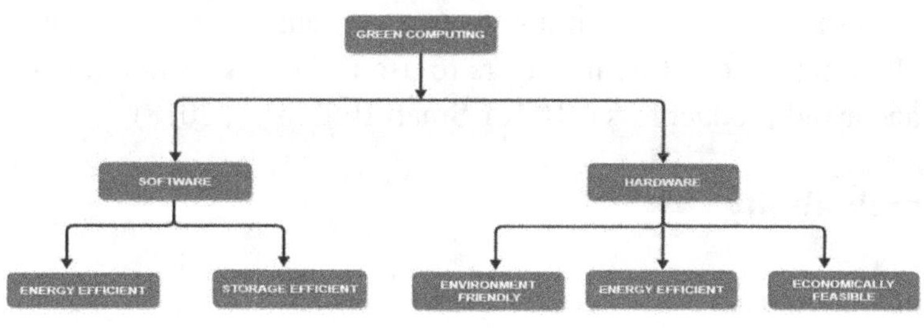

indicated in figure 2. According to Patel, Mehrotra & Soner (2015), the green computer can have two technical aspects:

1. The purpose of software technology is to create methods that can improve program, storage and energy efficiency.
2. In terms of hardware, along with being energy efficient, it should also be economically feasible and environmentally friendly.

THE CONCEPT OF GREEN BUILDINGS

A building can be termed as 'Green building' if every aspect of it is environment-friendly as well as resource-efficient right from its layout design to its maintenance throughout its life. Even though *sustainable development* and green building have a close relationship, but they are not the same. Over a period of time, the definition of green building has evolved due to the advancement in technology. (Robichaud & Anantatmula, 2011) stated the four pillars of the green building, i.e. minimized environmental impact, enhanced occupant's health conditions, the return on investment to builders and community, and assessment of longevity of the building during the planning and construction stage.

Smart Automation and Effect on the Environment

With the help of the 'Internet of Things', the green building can be transformed into a smart building as the purpose of the IoT is to boost the abilities of electronic devices to structure a smart environment. With the population growing at a higher rate, the demand for resources is also growing, as resources on earth are rapidly being depleted. To tackle this problem, the concept of green building is rising, which is smartly automated and capable enough to manage these issues that are being currently faced by mankind (Dandekar, cans, Magar, Tavhare, & Katkar, 2017). Enabling the green cloud computing in the green building via IoT would reduce the harmful effects caused by the construction of buildings, CO2 emissions, degradation of air and water quality, energy wastage. Making the use of IoT sensors in green building a method called 'Predictive maintenance' helps in detecting the exact time of maintenance which would eventually conserve the resources. In the new era of technology, heavy construction types of equipment are being replaced by sensors with the help of IoT which can be governed by indicators. Therefore, less energy usage means cost-cutting in terms of maintenance and eventually leads to better asset optimization. It allows installation managers to use technology solutions, improve construction equipment uptime and avoid product loss ("10 IoT Smart Building", 2018).

Concept of Sustainability

Sustainability for software and hardware developers and users has become increasingly important Due to the rapid growth in energy consumption in the last two decades (Radu, 2017), it is becoming an increasingly important consideration for building practitioners to make the structure economically feasible and ecologically sustainable. According to Sinha, Gupta, & Kutnar (2013), the following objectives should be met to achieve sustainability:

- Reducing matter and energy consumption.
- Material reusability and recyclability.
- Minimum environmental and energy impacts.
- Satisfaction of society.

The concept of green building and sustainable design are a growing engineering phenomenon with an unprecedented growth rate and acceptability. The fact that energy supplies and resources are diminishing, together with people's increased awareness of sustainability, helps drive this rapid growth of green buildings (Sinha, Gupta, & Kutnar, 2013).

Methods to Minimize Energy Consumption in Buildings

Since green buildings need a separate space to store the data and the required storage is provided by data center's and cloud storage which in turn generates an enormous amount of heat and energy and if not managed properly can result in overheating of the servers. The heat and energy produced by the servers could even result in the melting of paraphernalia. One of the crucial methods that can be followed in data center's to minimize energy consumption is to improve airflow management. In proposition to *Green House Data*, adjusting the computer cabling in order to efficiently exhaust heat, which will eventually reduce the possibility of overheating. Following this simple approach would minimize energy expenditures (Prashant, 2017). Computer spacing could also minimize the airflow. One of the methods designed by IBM is to arrange the server racks with appropriate distance between them in order to make the airflow system efficient (Prashant, 2017). Floor tile placement approach could also result in minimizing energy consumption (Prashant, 2017). According to the Data Trend, one of the foundational steps taken by the company is to set up the entity in locations which have a cooler climate. It would eventually drive down the cooling costs of companies as cooler climates will naturally help in cooling down the systems externally (Prashant, 2017).

The general trend shows that the servers are not performing up to its full efficiency and capacity. Integrating computing resources could significantly contribute to improving performance. Using better software might also increase the efficiency of the system. Amalgamating all the aforementioned methods could result in computing usage in the range of 30-80% (Prashant, 2017).

IOT ARCHITECTURE IN GREEN BUILDING

The IoT considers all the objects in the environment as part of the internet. The impact of IoT on the world is comparable to the effect caused by the internet and therefore it is considered a new revolution in human life. Providing services such as payment through mobile phones and transfer of money just a few clicks away through connecting various digital devices to one another and processing everything with the help of internet. IoT system of the green building amalgamates all the resources such as the environment, people, sensors and physical devices. An integrated system in a building provides real-time monitoring and control system (Wang, Qiu, Dong, & Wang, 2017). Integrating various functions such as the lighting of the building, smart windows whose glass adapts automatically to the changing conditions and various automated features into a single interactive system which in turn makes it a convenient choice for the user. This section describes the architecture in green building through IoT devices and the three driving factors of IoT are presented in figure 3. According to Abraham, (2016) the three following important factors drive the IoT forward:

- **Sensing Nodes:** A sensor is a type of information collection tool, which is in charge of gathering the relevant data in a specific environment, then carries information from the collected information with a specific electrical signal. These nodes contain distinctive identification and can be controlled independently through a distant command and organized topology.
- **Embedded Processing Nodes:** The heart of the IoT is embedded processing. Local processing potential for microcontrollers or microprocessors. *Microcontroller Unit (MCU)* can offer embedded processing in real time, which is a major necessity for most IoT applications. Cost-effectiveness,

Figure 3. IoT factors

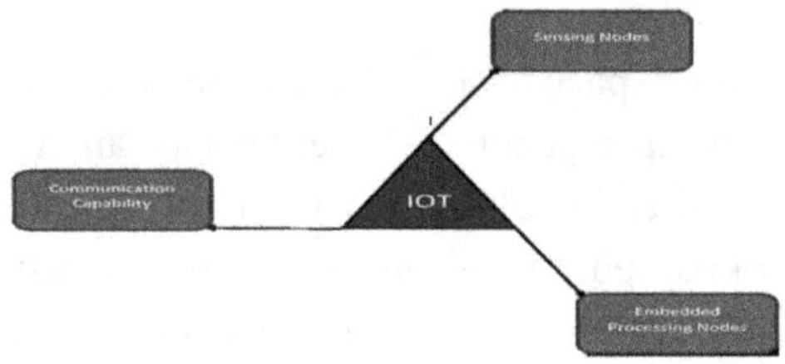

low power consumption, quality and reliability and security are some major utilities of embedded processing.

- **Communication Capability:** The communication node's primary responsibility is to transmit information collected by the sensing nodes to the targets recognized by the local embedded processing nodes.

IoT Devices

A huge number of devices connected to the internet mass-produce enormous data as well as information. cloud computing being a *pay-as-you-go service* provides access to the data storage from anywhere at any time with the use of IoT (B.B, Saluja, Sharma, Mittal, & Sharma, 2012). All the IoT devices require data storage and cloud complements the need for IoT devices efficiently. IoT is comprised of millions of devices that can detect, calculate, communicate and act according to the requirements (Jung, Ganasamband, & Mukerjee, 2012). Interconnecting different digital devices over the network has proven to be an inexhaustible source of information. The phenomena was made possible due to three developments. Firstly, miniaturization, where it is possible to obtain technology in the hands of the user anywhere, anytime. Secondly, forthcoming of mobile communications. Lastly, intelligence in application services that is converted to the vast amount of data generated through the IoT sensor network into functional information for different uses (Alcaraz, Najera, Lopez, & Roman, 2010). Smart building management systems can be successfully managed remotely from any part of the world. In addition, the indoor and outdoor environment is always sensed to provide optimal results for ventilation, lighting, fire and safety purposes. Another example is an ultra-sonic noise detection. Electric lines with cracks or holes produce ultra-sonic sound. This may not be noticed in the usual cases, but it is easy to detect the location for maintenance with IoT technology. Another area that IoT has completely transformed is how managers of commercial facilities can monitor, compute and accumulate data from each part of the building without facing any reach issue. Moreover, IoT sends the data at a very fast rate. IoT devices provide valuable insights to developers to react rapidly to emerging trends. IoT applications enable facility managers to carry out different experiments to check the optimization results. It also allows them to use IoT devices with a single panel to monitor building systems ("10 IoT Smart Building", 2018).

Architecture of the IoT Devices

In IoT, every device is assigned an IP address and is connected to the IP network using the dedicated hardware. The trading of data in such architecture could prove to be a costly affair. But using the cloud to manage and oversee IoT can be cost-effective as well as energy-efficient (B.B et al., 2012). According to B.B, Saluja, Sharma, Mittal, & Sharma, (2012) the following models are appropriate for IoT:

- IaaS, for business models of sensor & actuator and resource access models.
- PaaS offers IoT data and control access services.
- SaaS, application domain for monitoring services.

Since sensors are quite expensive, also the maintenance is costly, but with the integration of cloud in the respective process could effectively consume the resources with reduced costs on sharing basis. Sensors are controlled and maintained by cloud computing. The sensor cloud provider offers various services to the user involving the sensor from which the user selects an option according to its requirements. Through this way, the user can take benefit of the sensor services without actually incurring the buying cost (B.B et al., 2012). Some applications such as transport monitoring, including traffic signal control, congestion control, navigation, dynamic traffic lights, require a Wireless Sensor Network (WSN) and computing cum data-intensive applications benefit from cloud storage and computing power (B.B et al., 2012). Applications for military sensors are benefited with the infrastructure since it's safer to send cloud infrastructure sensor data on the Internet, the application of cloud sensor even spreads across the health care system as well as the weather forecast. IoT enables commercial homeowners to have energy-efficient buildings. It also affects the layout of the buildings and enables them to be environmentally friendly and efficient and effective in terms of resources ("10 IoT Smart Building", 2018).

CONCEPT OF GREEN CLOUD COMPUTING

Green cloud computing could greatly contribute to reducing the impact of computing on the environment. Although the generally perceived notion is that cloud computing consumes a large amount of energy, but it has a green lining. In cloud computing, comparatively lower carbon emissions are expected due to its highly optimized structure and energy saving construction. The backbone in making cloud computing energy efficient is "Virtualization", through which several organizations share a common infrastructure saving infrastructural cost as well as energy costs (Abdullah, 2014). Green cloud computing saves a lot of energy, time, money and therefore more and more organizations are seeking green cloud solutions, but a considerable number of corporations still prefers traditional data center's.

Cloud Computing Architecture

The architecture of Cloud Computing comprises of two broad sections which are connected over a network, typically Internet: -

- Front-end
- Back-end

The client's computer or computer network and the application needed to access the cloud computing system are part of the front end. It is important to note that not all cloud computing systems have the same user interface. For example, Web Browsers such as Chrome or Firefox, and services like web-based email programs. Other systems have distinct applications that allow network access to clients. On the back end, the "cloud" of the computing services is created by the variety of computers, server, and data storage systems. A cloud computing system can contain any type of data, program or information. Normally, every application will have its own exclusive server. To make sure that everything runs smoothly, a central server directs the system, monitoring traffic and client demands. Middleware, a special kind of software is used by the central server, and a set of rules called protocols are followed. The middleware software permits the communication of networked computers with each other. Since, the servers do not work with its full ability, the unconsumed computation power is wasted. Server virtualization, which is the technique that reduces the need for more physical machines and works on maximizing the output of individual servers to their full potential. A cloud computing company with plenty of clients is likely to have high demands for a lot of storage space. Hundreds of digital storage devices are required by some companies. To keep all the clients' information stored, a cloud computing system would require at least twice the number of storage devices than it usually needs. This is because, like all other computers, these devices occasionally break down. Backups of client information need to be created by cloud computing systems and saved on all other devices, which would ensure easy access to retrieve data that otherwise would be unreachable. This process of backup is known as redundancy (Strickland, 2016).

Requirements of Green Cloud Computing

Since computers are being widely used to increase computational speed as well as accuracy, the power consumed in the process is high and growing exponentially in the current scenario. It also leads to the greater emission of greenhouse gases for instance Carbon Dioxide (CO2) which is considered very harmful for the environment and natural resources. It was also observed that most of the traditional data center's use old technologies which possessed insufficient cooling capacities and lead to the overheating of the servers and ultimately leading to the wastage of energy and emission of CO2 (Negi, Kapoor, & Kapoor, 2015). Also, data center's are constructed keeping in mind the future demands which lead to the construction of huge infrastructure. The huge data center is not functional throughout the year, but is in demand only in the short peak season, which is not energy efficient as the huge infrastructure not only costs more but also result in the wastage of energy. The issue with the vast data center's is solved via cloud computing devices occasionally used, i.e. Pay-As-You-Use Service which extricates the infrastructure cost and high energy consumption, thereby, making cloud computing green and adding benefits to the environment. For example, if three companies require data storage then each one of them would attempt to build a data center for themselves individually, which would not only cost them a huge sum of money but also additional expenses like land acquisition and factors like scalability. Cloud computing makes this job hassle-free as it enables the three companies to use the same servers for data servers through "Virtualization" and also saves them the cost as it is a Pay-As-You-Use Service. All the aforementioned issues create demand for green cloud computing, which would not only be energy-efficient but will also protect the environment.

Virtualization

A research was carried out which compared the setup consisting of virtualization and setup containing no virtualization. It was found out that On-site servers not using virtualization release about 46 kg of CO2 per year. The figure is 2 kg if a person uses a public cloud that abides best practice (Scott, 2016).

Figure 4. Virtualization

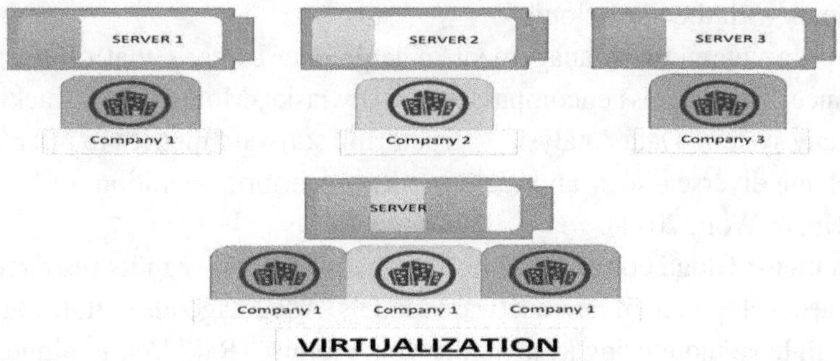

Virtualization plays a pivotal role in cloud computing. Since virtualization reduces the complexity of services offered by the cloud provider, which eventually results in cloud computing being economically feasible. Most companies use Hypervisor to manage different aspects of cloud computing. A hypervisor enables multiple operating systems to use single hardware and it is also known as a *virtual machine manager* (VMM). Each operating system behaves as if it is the only one using the host's processor, memory and other resources, but it is the hypervisor that is actually controlling everything and making sure no two operating systems obstruct each other. A hypervisor performs a certain task with great efficiency without compromising the computing speed. Virtualization makes it easier to deploy, manage, and deliver resources. The VM pool comprises multiple operating systems having different applications in it. The hypervisor controls the whole architecture and there is a secure partition between each OS environment which eliminates the security concerns. It is common to mistake cloud computing as virtualization as the system runs on the virtual server in each case. However, cloud computing is a service where virtualization is an integral part of the physical architecture of cloud computing. Virtualization becomes a necessary need to adopt cloud computing infrastructure (Nimje, Gaikwad, & Datir, 2013). Data center's consume a lot of energy and resources and emits CO_2 in large quantity making it harmful to the environment. However, all the aforementioned issues could be solved by virtualization in cloud computing, which basically aims to use the resources efficiently and thus reduces the wastage of energy and is environmentally-friendly. Hence, virtualization plays a key role in making cloud computing efficient. Virtualization is explained via an example, in figure 4.

Monitoring

Cloud monitoring is an essential task for both cloud service providers and cloud service consumers. It performs a major function in supervising and administrating the hardware and software structures; whereas, it also yields useful data regarding the application process such *Key Performance Indicators (KPI)* (Aceto, Botta, Donato, & Pescape, 2012). (Aceto, Botta, Donato, & Pescape, 2012) were on the views that cloud computing involves many activities that are benefitted through monitoring such as:

- **Capacity and Resource Planning:** Ahead of adopting cloud computing on a large scale, capacity and resource planning remains a big challenge for developers.

- **Capacity and Resource Management:** Monitoring system plays a vital role in the management of heavy systems with the likes cloud.
- **Data Center Management:** Management of large data center's that offer cloud services is of huge importance. This process encompasses two key tasks, Monitoring – tracking of the preferred resources on any system. Data Analysis – structuring software metrics to infer systems or applications for resolving diverse issues, and various other governing operations (Wang, Schwan, Talwar, Eisenhauer, Hu, & Wolf, 2011).
- **SLA Management:** Cloud computing offers immense flexibility in its resource management that encourages the development of new software models. Advantageous to the cloud applications, this new feature will have monitoring as its underlying premise (Rak, Venticinique, M'hr, Echevarria, & Ensal, 2011).
- **Billing**: As discussed earlier, "measured services" are offered to consumers via monitoring, which allows them to pay substantially for a metered component. In addition, however, monitoring helps consumers in verifying their effective use of cloud services as well as in making a price comparison across different providers (Li, Yang, Kandula, & Zhang). Thus, monitoring is crucial for the provider (or auditor) as well as the consumer.
- **Troubleshooting:** Since the cloud infrastructure is both intricate and enormous, it presents a major challenge for troubleshooting since the root issue would have to be explored throughout the system leading to both energy and time consumption. Therefore, monitoring is required in this aspect in order to identify the location of the problem and help the provider as well as the user in a difficult situation in terms of the complexities of the infrastructure.
- **Performance Management:** Armbrust argued that even though providers are careful in delegating all the services and tasks up to the level which can satisfy customers, but many a time some cloud nodes may not perform well as compared to the other nodes (Armbrust, et al., 2009).
- **Security Management**: For several reasons, the security of the cloud is extremely crucial. It is one of the most crucial barriers for the spread of cloud computing, particularly in view of certain types of applications and consumers (Chen, Paxson, & Katz, 2010).

Hazarika & Singh (2015) believed that for a successful cloud monitoring system, there are a few necessary properties that should be consistent with the infrastructure, like:

- **Scalability**: It means that the system should be capable enough to deal with a large number of resources. This property is essential because there are a diverse number of parameters to be monitored for many dissimilar resources. A scalable monitoring system should, therefore, be able to collect, transmit and analyze such data volumes efficiently without interfering with normal cloud operations.
- **Elasticity**: An elastic monitoring system can handle dynamic changes of monitored entities. It is also referred to as Dynamism (Clayman, Galis, & Mamatas, 2010).
- **Adaptability**: Clayman proclaimed that a monitoring system is considered as adaptable if it adapts to different computational and network loads so as not to be invasive (Hazarika, 2015). Owing to the technical complexities and dynamism of the cloud services, adaptability becomes a fundamental requirement.

- **Timeliness**: Getting results from monitoring on time is important as taking any preventive measures after it is too late is useless. Therefore, it becomes necessary to have monitoring on the basis of fixed time.
- **Automaticity**: Distributed resources can be managed by an autonomous monitoring system by automatically reacting to arbitrary changes and hiding the intrinsic intricacy of the suppliers and consumers.
- **Accuracy**: An accurate monitoring system provides accurate measures, i.e. these measures converge with the actual result.

Network Infrastructure

As far as the network level is concerned, energy efficiency can be accomplished in the network interface card which is considered to be a node level or switches and routers i.e. the infrastructure level. As per (Abdullah, 2014), "Green Networking" refers to the collaboration of energy understanding in the devices, design and network protocols, and it deals with the issues associated to energy efficiency at the network level.

GREEN CLOUD ARCHITECTURE IN GREEN BUILDING

From the study which aims to make cloud computing efficient, it is clear that although different cloud components are efficient in terms of power and performance, still, the collaborated infrastructure ceases to exist. The majority of cloud computing sustainability efforts have missed the network contribution. For example, if a file is quite large in size then operating it on the network will contribute to major energy consumption and in that case transmitting it locally than though cloud would prove to be a greener solution. Even though VM consolidation significantly lowers the active servers, but it puts a lot of load on other servers which might overheat due to the excessive load and therefore the heat distribution would cause a problem. Some realms majorly focus on the reordering of workload for efficient cooling rather than emphasizing on the virtualization. Adding to that, as cloud providers are profit-oriented, they only look for solutions which reduce the power consumption leading to Carbon emission without cutting down the profits. Hence, a consolidated architecture is required to enable green cloud computing and the aforementioned structure is explained in this section (Kumar & Buyya, 2012).

The green cloud architecture consists of a middleware *Green Broker* to which users submit requests and it manages and selects the greenest cloud provider available to facilitate the user's request. Users can request for three services, i.e. infrastructure, platform or software. Cloud providers can register their services in the public directory as "green offers" which can be accessed by the green broker. The green offers are green services, pricing and time to access them for the lowest carbon emissions. Green broker receives the current state of the energy parameters from the Carbon Emission Directory for the use of different cloud services. The Carbon Emission Directory keeps all data relating to cloud service energy efficiency. Green broker calculates the carbon emissions of all cloud providers providing the cloud service requested. A set of services that lead to lower carbon emissions are recognized and then selected. These services are then purchased on behalf of the users (Kumar & Buyya, 2012).

The green cloud framework has been designed to track the overall energy use of a user's request. The main function of the green broker from the user's point of view is to ensure that the minimum carbon emission cloud services are provided to the user along with managing the QoS requirements. The three services, i.e. SaaS, IaaS, PaaS can be accessed by the user through the cloud (Kumar and Buyya, 2012). The requirements of the three levels of cloud computing and the traditional data center are illustrated in figure 5.

IaaS Level

IaaS provider both provides an independent architecture to the users and facilitates other cloud services. Thus, it is integral to the whole green architecture. They use the latest technologies to provide the most energy-efficient infrastructure for IT and cooling systems. The use of virtualization and consolidation further reduces energy consumption by switching off unused servers. In order to evaluate each site of an IaaS provider in terms of being energy efficient, various energy meters and sensors are installed. This information is frequently posted on the Carbon Emission Directory by cloud providers. The cloud provider offers different green offers and pricing schemes to encourage users to use their services during off-peak or maximum hours of energy efficiency.

PaaS Level

This layer provider offers platform services in general for application development. The platform facilitates the development of applications that guarantee energy efficiency throughout the system. This can be achieved by including different energy profiling tools like JouleSort. In addition to the application development, cloud platforms expedite user applications deployment on Hybrid cloud. In this type, the platforms profile the application and decide which part of the application or the data should be processed at home and in the cloud in order to achieve maximum energy efficiency (Rivoire, Shah, Ranganathan, & Kozyrakis, 2007).

SaaS Level

This level is considered to be the least complex level as the whole software is provided by the SaaS providers and makes it easier for the user to use its service.

The SaaS providers deliver the software installed in their data center to the customer. Also, they provide the IaaS provider's resources. They evaluate different aspects of their software such as energy efficiency, implementation and deployment. Not only the SaaS providers offer data center's which have close proximity to the user to make it convenient, but also maintain a minimum number of user's confidential data facsimile on the energy efficient storage.

FURTHER CHALLENGES AND FUTURE RESEARCH DIRECTIONS

One of the biggest challenges lies in the traditional trend of building a data center for the organization's storage requirements rather than using cloud storage is to keep the data secure and private. In the changing era of technology, failing to keep the pace with latest technologies can have massive impacts on the existence of the organization. Using more data center's causes more carbon emission and eventually

Figure 5. Requirements of different levels of cloud

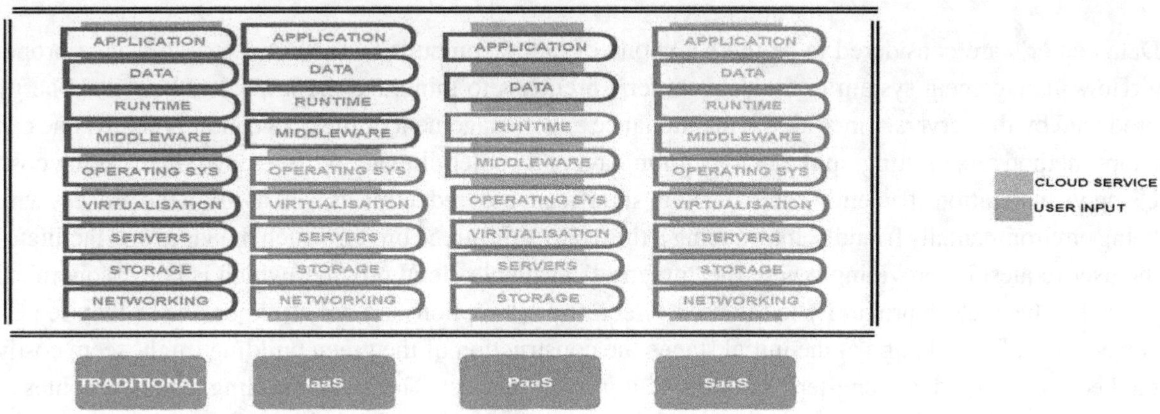

Figure 6. Requirements of different levels of cloud

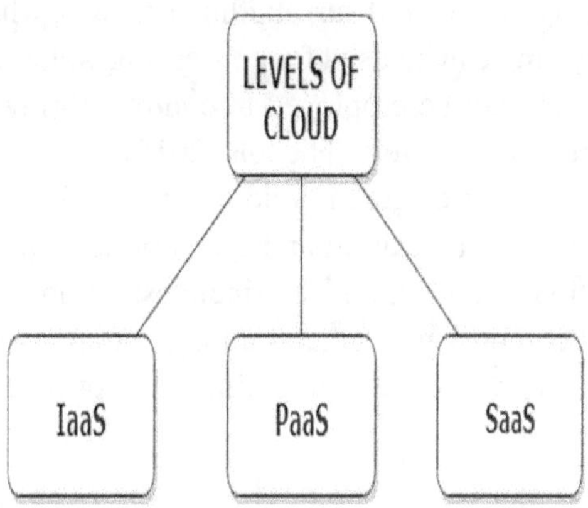

impacts the environment negatively. Switching to cloud from is not as easy as it seems and there are various factors that should be regarded.

Coming to the architectural side of cloud computing, it has broadly two main challenges: (1) technological and (2) systematic.

From the technological point of view making cloud "green" has its own challenges. For example, allocating resources efficiently and virtualization leading to the overheating of servers causes stumbling block to both users as well as providers. Unable to create efficient cooling inlets and outlet in can cause increased carbon emission.

Now, the systematic challenge is developing the system in a holistic way so as to make it efficient as well as eco-friendly. The cloud providers being profit oriented majorly focuses on minimizing the energy resources so as to reduce the cost and often ignores the environmental factor. Another obstacle that remains even after seeking a holistic development is satisfying the customer needs economically. Future research should focus on making the cloud storage efficient, sustainable, eco-friendly and finally economical through the use of the latest technologies.

SOLUTION AND CONCLUSION

Data center's are considered to be the backbone of cloud computing, it is necessary to have a proper airflow management system in the data center's premises to minimize the impact of heat and energy produced by the servers. Since all the maintenance and management of the data center is costly, one can adopt methods like setting up the in a location having a cooler climate to drive down the cooling cost. Using virtualization efficiently could result in significant cost reduction as well as align the process with being environmentally friendly. Introducing a third party, the green broker which manages and facilitates the user request by providing it with the "greenest" available cloud provider which is both economical as well as have close proximity to the user. Green friendly options are not as expensive as they're perceived to be. Even though at the initial stages the construction of the green building might seem costly but keeping in mind the long-term benefits of it justifies its cost. The green building also contributes to saving energy as well as water production. It lowers utility costs and provides for constructive output, such as the property value of a green building would increase over time as the fossil fuels are phased out. Similarly, the dependency on power plant is reduced, leading to an increase in efficiency, thus having a positive impact on the local infrastructure and surrounding area as a whole. Systemized environmental measures such as rainwater recycling can be used for conserving water in a better way. Along with this, green water roofs and solar panels can be employed to control stormwater management and produce clean energy to reduce utility bills respectively (Nichols, 2016).

This chapter deals with the concept of green building and the incorporation of green cloud architecture in it utilizing the notion of virtualization and green broker. In this chapter, green services are introduced whose implementation follows the ICT principles. Although this chapter has not included all the horizons of green cloud computing in green building, attempts to describe the importance and requirement of green building and green cloud computing. Moreover, it explains the green cloud architecture in green building.

FUTURE SCOPE

In future researchers should focus on the applicable aspect of cloud computing in the green building. Even though the major focus point of the research includes increasing energy efficiency and reducing environmental damage in the architecture of cloud computing but apart from that it should also be noted that reducing costs and making it a feasible solution is equally important.

REFERENCES

Abdullah, A. (2014). Green cloud computing: The need of the hour. *International Journal of Research in Advent Technology*, *2*(1), 316–321.

Abraham, S. (2016). Internet of Things (IoT) with Cloud Computing and Machine-to-Machine (M2M) Communication. *International Journal of Emerging Trends in Science and Technology, 3*(9), 4654-4661.

Aceto, G., Botta, A., Donato, W., & Pescape, A. (2012). Cloud Monitoring: definitions, issues and future directions. In *proceedings of the IEEE International Conference on Cloud Networking, CLOUDNET*. IEEE. 10.1109/CloudNet.2012.6483656

Alcaraz, C., Najera, P., Lopez, J., & Roman, R. (2010). Wireless Sensor Networks and the Internet of Things: Do We Need a Complete Integration? In *Proceedings of the 1st International Workshop on the Security of the Internet of Things*. Tokyo, Japan: IEEE.

Armbrust, M., Fox, A., Griffith, R., Joseph, A. D., Katz, R. H., Konwinski, A., & Zaharia, M. (2009). *Above the Clouds: A Berkeley View of Cloud Computing* (Rep. No. UCB/EECS-2009-28). UC Berkeley Reliable Adaptive Distributed Systems Laboratory. Retrieved from https://www2.eecs.berkeley.edu/Pubs/TechRpts/2009/EECS-2009-28.pdf

B.B., P.R., Saluja, P., Sharma, N., Mittal, A., & Sharma, S.V. (2012). Cloud computing for Internet of Things & sensing based applications. In *Proceedings of the International Conference on Sensing Technology*. Bangalore: ICST.

Berral, J. L., Goiri, I., Nguyen, T. D., Gavalda, R., Torres, J., & Bianchini, R. (2014). Building Green Cloud Services at Low Cost. In *Proceedings of the2014 IEEE 34th International Conference on Distributed Computing System.* Madrid, Spain: IEEE. 10.1109/ICDCS.2014.53

Bostic, K. (2013, March 21). *Apple data centers now 100 percent renewable-powered, but emission rise 34 percent.* Retrieved from https://appleinsider.com/articles/13/03/21/apple-claims-data-centers-100-renewable-powered-75-for-corporate-facilities

Chen, Y., Paxson, V., & Katz, R. H. (2010). *What's New About Cloud Computing Security? (Rep. No. UCB/EECS-2010-5).* Berkeley, CA: EECS Department University of California. Retrieved from https://www2.eecs.berkeley.edu/Pubs/TechRpts/2010/EECS-2010-5.html

Clayman, S., Galis, A., & Mamatas, L. (2010). Monitoring Virtual Networks with Lattice. In *Proceedings of the 2010 IEEE/IFIP Network Operations and Management Symposium Workshops*. Osaka, Japan: IEEE. 10.1109/NOMSW.2010.5486569

Dandekar, A., Kanse, R., Magar, S., Tavhare, S., & Katkar, A. (2017). IoT Based Green Building. *International Journal of Computer Science and Mobile Computing*, *6*(4), 439–443.

Data Center Knowledge. (n.d.). *Wind-Powered Data Centers.* Retrieved from https://www.datacenterknowledge.com/wind-powered-data-centers

Department for Business, Energy & Industrial Strategy and Environment Agency. (2015, July 15). *CRC Energy Efficiency Scheme*. Retrieved from https://www.gov.uk/government/collections/crc-energy-efficiency-scheme#history

Forrest, W., Kaplan, J. M., & Kindler, N. (2008). *Data centers: How to cut carbon emissions and costs*. Retrieved from McKinsey & Company Website: ftp://public.dhe.ibm.com/software/uk/itsolutions/optimiseit/greenhub/carbon-managment/pov-mckinsey-report.pdf

Hazarika, B., & Singh, T.J. (2015). Survey Paper on Cloud Computing & Cloud Monitoring: Basics. *SSRG International Journal of Computer Science and Engineering, 2*(1), 10-15.

10 . IoT Smart Building Trends to Look Out For in 2018. (2018). [Web log post]. Retrieved from https://blog.senseware.co/2018/01/08/10-iot-smart-building-trends-look-2018

Jung, G., Ganasambandam, N., & Mukerjee, T. (2012). Synchronous Parallel Processing of Big-Data Analytics Services to Optimize Performance in Federated Clouds. In *Proceedings of the 2012 IEEE Fifth International Conference on Cloud Computing*. Honolulu, HI: IEEE. 10.1109/CLOUD.2012.108

Kumar, S., & Buyya, R. (2012). Green Cloud Computing and Environmental Sustainability. In S. Murugesan & G. R. Gangadharan (Eds.), *Harnessing Green It: Principles and Practices* (pp. 315–339). IEEE. doi:10.1002/9781118305393.ch16

Le, K., Bianchini, R., Nguyen, T. D., Bilgir, O., & Martonosi, M. (2010). Capping the brown energy consumption of Internet services at low cost. In *Proceedings of theInternational Conference on Green Computing*. Chicago, IL: IEEE. 10.1109/GREENCOMP.2010.5598305

Li, A., Yang, X., Kandula, S., & Zhang, M. (2010). CloudCmp: Comparing Public Cloud Providers. In Proceedings of the 2010 Internet Measurement Conference. Melbourne, Australia: Association for Computing Machinery, Inc.

Miller, R. (2014, October 14). *Super-Sizing Solar Power for Data Centers*. Retrieved from https://www.datacenterknowledge.com/archives/2014/10/14/super-sizing-solar-power-for-data-centersl

Mohamed, A. (2018). *A history of cloud computing*. Retrieved from https://www.computerweekly.com/feature/A-history-of-cloud-computing

Nayyar, A. (2011). Interoperability of cloud computing with web services. *International Journal of ElectroComputational World & Knowledge Interface, 1*(1).

Nayyar, A. (2011). Private Virtual Infrastructure (PVI) Model for Cloud Computing. *International Journal Software Engineering Research and Practices, 1*(1), 10–14.

Negi, G. P., Kapoor, H., & Kapoor, A. (2015). Green Computing: A demand of Time. *International Journal of Engineering Science, 14*(05), 34–36.

Nichols, M. (2016, April 22). *Green building challenges*. Retrieved from https://usgreentechnology.com/green-building-challenges

Nimje, A. R., Gaikwad, V. T., & Datir, H. N. (2013). Green Cloud Computing: A Virtualized Security Framework for Green Cloud Computing. *International Journal of Advanced Research in Computer Science and Software Engineering*, *3*(04), 642–646.

Patel, Y. S., Mehrotra, N., & Soner, S. (2015). Green cloud computing: A review on Green IT areas for cloud computing environment. In *Proceedings of the 2015 International Conference on Futuristic Trends on Computational Analysis and Knowledge Management (ABLAZE)*. Noida: IEEE. 10.1109/ABLAZE.2015.7155006

Prashant, D. (2017). Greening the Cloud Computing. *International Journal of Engineering and Computer Science*, *3*(05), 5811–5814.

Radu, L.-D. (2017). Green Cloud Computing: A Literature Survey. *Symmetry*, *9*(12), 295. doi:10.3390ym9120295

Rak, M., Venticinque, S., M'hr, T., & Echevarria, G., & Ensal, Gorka. (2011). Cloud Application Monitoring: The mOSAIC Approach. In *Proceedings of the 2011 IEEE Third International Conference on Cloud Computing Technology and Science*. Athens: IEEE. 10.1109/CloudCom.2011.117

Rivoire, S., Shah, M. A., Ranganathan, P., & Kozyrakis, C. (2007). JouleSort: A balanced energy-efficiency benchmark. In *Proceedings of the the ACM SIGMOD International Conference on Management of Data*. Beijing, China: IEEE. 10.1145/1247480.1247522

Robichaud, L. B., & Anantatmula, V. S. (2011). Greening Project Management Practices for Sustainable Construction. *Journal of Management Engineering*, *27*(1), 27–57. doi:10.1061/(ASCE)ME.1943-5479.0000030

Saha, B. (2018). Green Computing: Current Research Trends. *International Journal on Computer Science and Engineering*, *6*(03), 467–469.

Scott, G. (2016). *How Green is Cloud Computing*. Retrieved from https://www.ewastesecurity.com/how-green-is-cloud-computing-new-study/

Sinha, A., Gupta, R., & Kutnar, A. (2013). Sustainable Development and Green Buildings. *Drvna Industrija*, *64*(01), 45–53. doi:10.5552/drind.2013.1205

Solanki, A., & Nayyar, A. (2019). Green Internet of Things (G-IoT): ICT Technologies, Principles, Applications, Projects, and Challenges. In Handbook of Research on Big Data and the IoT (pp. 379-405). IGI Global.

Strickland, J. (2016). *How cloud computing works*. Retrieved from https://computer.howstuffworks.com/cloud-computing/cloud-computing1.htm

Uddin, M., & Abdul Rahman, A. (2011). Carbon Sustainability Framework to Reduce CO2 Emissions in Data Centers. *International Journal of Green Economics*, *5*(4), 353–369. doi:10.1504/IJGE.2011.044619

Wang, C., Schwan, K., Talwar, V., Eisenhauer, G., Hu, L., & Wolf, M. (2011). A flexible architecture integrating monitoring and analytics for managing large-scale data centers. In Proceedings of the ACM international conference on Autonomic computing. Karlsruhe, Germany: IEEE. doi:10.1145/1998582.1998605

Wang, M., Qiu, S., Dong, H., & Wang, Y. (2017). Design an IoT-based building management cloud platform for green buildings. In *Proceedings of the 2017 Chinese Automation Congress (CAC)*. Jinan, China: IEEE. 10.1109/CAC.2017.8243793

What is green building? (2019). Retrieved from https://www.worldgbc.org/what-green-building

KEY TERMS AND DEFINITIONS

Green Broker: It is the link between the green cloud architecture and the User. It facilitates the user request and provides the most suitable green cloud provider to the user according to its requirements.

Green House Data: It is a data center service provides and its headquarters is based in Wyoming, United States. It offers various services such as IaaS, colocation, and VMware cloud hosting.

Key Performance Indicator (KPI): It is a quantity which can be measured, and it is used to analyze an organization based on its key objectives and it is also used to evaluate a company's performance.

Microcontroller Unit (MCU): It is a small integrated circuit which is designed to perform a certain specific task installed in a system. It consists of a processor, memory and input/output.

Pay-as-You-Go Service: It is a type of service in which the payment takes place only when you use the service and till the time it is being used. So, the money is not paid on an advanced basis and hence saves a lot of capital.

Quality of Service (QoS): It provides a suitable environment for the data traffic over the network and controls the network resources. It expedites the data transfer and exchange of packets and reduces the packet loss.

Sustainable Development: It is economic development through which the system is sustained for the future and at the same time it does not depletes the natural resources or nature.

Virtual Machine Manager: It is a software which controls resources of virtual machines and provides services to the cloud. It is also called VMM or hypervisor.

Chapter 9
IoT-Based Green Building:
Towards an Energy-Efficient Future

Rachna Jain
Bharati Vidyapeeth's College of Engineering, India

Vaibhav Goel
Bharati Vidyapeeth's College of Engineering, India

Jusleen Kaur Rekhi
Bharati Vidyapeeth's College of Engineering, India

Jafar A. Alzubi
https://orcid.org/0000-0001-6724-1421
Al-Balqa Applied University, Jordan

ABSTRACT

Among the various domains of IoT, one domain that is highly emerging in recent years is the application of IoT in green buildings. With the advent of IoT, the concept of green buildings has taken an even broader perspective. Incorporating intelligence into the current building management systems could revolutionize the buildings in terms of energy efficiency. The chapter explores some sound benefits of integrating IoT into a green building. It offers insight into the various technologies used in green construction, followed by some IoT-based architectures. Some machine learning algorithms that can be used to boost the efficiency of IoT devices are also discussed. Finally, the chapter dives into the future of IoT-enabled green buildings, and explores the challenges in achieving zero-energy buildings, while addressing the questions it raises. It focuses on how a green building, together with the internet of things, may lead to zero-energy buildings, thus carving our path towards a secure and energy-efficient future.

DOI: 10.4018/978-1-5225-9754-4.ch009

INTRODUCTION

Minimizing the massive energy consumption in the industrial and residential sector is a big challenge worldwide. Increasing the energy demand patterns of cities are one of the major challenges that the globe has encountered in recent years. This ever-increasing demand for energy, when coupled with the limited energy resources, constitutes the energy problems which have to be resolved as soon as possible, in order to prevent the expected extinction of energy resources in the near future.

When we dig deeper into finding the causes of such enormous energy demand of cities, it is found that buildings, particularly, are the biggest electricity consumers, accounting for more than 60% of the total global energy consumption. In other words, buildings are solely responsible for more than 60% energy consumption, which is a lot to consider (Tushar et al., 2018).

The life cycle of a building, in general, comprises 3 phases: construction, operation, and deconstruction. Each phase is responsible for hurting the environment in one or the other way. For instance, the factories that produce the materials used in the construction of a building produce damaging CO_2 emissions; according to the US Green Building Council (USGBC), buildings are the biggest energy consumers, accounting for around 41% of the world's energy use; the destruction of buildings (in the interest of renovation) also contribute to a large amount of hazardous waste.

Recent years have seen a tremendous growth in the number of buildings being built, and consequently their disastrous effects on the environment. Thus, having a sustainable and efficient building is more of a necessity today than anything else.

What Is a Green Building?

The United States Environmental Protection Agency (U.S. EPA) defines green building as *"the practice of creating structures and using processes that are environmentally responsible and resource-efficient throughout a building's life-cycle from siting to design, construction, operation, maintenance, renovation and deconstruction. This practice expands and complements the classical building design concerns of economy, utility, durability, and comfort. Green building is also known as a sustainable or high-performance building"*.

The most widely accepted and pronounced benefit green homes have on the environment is, without a doubt, energy efficiency. Other benefits include, but are not limited to, massive reduction in carbon emissions, improved indoor air quality, reduction in waste sent to landfills, and protection of natural resources, as the green construction involves the use of renewable and biodegradable resources. Moreover, green building doesn't just have an outstanding result on the environment. It is also advantageous to the occupants of the buildings.

However, certain barriers or obstacles are present for green buildings, such as higher inceptive costs, lack of incentives and political assist, absence of public awareness, and the perception that green is for the high end sector only, as per the Dodge Data and Analytics 2016 (Jadhav, 2016).

But, the construction of a green building should be looked upon as an investment. An investment that will save money (in the future), and at the same time preserve the environment and prevent depletion of natural resources. That is a win-win situation for everyone.

Figure 1. Internet of Things (IoT)

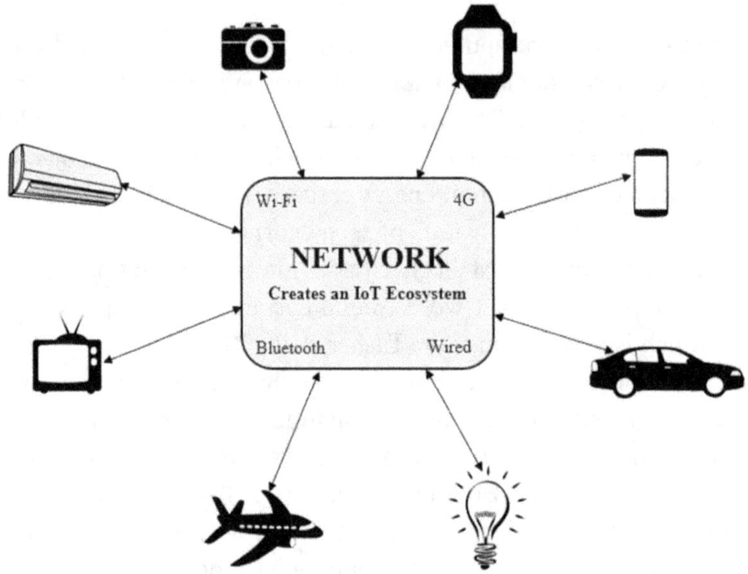

What Is IoT?

IoT is short for Internet of Things. It is an emerging concept that aims to connect millions of devices with each other (Arshad, Zahoor, Shah, Wahid, & Yu, 2017). It may be defined as a giant network of interconnected devices. The devices communicate amongst themselves using the Internet, with or without human interference. The Internet of Things ecosystem consists of a number of intelligent devices, smart objects, smartphones, tablets, etc. It generally makes use of Quick Response (QR) codes, Radio-frequency Identification (RFID), sensors and wireless technology to achieve intercommunication between devices (S. Singh & N. Singh, 2015).

In simpler words, Internet of Things is a computing concept that enables everyday appliances to connect to the Internet and communicate with other appliances, so as to perform a series of intelligent actions or operations. Fig.1. shows a very basic IoT network where a number of devices like car, smartphone, camera, wearable, etc. are connected to the network, through wired or wireless media, in order to accomplish intercommunication between the devices.

The next section describes the need of implementing the concept of Internet of Things to the traditional green buildings.

Need of IoT in Green Buildings

Now, we know what is meant by a green building, and what IoT is. Let us try to combine both.

Imagine if we could enhance the energy efficiency capabilities of a green building, and at the same time providing the users with various customization facilities, thus improving their comfort and productivity. This can be achieved through implementing the Internet of Things within a green building. Among the various domains of the Internet of Things, one domain that is highly emerging in the recent years is the application of IoT in Green Buildings. With the advent of IoT, the concept of green building has taken an even broader perspective. Incorporating intelligence into the current Building Management Systems could revolutionize the buildings in terms of energy efficiency and a bunch of other aspects, and IoT has emerged as a promising tool to transform this integration from a mere concept to reality.

The IoT for buildings can be viewed of as a giant network where all the devices (things) present in the building, ranging from the lighting systems, to the daily utility appliances like alarm clock, coffee maker, etc., to the Heating, Ventilation and Air Conditioning (HVAC) systems, are connected to a central hub, that enables them to share and exchange data, in order to monitor and manage the energy consumption of the building in a cost-effective manner, and at the same time maximizing the residents' comfort and productivity.

An advanced IoT system is implemented to design a smart energy management system in the buildings, by extending existing approaches and integrating cross-domain data, including the building's data, energy produced and the energy consumed, energy costs and the users' behavior, to produce daily efficient energy plans for the users. The wireless flexibility is an attractive reason behind the popularity of IoT in Green Buildings, which constitutes the main factor for energy efficiency and environmental benefits of the buildings. IoT can provide its owners with timely information to make intelligent decisions regarding how to operate and maintain the technology used. Moreover, computerized networks are employed in buildings to monitor life safety, security, lightning systems and elevators. These networks help in optimizing the performance and thereby leading to a cost-effective building operation and an improved level of comfort for the occupants. All in all, this integration of IoT in green building will help in reducing the energy consumption of the building further to a great extent. Installation of IoT may look pricey upfront, but it is an investment that will save huge energy costs over time, as well as contributes to the occupants' comfort and productivity.

STATE OF THE ART

History of Green Buildings

The appearance of Green Buildings emerged as an abode that integrates the automation and intelligence, so as to enable the availability of some necessary and effective services for the inhabitants, thereby ensuring a sense of security and comfort inside the house, and at the same time encouraging efficient and wise usage of energy. IoT came into the picture as it provides adaptable and trustworthy wireless technologies to constitute an important element in promoting well-organized use of energy, thus facilitating environmental management of the Green Buildings.

The concept of Green Building in the United Stated came through practices that involved the use of renewable materials and solar design. Backdate in the golden age, the Anasazi in the Southwest built an entire village in a way that all the houses received Sun energy even in the winters. The need for more energy efficient and environment friendly building practices ignited the beginning of the Green Building Movement back in the 1960s and 1970s.

Eventually, various rating systems were developed in different countries as the green building concept was now famous worldwide. Green building rating systems like LEED (Leadership in Energy and Environmental Design, The United States and Canada), BREEAM (Building Research Establishment Environmental Assessment Method, United Kingdom), etc. aid the end-users in determining the building's degree of environmental performance. It allocates credits to parameters like site's placement, maintenance, preservation of resources such as building materials, energy, and water, and occupants' health and comfort. The number of credits given to a particular parameter usually decides the level of attainment of that parameter by the building.

The Indian Green Building Council (IGBC) had been established in 2001. It became a fragment of the Confederation of Indian Industry (CII). The IGBC aims at enabling an environment that is sustainable for all and facilitating India in becoming one of the global leaders in the sustainable built environment by 2025.

Recent Developments

After years of use, the buildings deteriorate thermally and its indoor and outdoor starts wearing out. As a result, the buildings fall short of expectations. But IoT provides new opportunities by enforcing intelligence into the building management system (Manic, Wijayasekara, Amarasinghe, & Rodriguez-Andina, 2016). Though slow, the concept of green buildings has travelled a long journey from employing basic methods like solar panels, green insulation, cool roofs, etc., to major technological advancements using IoT, thus transforming a "just green" building to a "green smart" building. Applications of IoT in green buildings, till now, involves the following categories:-

- A **Building Management System (BMS):** Also known as a Building Automation System (BAS), is a computer-based control system, which is usually fixed in buildings to control and monitor the building's mechanical, automatic and electrical equipment, equipment that provides services of ventilation, lighting, power systems management, fire systems, and also security systems. A BMS is made up both software and hardware; the software program, usually configured in a graded manner, using protocols such as C-Bus, Profibus, etc. BMS is usually based on IoT & Machine Learning techniques. It helps to quote the building occupancy information through simple and affordable IoT sensors and studies the impact of human activities on the energy usage of a building, which is used as a basis for implementing efficient building management practices (Minoli, Sohraby, & Occhiogrosso, 2017). The BMS basically works by collecting a large amount of data from various subsystems within the buildings which can then be analyzed using different types of signal processing tools, thereby helping in monitoring and identifying different causes and extent of energy and resources consumption (Pan et al., 2015).
- **Human Detection System:** The human detection system for an efficient BMS (building management system) uses some machine learning techniques such as parametric and non-parametric algorithms. These algorithms include background subtraction models and Gaussian processes (Li et al., 2017) that can be used along with a camera for counting the number of heads, (i.e., the number of persons currently present in the building). OpenCV library can be used for implementing these algorithms. The use of machine learning techniques in IoT-based systems is discussed in detail in later sections.

Traditional vs. IoT Enabled Green Buildings

Apart from minimizing energy consumption, there is a rich profusion of other benefits that a green building possesses over a traditional building. Some of them are discussed below in brief.

Environmental Sensitivity

The life cycle of a building, in general, comprises 3 phases: construction, operation, and deconstruction. Each phase is responsible for hurting the environment in one or the other way. For instance, the factories that produce the materials used in the construction of a building produce damaging CO2 emissions; according to the US Green Building Council (USGBC), buildings are the biggest energy consumers, accounting for around 41% of the world's energy use; the destruction of buildings (in the interest of renovation) also contribute to a large amount of hazardous waste. There is tremendous growth in the number of buildings being built and consequently their catastrophic effects on the environment. Buildings usually contribute to atmospheric change and also affects health through the materials used and even the decisions about sites, efficient use of electricity, water and landscape surroundings (Younger, Morrow-Almeida, Vindigni, & Dannenberg, 2008). By the efficient use of non-renewable resources, green buildings prove to be very sensitive towards the environment. The use of solar panels, rainwater harvesting techniques makes these buildings, energy and water efficient. These buildings lead to the reduction in consumption of fossil fuels and less emission of atmospheric pollutants such as CO2, SO2, and NO2. Utilization of non-toxic, biodegradable materials in the construction helps in saving the environment.

Cost

The initial costs incurred in the construction of a green building are similar or sometimes more than a conventional building, owing to the requirement of special equipment in green building, such as solar panels, smart appliances, materials for insulation, cool roofs, etc. But, the green buildings definitely shine in the long run, as the energy costs will reduce by a huge margin, and the cost of maintenance of a green building is minimal when compared to a traditional one.

Water Efficiency

Water is the most essential element for human life. Although many might think that water is a limitless resource, the quantity of potable water is exhausting rapidly, leading to the scarcity of water in various regions of the world (Sheth, K.N., 2017). Thus, efficient use of water is an unavoidable necessity today. Traditional buildings consume a lot of water in their lifetime. But, that is not the case with green buildings. They minimize the usage of water and at the same time prevent it from getting wasted, by using technologies like rainwater harvesting, reusing gray water for flushing the toilets, etc.

Comfortable Indoor Environment

The indoor environment is far more comfortable and secure in green buildings as compared to the conventional ones. The indoor air pollutants like dust, various kinds of irritants, etc. sometimes make the indoor air more polluted than outside (this is a major case in metropolitan cities). The green buildings provide the inhabitants with various customization facilities, thus improving their comfort and productivity. This can be attained by implementing the Internet of Things (IoT) within a green building. Moreover, computerized networks are employed in buildings to monitor life safety, security, lighting systems, and elevators. These networks help in optimizing the performance and thereby leading to cost-effective building operation and an improved level of comfort for the occupants.

Materials Required in the Construction

Green buildings are built with non-toxic materials that are not harmful to the environment in any way. These materials include earthy materials, wood, bamboo, SIPs, insulated concrete forms, cordwood, straw, earth bags, slate / stone roofing steel, thatch, etc. When compared to the materials used in the construction of conventional buildings, these materials emit far lesser air pollutants and are also renewable.

BENEFITS OF INTEGRATING IOT

We have studied in the previous section how a green building possesses a bunch of benefits over a traditional building. Now, what if we could enhance the green building even further by integrating IoT within a green building? That's what this section is all about. It discusses all the benefits that the process of integrating IoT may yield (Kats, 2003).

Energy-Savings

IoT-based green buildings gained popularity due to their ability to reduce the power consumption. It is basically done by various IoT systems, embedded within the building, such as counting of heads, automatic switching on/off of lights, and many more. For example, an IoT sensor checks if there is a person in a room or not. If it is empty, then the system automatically switches off the lights of the room, thus saving energy. These buildings also became the part of ENERGY STAR program, a program initiated by the U.S. Environmental Protection Agency.

Automation

Using IoT, we can transform a green building into a smart building that is intelligent enough to automate a number of tasks, thus providing convenience to the occupants. For example, the coffee maker starts brewing coffee as soon as the alarm rings, or the air conditioner is switched on half an hour before the occupants arrive. This is achieved by communication between the various IoT enabled devices. It saves time as well as provide a lot of comfort and convenience to the occupants.

Personalization

Another great benefit of integrating IoT in buildings is personalization. Personalization means organizing things and actions according to one's own taste. It may include, for example, setting the desirable light color, the temperature of the air conditioner, the fan speed, etc.. This helps in meeting one's needs efficiently, thereby maximizing comfort, satisfaction, and productivity.

Business Opportunities

The increasing demand of IoT-enabled green buildings has shifted the focus of both the government agencies and the builders towards the green building. The strong craving for minimizing energy consumption, along with maximized comfort and personalization levels, assigns this industry a strong business opportunity (Minoli, Sohraby, &Occhiogrosso, 2017). The green buildings are naturally built, thus they provide a huge return on investment rates. These buildings are usually sold at high prices. The major attraction for energy-conscious house buyers is the Annual Cost Comparison Chart which gives the yearly reports of the cost as compared to the traditional buildings. The indoor quality and comforts are also major reasons for such a large number of buyers.

TECHNOLOGIES USED FOR GREEN CONSTRUCTION

By Green Construction, we basically mean the use of resource-efficient and environmentally friendly processes in construction, so as to ensure the lifetime sustainability of the building, with the least harm to the environment and the occupants' health. Sustainability refers to minimizing the usage of natural resources and energy, thereby lowering generation of waste as well as pollution, as the byproducts of construction (Sandanasamy, Govindarajane, &Sundararajan, 2011).

Some of the widely used technologies for green construction of buildings are listed below.

Biodegradable Materials

Majority of the traditional materials used in construction lead to accumulation of hazardous waste products and toxic chemicals that usually take hundreds of years to degrade. Moreover, even after they degrade, it still harms and contaminates the environment. Thus, the use of biodegradable materials in construction is an unavoidable necessity that needs to be put into action, thereby ensuring the building's sustainability.

Biodegradable materials degrade naturally, without contaminating the soil. They break down easily without releasing any toxic chemicals, and therefore, aid in a reduction in the negative impacts of construction materials on the environment (Wang, Chen, Lin, & Zhang, 2011).

An excellent example of a biodegradable material is biodegradable paint, also known as organic paint, which is created using milk protein, lime and mineral pigments. Another great example is hemp, which is used heavily (outside United States) in construction for everything from the foundation of the building to the insulation systems.

Figure 2. Biodegradable paint

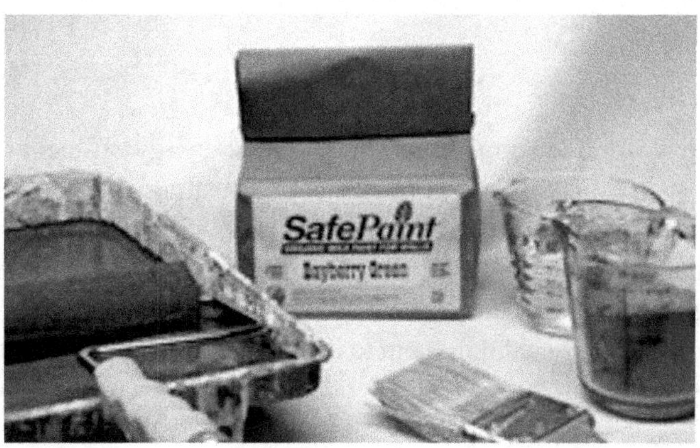

Figure 3. Solar panels

Solar Power

Conversion of light from the sun rays into electricity is a new trend that is gaining popularity in the recent decades. Solar power is a sustainable technology used in green construction.

There are basically two types of solar power:

1. **Active Solar Power:** Active solar power refers to the use of functional solar panels that absorbs the sun's radiation, for heating and electricity purposes, thereby minimizing the gas or electricity consumption.
2. **Passive Solar Power:** Passive solar power is based on using the radiations emitted by the sun to heat a home. To achieve this, few actions like placing the windows strategically and using surfaces that can absorb the heat, need to be carried out. The windows aid the solar energy to enter the house, and the heat-absorbing surfaces retain the heat entered.

Solar energy can also be used to heat up water. An array of about 24 solar water heaters can deliver up to 2000 litres of hot water every day. Moreover, a series of photovoltaic panels can generate up to 10.7 kW of energy, which can be fed into a 900Ah/240V battery bank (Mohanty, Skandhaprasaad, &Samal, 2010).

The installation of solar panels might look costly up front, but from the long-term perspective, it saves a lot on energy bills, as well as aids in reducing the greenhouse gas emissions (GHGs) from non-renewable energy resources such as fossil fuels which would be otherwise used as traditional sources of energy.

Smart Appliances

The appliances in today's world are smarter and much more energy efficient than ever. For example, LG's smart appliances such as its SmartGrid refrigerators, washing machines and dishwashers, all are connected to a smart meter for maximized energy efficiency. Smart meters, refer to the meters that provide useful power data by gathering real-time data as a result of their communication with the IoT devices, and analyzing that data. The data generated by those meters are used by the LG's appliances to calculate the energy costs, thus enabling them to decide automatically when to run and when not to, depending on the energy rates.

Though the smart appliances are one small part out of many of our environmentally friendly elements of construction, their energy efficiency capabilities make them a fantastic and wonderful green technology that is oriented towards establishing zero-energy homes.

Cool Roofs

Cool Roofs are a green design technology that minimizes the dependence of homes on air conditioning. They are designed to offer increased solar reflectance and reduced heat absorption and thermal emittance, thus keeping the homes at standard room temperatures. In simpler words, the cool roofs reflect much more solar radiation compared to a traditional shingle roof. They also trap the air inside, thus preventing the warm or cold air from escaping through the roof. Thus, a cool roof aids in improving the interior temperature of a building.

Cool roofs are fabricated from a variety of materials, such as certain reflective paints, cool roof shingles and tiles. A great environmental reason to trust the cool roofs is that their reflectance helps to minimize the heat-island phenomenon. It is a phenomenon that occurs in urban and suburban regions, causing exceptionally higher temperatures than the surrounding rural areas.

Green Insulation

Green insulation is a green technology that includes insulating the walls of a home using certain stuff that does not need to look nice or attractive, as it is essentially a wall-filler at the end. Green insulation eliminates the need of using high-end materials that are usually made from non-renewable resources.

Because it is filled in the walls, it will always remain out of sight. Then why not make use of any old junk as the insulating material? That is the primary idea behind the green insulation technique. It utilizes recycled material to insulate the walls. There are different types of green insulation; n materials, some of which are discussed below:

- **Cotton Insulation:** Uses cotton as the insulating material, which is both cheap and safe to the environment.
- **Soft Blue Insulation:** It is made out of recycled denim. Ever could you envision that your old torn jeans could also be used to insulate your house? Cool, isn't it?
- Cellulose insulation. It takes recycled newspaper as the insulator. Blow-in-cellulose is one of the most common forms of recycled papers. We can simply spray it onto walls instead of laying it out in sheets.
- **Fiberglass Insulation:** Recycled glass is used as the "mystery" material here. But, compared to the cellulose insulation, fiberglass has some limitations. One of them is that melting the glass to form the fiberglass insulation is much more power consuming when compared to generating cellulose insulation using paper. Moreover, the fiberglass only consists of 30-40% recycled material, in contrast to 75-85% in the case of cellulose insulation. Thus, cellulose and cotton are definitely preferred over other types when it comes to green insulation.

Figure 4. A wall with green insulation

IOT-BASED ARCHITECTURES FOR GREEN BUILDINGS

A large and diverse variety of IoT based architectures for green buildings have appeared or been proposed in the past. But, there still isn't any framework yet, that has been accepted industry-wide (Tiburski, Amaral, Matos, & Hessel, 2015). This section provides an overview or a critical literature review of some of the IoT models proposed in the recent years along with their respective architectures that could be integrated into green buildings in the interest of energy efficiency.

- Marinakis&Doukas (2018) proposed a system that extends the existing approach by integrating cross-domain data, like weather data, behavior of end-users, decentralized sensor-based data, energy price data, and energy production data. Based on this collected data, it generates timely action plans for the occupants, with personalized details or particulars. Short-term predictions are produced of the users' behavior and their respective patterns of energy usage (Marinakis & Doukas, 2018).

The proposed system's internal architecture, comprises the following components:

- 5 modules for data capturing. They collect data like energy prices, building sensors' data, energy production, etc. from various sources.
- The semantic framework. It's a communication system. The main job of this framework is to integrate the data collected from a number of sources and domains. It uses Semantic Web technologies in order to perform this task.
- The action engine. It is an integrated system that predicts the energy behavior of the building, analyzes it, and suggests appropriate actions to be taken to boost the energy efficiency.

The authors implemented a system that was web-based, using the aforementioned internal architecture. The system possesses one significant feature known as "immediate, virtual distribution" of the energy consumption in building on the Internet. Hence, the occupants are persistently updated about the energy utilization as well as other parameters such as GHG emissions, monthly energy costs, etc. just by using the website.

The first level includes the collection of data, followed by analyzing the collected data, and present the data amongst 4 groups of indicators. This facilitates the energy management. The values of these indicators are based upon virtual sensors called synthetic sensors, the data of whom is obtained from existing sensors.

At the second level, the system gives suggestions for minimizing the energy consumption, based on the data collected and analyzed at the first level. It also depends on factors like occupancy, indoor thermal comfort, cooling and heating technical systems, etc.

Thus, this system enables the end-users to know about the total energy consumption, as well as the contribution of each specific occupant. It further provides recommended actions that are personalized for every end-user, to reduce energy consumption, accompanied by an approximation of their possible impacts upon the usage of energy and occupant comfort.

- Arrowhead Framework (Delsing, J.) is an initiative taken by the European Union (EU) that was endeavored to achieve collaborative automation of interconnected appliances. The plan is to use the Transmission Control Protocol/Internet Protocol (TCP/IP) everywhere. This framework uses an approach where IoTs are seen as abstracted services. Therefore, it is aimed at worldwide interoperability over a number of Service Oriented Architecture (SOA) based IoT environments, and the applications revolving around smart energy, smart buildings, mobility, and smart production.
- Chasta et al. proposed a system that further controls the active systems in a building viz. lighting, air conditioning, and safety systems such as gas and fire alarms (Chasta, Singh, Gehlot, Mishra, & Choudhury, 2016). Various IoT sensors such as LDR, PIT, LM35, etc. are attached to the Arduino board. LDR stands for Light Dependent Resistor. It is a sensor that is connected for measuring degree of intensity of light in the area under consideration. PIR stands for Passive Infrared. It is another sensor that is employed for determining the occupancy of the room, i.e. the number of persons currently present in the room. Fire and gas sensors are connected for emergency alert. For detecting the temperature of the room, LM35 sensor is applied. Now, the idea is to design a system that will control the lighting systems (on/off or dimming the intensity) and also air conditioning system autonomously based on the outputs of LDR and PIR sensors. Also, the system will alert with a hooter in case of fire or gas leakage.

The transmitter or sensing unit is shown in Fig.5. The main control module is Arduino UNO board. Different sensors, along with the RF modem and the power supply unit, are connected to the inputs of the controller to feed data to the controller unit. The RF modem wirelessly transmits data to the receiver section, which is shown in Fig.6. Arduino board, control circuits, & the RF modem are the primary elements in the receiving unit. This receiver module accepts the data sent through transmitter module. The controller processes this data, and accordingly, various control circuits are activated. The lighting systems are controlled through the dimmer circuit, on the basis of the data from the LDR sensor in transmitter module. The alarm hooter is controlled by the relay device corresponding to data inputs from the fire and gas sensors. Air conditioner (AC) is controlled by the relay circuit to switch on/off the device.

Figure 5. Block diagram for sensing (transmitter) unit

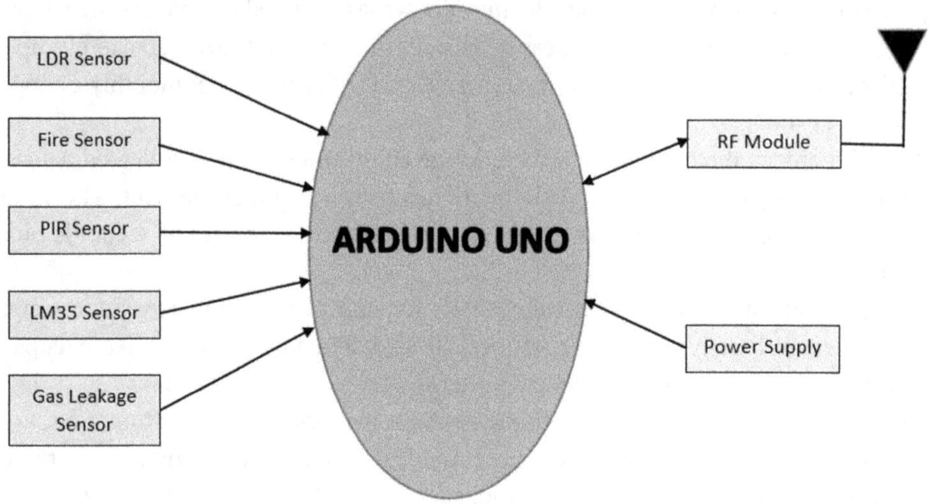

Figure 6. Block diagram for receiving unit

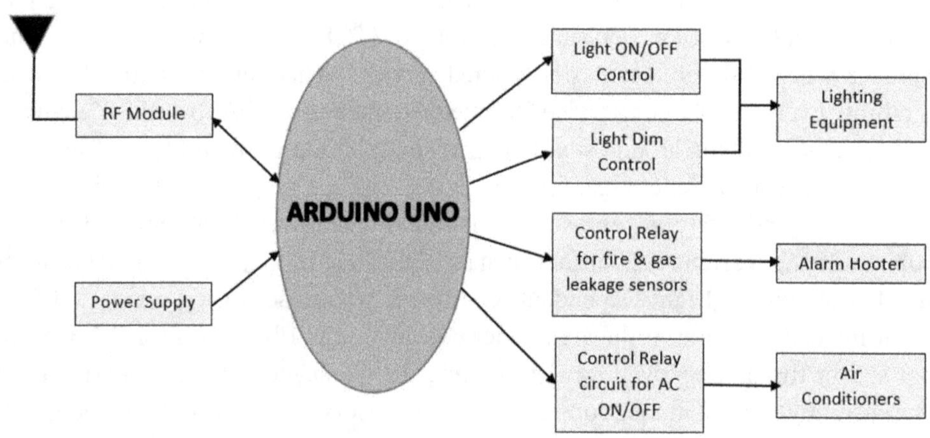

Thus, this system is capable of controlling the electrical equipment in the building, along with providing some safety features too. This demonstrates that the wireless networks are the most capable and inexpensive sensor networks today that are easy to deploy as well. This minimizes the energy consumption and wastage due to human laziness.

- The European Telecommunications Standards Institute (ETSI) developed a high level architecture for Machine-to-Machine (M2M) (ETSI, 2011). While it's a widely accepted framework, it's essentially concentrated on M2M. It describes a resource-based architecture that could facilitate the interchange of events and data or interaction amongst devices. It involves communication over

Figure 7. ETSI high level architecture for M2M

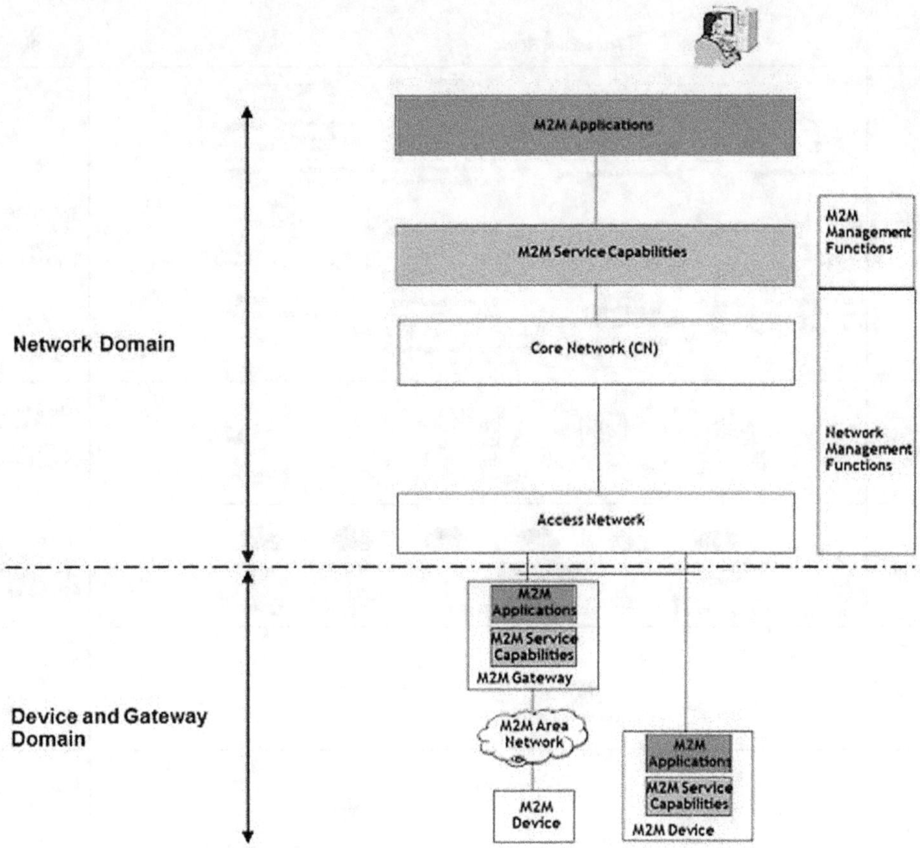

networks, without needing any kind of human assistance. This is the main goal of this architecture. ETSI felt that this architecture should have the ability of supporting various security services, such as authorization and access control, sensitive data handling, security association management, security administration and identity protection (Minoli et al., 2017).

- Minoli et al. proposed an IoT-based architecture consisting of seven layers. They called it the Open Systems IoT Reference Model (OSiRM) (Minoli, Sohraby, &Kouns, 2017). This model calls special attention to the concept of security and its significance in IoT devices. To construct the needed capabilities on a well-defined baseline, the security architectures of IoT devices essentially rely on the accessibility of the elementary ICT (Information and Communication Technology) architecture. Few of the proposed IoT-based architectures do consider the concept of security, but most of them view security as a homogenous vertical stack. Security mechanisms that support availability, integrity, and confidentiality, are required at all the architecture's layers, for them to be really effective. Such capabilities are incorporated in the OSiRM.

OSiRM supports three security mechanisms standard to and existing solitarily at each layer:

Figure 8. The Open systems iot reference model (OSiRM)
Source: Minoli et al., 2017

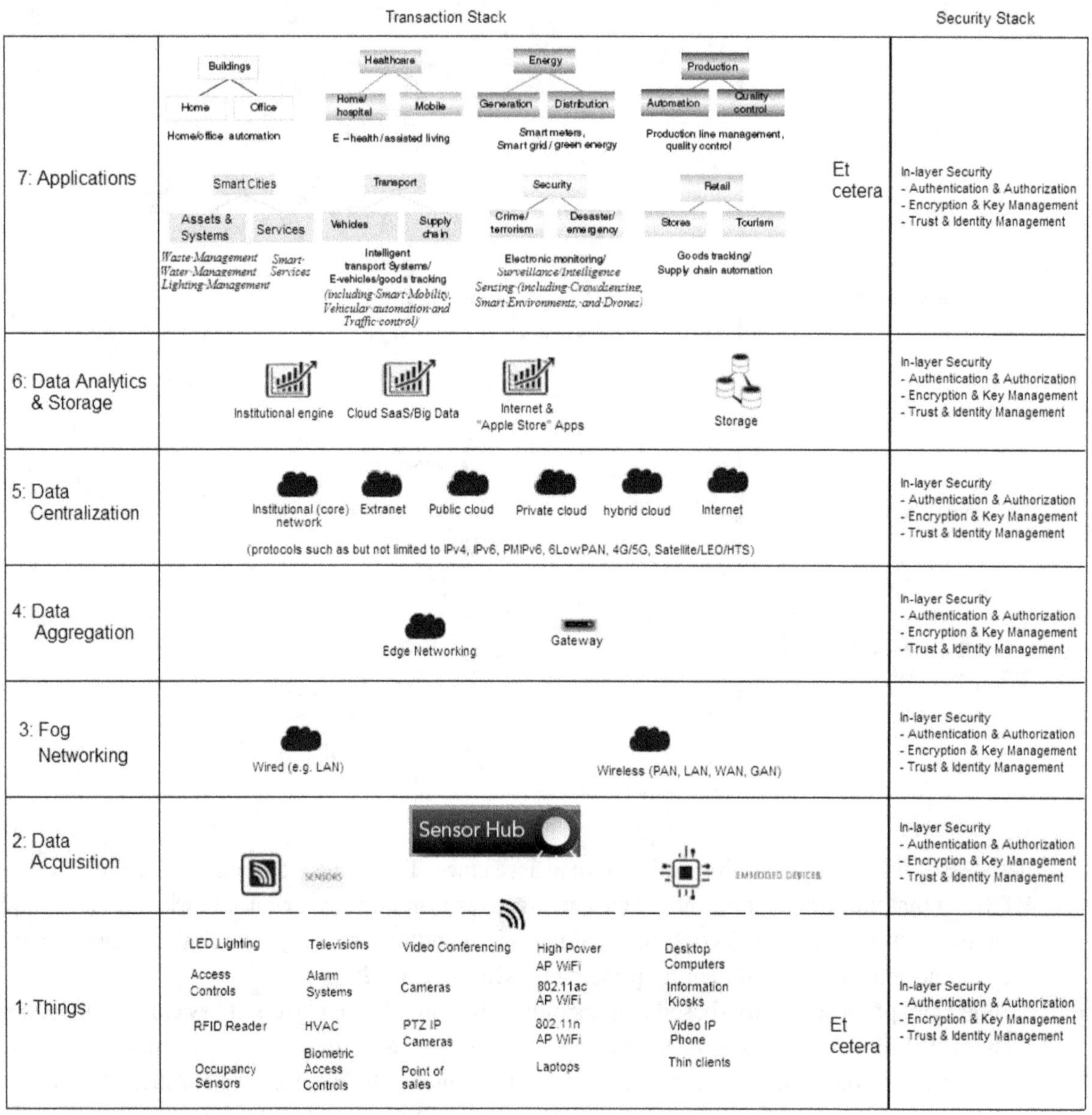

1. Authorization and Authentication.
2. Encryption and Key Management.
3. Trust and Identity Management.

The OSiRM can even accommodate other realms and mechanisms if required. Moreover, there will be optimized differences in this OSiRM model for the given security function at various layers, and specializations that may be needed or may occur based on the type of application.

INTRODUCTION OF MACHINE LEARNING IN IOT SYSTEMS

Like the Internet of Things, another technology that is facing massive growth and gaining widespread recognition around the globe in the current decade is Artificial Intelligence and Machine Learning. In simple words, IoT is a network of interconnected devices that communicate with each other and generate useful data, whereas, machine learning is a technology that is used to analyze data and infer certain conclusions, based on which some actions can be taken.

Looking at the success of both the technologies, some questions may arise in the minds of readers, like:

- What if we could merge these two technologies together?
- Is this merging possible?
- If yes, how can it be attained?
- What would be the consequences of this merging?

Yes, all these questions are valid, and we shall try to answer all of these along with some other questions as well in this section. Let us start by having a proper definition of machine learning.

Wikipedia defines Machine learning (ML) as *"the scientific study of algorithms and statistical models that use computer systems to progressively improve their performance on a specific task"*. In simpler words, Machine Learning is a branch of Artificial Intelligence (AI) that injects in computer systems the capability to automatically learn from their experience and take autonomous decisions, without being explicitly programmed, i.e., embedding intelligence into a machine. The most important and the most basic thing for a machine learning algorithm is the data on which it will be trained. And that is what an IoT device does- generates huge amount of useful data. That's where the concept of integration of Machine Learning algorithms in IoT devices came into the picture.

Machine learning has a crucial contribution in the success of IoT devices. It is used for handling the massive quantity of data that those devices generate. It can be applied to both the IoT-based systems and IoT devices. In IoT-based systems, it operates on the collection of data generated by different connected IoT devices, and accordingly generate action plans for the devices. In case of IoT devices, it provides them with a brain to learn and take intelligent decisions. This is called as "embedded intelligence" by some scholars. To apply Machine Learning to IoT systems (or devices), we need to first select the algorithm which would be best suited to the particular data under study, since there is a lot of variation in the data as it is being generated through different sources with a variety of data types. Thus, searching for the best ML model that would fit the data is amongst the biggest hurdles in the way of applying machine learning to IoT devices for fruitful analysis of IoT data. This issue has created various opportunities in extending newer developments.

The articles on IoT data analysis suggests that there exist a total of 8 major sets of algorithms applicable to the IoT data. These algorithms are classified on the basis of their structural resemblance and the type and quantity of data that can be handled and processed by them in a practical time, respectively (Mahdavinejad et al., 2018). Those eight groups, along with their brief description, are listed below:

Bayesian Statistics

Bayesian methods of machine learning adapt the concept of probability distribution in order to learn uncertain concepts efficiently, without over-fitting. The basic idea behind this method is to use the contemporary knowledge to update the prior beliefs into the posterior beliefs.

K-Nearest Neighbors (KNN)

The KNN is a supervised machine learning algorithm that classifies a new unseen data point (also called a query point) based on the values of the k data points nearest to it in the feature space. Fundamentally, it classifies k types of clusters such that the separation inside is least possible. Thus, it's a classification algorithm.

Neural Networks

Neural networks (or more specifically, Artificial Neural Networks) are computing systems modelled on the biological neural networks which makes up the human brain. The neural network itself is not an algorithm, but rather a framework for different machine learning algorithms to work together and process complex data inputs. Being compact models, neural networks are fast in processing new data. But, they generally require a large amount of mathematical computations to be processed (Mahdavinejad et al., 2018).

Support Vector Machines

SVM is a supervised ML classification algorithm. It classifies data values by making use of labeled data samples. It aims to discover a plane which will separate the two groups of the data by as wide as possible margins. The new data point is classified depending on which side of the plane it falls.

Decision Trees

It is another classification algorithm that predicts the data labels by iterating the input through a tree known as the decision tree.

Principle Component Analysis

PCA is a technique of compressing data by extracting essential information from the data and representing it with a collection of new variables known as principal components. This is done by orthogonally projecting the data points on an L dimensional subspace. This subspace is known as the principal subspace.

K-Means Algorithm

K-means is a clustering algorithm that revolves around the objective of clustering massive chunks of unlabeled data into a set of K clusters on the basis of similarities in data points. The separation between the data points is used to calculate similarity. This algorithm is employed extensively to solve node clustering problem. The reason is simple implementation and linear complexity.

Reinforcement Learning

It refers to a type of Machine Learning where a reinforcement agent decides the sequence of steps to be taken to perform a given task by learning from experience, in order to maximize reward.

These were some of the Machine Learning techniques which, if applied appropriately, have the ability to make the IoT devices as smart and intelligent as humans, and thus having excellent productivity and efficiency.

FUTURE OF IOT ENABLED GREEN BUILDINGS

The Internet of Things has always amazed the world when it comes to exploring new technologies that can inject, in any object, a sense of smartness and intelligence. Thus, it would be no surprise if IoT could revolutionize the green buildings in the years to come.

Some of the improvements or add-ons that can be implemented in the current Green Buildings to make them even more efficient may include the following:

Foretelling the Maintenance

Wouldn't it be great if we could predict in advance, the time at which the maintenance activities must be performed to ensure the flawless working of all the technologies that were embedded inside the building to make it green? It would take the building to the next level in terms of efficiency. The predictive and preventive maintenance can be very useful. It will be of great importance if we could use IoT sensors and improve the hardware devices which could help in predicting the time for maintenance. This will ensure that there will be no unexpected breakdown of the system within the building. It might not be easy to implement at this time, but looking at the rapid pace of developments in IoT, it'll surely be possible in the near future, making the green buildings more promising than ever.

Zero Energy Buildings

Green Buildings, together with the Internet of Things, may lead to zero-energy buildings, thus carving our path towards a secure and energy efficient future. A zero-energy building may be defined as a commercial or residential building with largely decreased energy needs (Jadhav, 2016). It can be seen as an interaction between buildings and energy needs and sources (Sartori, Napolitano, & Voss, 2012). It can be attained by using inexhaustible energy sources, and balanced and efficient use of energy. The development of zero energy buildings can be viewed as a goal which, if accomplished, will result to be very fruitful. As the cost of solar photovoltaic (PV) cells has dropped exponentially, the adoption of these cells for on-site energy generation has increased manifold. The other technologies for on-site generation of renewable energy may include wind turbines, solar hot water collectors, etc. (Torcellini, Pless, Deru,

& Crawley, 2006). To meet the ever-increasing demand of energy and to make buildings energy efficient and reliable, these above mentioned techniques can be used in the future (Devabhaktuni et al., 2013).

IoT for Cleaner Air

IoT devices with sensors that can check the level of CO2 present in the air can be installed in each corner of the building in order to ensure that the air is clean and breathable, thereby protecting the health of the occupants as well as the environment.

Efficient Construction

The buildings made for commercial use are prone to alterations, and hence it can lead to a huge amount of destruction waste, which can be harmful to the environment as well as the workers. So, the idea of prefab homes can be involved to save energy, overall construction costs, and maintenance costs. Outdated construction materials and equipment used for construction can be replaced with more efficient ones (Kibert, 2016).

Cloud Computing

The Cloud-centric vision which is the backbone of IoT comprises a flexible yet very easily accessible architecture that can enable different users to interconnect through the IoT's framework (Bandyopadhyay & Sen, 2011). Each of the users can interact with IoT framework according to their own requirements or needs. This can threaten the IoT-enabled green buildings (Nayyar, 2011). Thus, the framework comprises provisions to meet different requirements for privacy, security, data ownership, information sharing, etc. (Nayyar, 2011). Some IoT specific challenges are privacy, data analytics, participatory sensing, and Cloud computing apart from the WSN challenges that includes energy efficiency, architecture, protocols, security, quality of service, etc.

Secure and Programmable Networks and the Main Concern of Privacy

As we are considering that all the IoT sensor based devices will be connected to each other, it means that they will be connected over a network. The information from various devices is made available across all platforms to get a common operating picture (Gubbi, Buyya, Marusic, &Palaniswami, 2013). Wherever network comes, along with it comes the major concern of security. Privacy can be invaded and there can be serious threats to the security of the occupants of the IoT enabled green building. Cryptography, being the first soldier against data corruption, can be of great use. Unauthorized users tracking the activities of the buildings can be vulnerable. A smart green building, usually includes excellent facilities such as automatic switching off of lights, opening of the garage, etc., which, if unauthorized users get access to can become a shortcoming of green building. But these problems can be solved by some more research in the field of cryptography. There exist a variety of ways in which the system could be attacked, such as injecting fallacious information into the network; impairing the network availability; accessing confidential data; and many more.

CHALLENGES TO IOT INTEGRATION

We have seen in the previous sections how the Internet of Things can enhance the energy efficiency capabilities of a green building by incorporating a sense of intelligence into the equipment used in the building. We have also studied the other benefits of integrating IoT into a green building, like easy maintenance procedures, lower perceived costs over lifetime, personalization capabilities, and business opportunities, etc. We have seen how IoT can revolutionize the future of buildings around the globe. But, all this comes at a cost. The industry has realized that the wider introduction of IoT techniques into this green building sphere has to face some challenges that include industry-wide standardization, and most importantly cyber security challenges. The IoTSF (Internet of Things Security Foundation) claims that fewer than 10 percent of all the IoT products in the markets all over the world are fabricated with sufficient security. This matter is well communicated (Fuqaha et al., 2015; Granjal et al., 2015; Lai et al., 2015). For example, one major challenge to the IoT devices is that the huge amount of data that these devices generate may get stolen or used inappropriately, which might affect the confidentiality of the building or the occupants (Solanki & Nayyar, 2019). Hence, these IoT security challenges need to be resolved. Some of the challenges that have to be addressed by the green building industry as well as by the stakeholders (Minoli et al., 2017) include:

- **Intrinsic IoT Security (IoTSec) Issues:** In the IoT environment, there exists a large attack surface that houses the software, the physical, and the network attack surface within itself. Specifically, all the software code in the IoT environment has exploitable vulnerabilities. Every communication link endpoint is the part of the network attack surface. If the device is not protected at the physical level, it can be accessed via physical attacks, and consequently the whole IoT ecosystem will be attacked.
- **Device Mobility**: The device mobility, like roaming on open networks, needs a Mobility Management Mechanism (MMM). This mechanism gobbles up precious computing resources. The device may even be placed on a foreign network having unsure security status.
- **Low Complexity Devices:** This situation limits the extent of on-board computing power that is needed for certain operations, like firewalling, encryption, deep packet analysis, etc.
- **Use of Vulnerable Gateways/Concentration Points:** Gateways are used by a number of IoT configurations as a means of connecting IoT devices to a bigger network. Such gateways represent intensive points of attack, known as points of concentration (or concentration points). Like gateways, there are other concentration points too, such as data repositories that represent target points of attack.
- **Accessibility:** The devices may be in an open environment, like the corridor of a building, or the hall area, which makes them susceptible to physical tampering, or theft.
- **System Size:** There are different sizes of systems, some may be containing less number of IoT devices, while some containing very large numbers. Thus, scalable solutions are required, as the existing approaches to security issues may not scale up well.
- **Always Connected/Always on Mode of Operation:** IoT devices are almost always active systems, meaning that they are always connected, or in always on mode of operation. Hence, they are very much prone to cybersecurity attacks. To prevent this, we may need periodic re-authentication of the devices.

- **Lack of Agreed-Upon End-To-End Standards:** As most of the IoT systems are employed in the short-term, they are mostly vendor-specific. Thus, a wide-ranging, comprehensive end-to-end standard hasn't been developed or being implemented, which results in limiting the usage of security solutions.
- **Lack of Agreed-Upon End-To-End Architecture:** As most IoT systems are employed in the short-term, they usually doesn't stick to an agreed end-to-end architecture, which would otherwise enable the function standardization/simplicity and the capability of integrating security systems from a variety of vendors. This leads to a fragmented functional environment.

The above mentioned challenges are common in the whole IoT environment. But, few of them are specific to the green buildings than others, including IoTSec issues, absence of agreed-upon end-to-end architectures, and the fragmented situation of this industry. These challenges need to be resolved quickly in order to eradicate the barriers in the path for achieving zero-energy buildings.

CONCLUSION

This chapter has provided a broad-scale elucidation of the application of IoT in Green Buildings. A Green Building is a commercial or non-commercial building which is sensitive towards environment and uses the resources efficiently. IoT when embedded with these buildings make them even more efficient, reliable, and advanced. The IoT-enabled Green Buildings are beneficial in various aspects such as more efficient use of energy resources, affluent and ever booming new business opportunities as well as environmental sensitivity. Thus, the IoT-based Green Buildings are anticipated to evolve swiftly in the near future. Governments across the globe have shifted their attention from traditional to green smart buildings, for the buildings are massive energy guzzlers.

The technologies used in the construction of these buildings includes several different approaches that have been briefly covered in the chapter. As we know that IoT is an ideal emerging technology to influence this domain of smart and environmentally-safe building, some of the proposed IoT architectures and techniques have been discussed in brief. The influence of technologies like IoT, IP (both IPv4 and IPv6), PoE (Power over Ethernet) will enhance the energy-efficiency, functionality, capabilities, and cost-effectiveness of buildings (Minoli et al., 2017).

Moreover, the authors have explained the use of ML (Machine Learning) algorithms to enhance the capabilities by embedding intelligence in IoT devices which can be implemented in Green Buildings for better action-taking decisions and thus can provide a drastic improvement in efficiency. It further explains the various possible challenges or threats that are currently hindering the performance of IoT-enabled Green Buildings.

The trends of IoT in efficiently maintaining commercial buildings are on boom and we would see that increasing even further in the upcoming years. Finally, the authors have provided important trends that are difficult to implement now, but can be implemented in the future, seeing the pace with which the technology is growing and emerging. These will be extremely useful for building and facility managers to achieve their performance goals, hoping to finally achieve zero-energy buildings in the years to come.

REFERENCES

Al-Fuqaha, A., Mohammadi, M., & Guizani, M. (2015). Internet of Things: A Survey on Enabling Technologies, Protocols, and Applications. IEEE Communication Surveys & Tutorials, 17(4).

Arshad, R., Zahoor, S., & Shah, M. A. (2017). *Green IoT: An Investigation on Energy Saving Practices for 2020 and Beyond*. IEEE.

Bandyopadhyay, D., & Sen, J. (2011). Internet of things: Applications and challenges in technology and standardization. *Wireless Personal Communications*, *58*(1), 49–69. doi:10.100711277-011-0288-5

Chasta, R., Singh, R., Gehlot, A., & et al, . (2016). A SMART Building Automation System. *International Journal of Smart Home*, *10*(9), 217–224. doi:10.14257/ijsh.2016.10.9.20

Delsing, J. (2015). Automation Systems from IoT-Arrowhead Framework: concepts and basic architecture. *Embedded Conference Scandinavia*.

Devabhaktuni, V., Alam, M., & Depuru, S. S. S. R. (2013). Solar energy: Trends and enabling technologies. *Renewable & Sustainable Energy Reviews*, *19*, 555–564. doi:10.1016/j.rser.2012.11.024

ETSI. (2011, October). *102 690: Machine-to-Machine communications (M2M). Functional architecture*. ETSI.

Granjal, J., Monteiro, E., & Silva, S. J. (2015). Security for the Internet of Things: A Survey of Existing Protocols and Open Research Issues. IEEE Communication Surveys & Tutorials, 17(3).

Gubbi, J., Buyya, R., Marusic, S., & Palaniswami, M. (2013). Internet of Things (IoT): A vision, architectural elements, and future directions. *Future Generation Computer Systems*, *29*(7), 1645–1660. doi:10.1016/j.future.2013.01.010

Jadhav, N.Y. (2016). *Green and Smart Building Trends*. Green Energy and Technology. Springer Science+Business Media Singapore.

Kats, G. (2003). *Green building costs and financial benefits*. Boston, MA: Massachusetts Technology Collaborative.

Kibert, C. J. (2016). *Sustainable construction: green building design and delivery*. Hoboken, NJ: John Wiley & Sons.

Lai, C., Lu, R., & Zheng, D. (2015). Toward Secure Large-Scale Machine-to-Machine Communications in 3GPP Networks: Challenges and Solutions. *IEEE Communications Magazine Communications Standards Supplement*.

Li, W., Lu, Y., & Sun, J. (2017, December). People counting based on improved gauss process regression. In *Security, Pattern Analysis, and Cybernetics (SPAC), 2017 International Conference on* (pp. 603-608). Shenzhen, China. IEEE.

Mahdavinejad, M. S., Rezvan, M., & Barekatain, M. (2018). Machine learning for internet of things data analysis: A survey. *Digital Communications and Networks*, *4*(3), 161–175. doi:10.1016/j.dcan.2017.10.002

Manic, M., Wijayasekara, D., Amarasinghe, K., & Rodriguez-Andina, J. J. (2016). Building energy management systems: The age of intelligent and adaptive buildings. *IEEE Industrial Electronics Magazine*, *10*(1), 25–39. doi:10.1109/MIE.2015.2513749

Marinakis, V., & Doukas, H. (2018). *An Advanced IoT-based System for Intelligent Energy Management in Buildings*. Athens, Greece: School of Electrical & Computer Engineering, National Technical University of Athens. doi:10.339018020610

Minoli, D., Sohraby, K., & Kouns, J. (2017). IoT Security (IoTSec) Considerations, Requirements, and Architectures. *14th Annual IEEE Consumer Communications & Networking Conference (CCNC 2017)*.

Minoli, D., Sohraby, K., & Occhiogrosso, B. (2017). IoT considerations, requirements, and architectures for smart buildings—Energy optimization and next-generation building management systems. *IEEE Internet of Things Journal*, *4*(1), 269–283.

Mohanty, S., Skandhaprasaad, A. L., & Samal, S. S. (2010). *Green technology in construction. In Recent Advances in Space Technology Services and Climate Change 2010 (RSTS & CC-2010)*. IEEE.

Nayyar, A. (2011). Interoperability of cloud computing with web services. *International Journal of ElectroComputational World & Knowledge Interface*, *1*(1).

Nayyar, A. (2011). Private Virtual Infrastructure (PVI) Model for Cloud Computing. *International Journal of Software Engineering Research and Practices*, *1*(1), 10–14.

Pan, J., Jain, R., & Paul, S. (2015). An internet of things framework for smart energy in buildings: Designs, prototype, and experiments. *IEEE Internet of Things Journal*, *2*(6), 527–537. doi:10.1109/JIOT.2015.2413397

Sandanasamy, D., Govindarajane, S., & Sundararajan, T. (2011). Energy-efficient materials: Route to sustainability in green buildings. In *International Conference on Green Technology and Environmental Conservation (GTEC-2011)*. IEEE. 10.1109/GTEC.2011.6167702

Sartori, I., Napolitano, A., & Voss, K. (2012). Net zero energy buildings: A consistent definition framework. *Energy and Building*, *48*, 220–232. doi:10.1016/j.enbuild.2012.01.032

Sheth, K.N. (2017). Water efficient technologies for green buildings. *International Journal of Engineering Innovation and Scientific Research, 1*(3), 5-10.

Singh, S., & Singh, N. (2015, October). Internet of Things (IoT): Security Challenges, Business Opportunities & Reference Architecture for E-commerce. In *International Conference on Green Computing and Internet of Things (ICGCIoT)* (pp. 1577-1581). IEEE.

Solanki, A., & Nayyar, A. (2019). Green Internet of Things (G-IoT): ICT Technologies, Principles, Applications, Projects, and Challenges. In Handbook of Research on Big Data and the IoT (pp. 379-405). Hershey, PA: IGI Global.

Tiburski, R. T., Amaral, L. A., de Matos, E., & Hessel, F. (2015, December). The Importance of a Standard Security Architecture for SOA -Based IoT Middleware. *IEEE Communications Magazine*, *53*(12), 20–26. doi:10.1109/MCOM.2015.7355580

Torcellini, P., Pless, S., Deru, M., & Crawley, D. (2006). *Zero energy buildings: A critical look at the definition*. National Renewable Energy Laboratory and Department of Energy.

Tushar, W., Wijerathne, N., & Li, W. T. (2018, May). IoT for Green Building Management. *IEEE Signal Processing Magazine*.

Wang, X., Chen, Y., Lin, Q., & Zhang, W. (2011). The thinking of promotion and using green building materials. In *Proceedings of 2011 International Conference on Electronic & Mechanical Engineering and Information Technology*. IEEE. 10.1109/EMEIT.2011.6023110

Younger, M., Morrow-Almeida, H. R., Vindigni, S. M., & Dannenberg, A. L. (2008). The built environment, climate change, and health: Opportunities for co-benefits. *American Journal of Preventive Medicine*, *35*(5), 517–526. doi:10.1016/j.amepre.2008.08.017 PMID:18929978

Chapter 10
Renewable Energy Distribution and Management in Green Buildings

Suman Kumar
Delhi Technological University, India

Gurjit Kaur
Delhi Technological University, India

ABSTRACT

There are various environmental problems (i.e., "global warming," air and water pollution), which need to be prevented. Construction of buildings plays a significant role in pollution. To reduce the harmful effects in constructing buildings, it is necessary to move on to sustainable architecture. In this chapter, different advantages and standards for green buildings will be discussed. Different organizations are contributing towards a green environment. There are even different sensors that are able to detect wastage of energy and can predict the requirement of energy. Machine learning, a hot topic these days, can also play its role in demand prediction. In this chapter, role of network communication and sensing to optimize the energy of green buildings and machine learning-based demand prediction to optimize the energy of green buildings are discussed. Further predicting energy harvesting from weather forecasts, return on investment of green buildings, and potential benefits of energy-efficient green buildings are also discussed.

INTRODUCTION

Communities growth has a huge impact on the environment (Saunders, 2008).. Building any residential house or office actually affect the environment. The big question is that can buildings be made able to let pollutants spewed out which are harmful to nature. If people move to an architecture which is sustainable in nature, i.e.green buildings, then one can save the environment and also reduce the cost. Sustainable architecture seeks to reduce the harmful components and can reduce the environmental problems (Baird, Leaman & Thompson, 2012).. Sustainable built architecture is also termed as green buildings which are

DOI: 10.4018/978-1-5225-9754-4.ch010

environmentally friendly and resource-efficient, whether it is to select the location of the demolition of these buildings after ending of their life cycle (Piomo, Atienza & Rosing, 2009).

The buildings which are used for living and office spaces should use natural resources for operation. The energy is consumed mainly in buildings while construction and later in electrical applications like lighting, fan or air-conditioning systems. While, these amenities like lighting, fan, air conditioning, microwave oven, water heating provide comfort to these building occupants, while consumes huge amount of energy and pollution gets added (Ganji, Budzisz,, Debel, Meo, Ricca & Wolisz, 2015). Further, occupant performs various activities which generate solid and water waste. Building green is too much efficient. It creates buildings which optimize using local ecology, local materials and most important thing is that these buildings are built so that it can reduce requirements of water and power.

Thus, keeping these things in mind, it has been realized that the traditional architecture used was very green. TERI has estimated that, if Indian urban area buildings, were made compulsory to adopt green building concepts, India could save more than 8,400 megawatts of power, which will be enough for lighting of 550,000 homes every year. To run the HVAC system efficiently, energy-efficient windows can be used, which locks the air coming in and block air which is outside from entering. A double entrance door can also be created to control climate system. Outdoor air temperatures affect the indoor. Here one door is used which leads to the entrance of a small room. This room is climate-controlled having other doors which leads into other parts of the building. Untreated air can be prevented from rushing in and rushing out using doors in the building every time the door gets opened. Unoccupied areas which get lightened can cause wastage of huge energy. The solution of these common problems while construction is installation of motion-activated lighting. These sensors work on a timer to provide its efficient nature. This is very useful for buildings having areas staffed 24 hours and areas which are only occupied while part of the day. Traditional switches get replaced through these automatic lights, so here is no risk of lights on by someone accidentally in unused area, and thus it ensures the lights are only running during needful.

There are a number of researchers which are working on these types of technologies. The example is the Recursive Deterministic Perceptron (RDP) neural network model. To detect and diagnose the faulty building energy consumption "Weather-Conditioned Moving Average (WCMA)" model can be used. This is a solar energy prediction algorithm based on different Exponentially Weighted Moving-Average (EWMA) estimation method. The amount of energy can be accurately estimated using WCMA algorithm. Strategically windows can be placed that skylight eliminates need for electrical lighting while the day. Figure 1 given below shows the green building with different energy utilizes and other resource utilization. Insulation of good quality reduces regulation costs of temperature in both season summer and winter (Fowler, Rauch, Henderson & Kora, 2010). . One can reduce emissions and consumption of water and thus contributes to make our ecological footprint smaller.

The sun offers huge solar energy potential which is carbon neutral and charge free. Technology for renewable sources of energy and commercially viable can cover completely the needs of building energy consumption. Figure 1 shows that the energy star appliances which can be used to reduce energy uses. Thus solar energy can play an important role in supplying power to green buildings in northern latitudes. One can also use photovoltaic (PV) and wind turbine systems. Many small on-site energy harvesting deployments at individual buildings can be used to generate electricity using Distributed Generation (DG). DG has potential to make more efficient generation by reducing transmission and distribution losses, carbon emissions, and demand peaks. Many countries like India are facing many issues. There are several initiatives are taken by the governments to frame some standards. Some of them are discussed as below:

Figure 1. Green building and energy emission

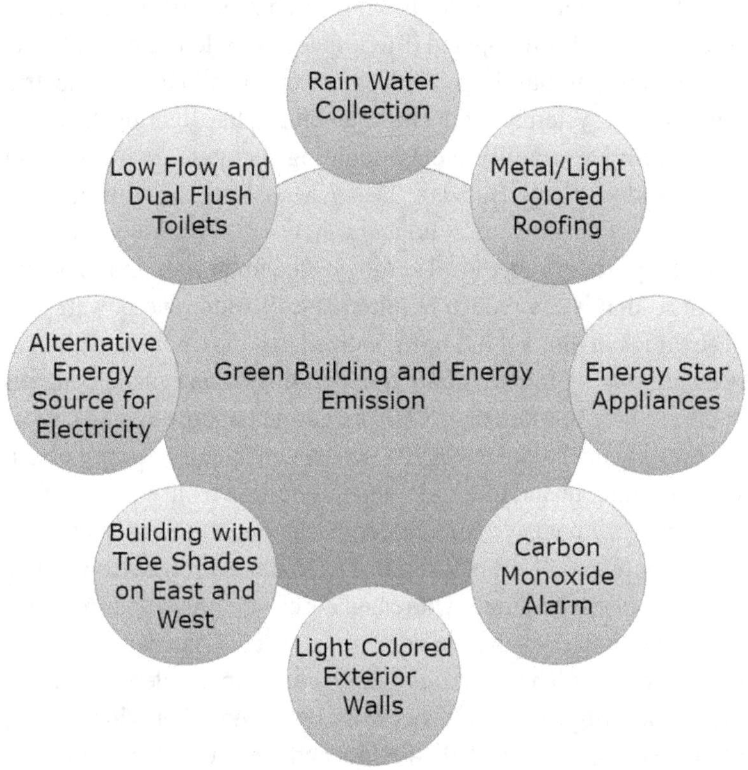

1. **LEED-India:** The Leadership in Energy & Environmental Design (LEED) is an international recognized certification system for the green buildings. The LEED-India Green Building Rating System is an international benchmark for the design, construction and operation of high performance green buildings (Kubba, 2009).
2. **IGBC Rating:** The Indian Green Building Council (IGBC) is a division of the Confederation of Indian Industry that works intimately with the administration and goes for economically manufactured condition. It offers four dimensions of rating for new structures that is legitimate for a long time: Certified, Silver, Gold, and Platinum. Aside from new structure confirmation, the IGBC Green Existing Building O&M Rating System offered for applying maintainable ideas for existing structures.
3. **Honey Bee ECBC:** The Energy Conservation Building Code (ECBC) was set up by the Indian Bureau of Energy Efficiency (BEE) to set vitality effectiveness benchmarks for plan and the development of structures.
4. **TERI GRIHAA:** The Green Rating for integrated habitat assessment is a national rating framework for green structures that is received while planning and assessing new structures.

Renewable energy is CO_2-free thus it can be used as an electrical energy source. In figure 2 it can be seen that in the long run to protect our nature and maintain environmental balance one need to switch to green buildings. Figure 2 represents the comparison between standard building and green building. It shows the comparison of different construction types and harmful impacts. These renewable energy

Figure 2. Comparison between standard building and green buildings

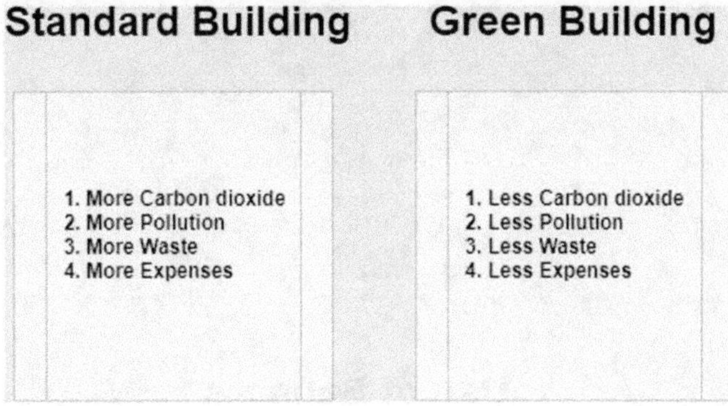

resources can be stored or different more technologies for stable electricity generations, likewise combined heat and power generation (CHP). Co-generation or combined heat and power (CHP) it is a system which generates energy and produce it in the form of both heat and electricity from a single primary source to satisfy the energy needs using existing power grid and with the new-renewable energy smart energy distribution and management system (SEDMS) is connected.

ROLE OF NETWORK COMMUNICATION AND SENSING TO OPTIMIZE THE ENERGY OF GREEN BUILDINGS

Wireless Sensor Networks (WSNs) are used in building to control energy and Building Management (BM). WSN can sense and communicate and is a key technology for accessing the data in real time. This data and process BM and WSN need to be integrated and thus information will be used for building management. Fig. 3 shows different sensors to interpret energy requirements and uses. A central platform required for data distribution and processing to integrate management. Using this, one can monitor the building energy uses continuously and facilitate BM functions using distributed databases and integrated BMS.

WSNs can be configured and sensor can function using less energy and long duration. It is playing a major role in various fields to process the data, i.e. bio-medical, environmental. WSN consists of two types of devices and those are sensor nodes and gateways. Usually five components are used to make sensor nodes, i.e. a microprocessor, a sensor board, peripherals, a radio board and power source. These five are contributing to functioning and processing of the sensor nodes from end to end. It's core component is a microprocessor, and coordinates all the node functions, which performs data processing tasks at the node level. Figure 3 represents the various sensor which can be installed in the building. The most crucial component is the radio board which links all the nodes together and thus forms an integrated network. Batteries are considered as a power source. In the design process, power consumption and optimization is a cumbersome task and batteries cannot be replaced for low voltage regulation. In fig. 3 it is shown that a wind turbine and solar panel are on rooftops and sensors can detect temperature in buildings. Wireless sensors are used which can control and detect temperature.

Figure 3. Sensors in the buildings

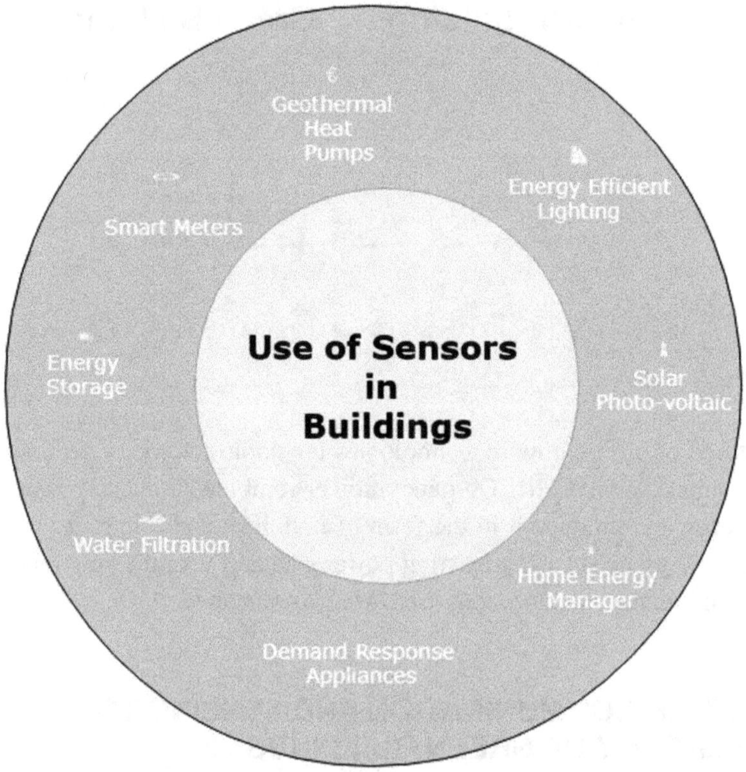

MACHINE LEARNING BASED DEMAND PREDICTION TO OPTIMIZE THE ENERGY OF GREEN BUILDINGS

User requirements are high priority in these buildings and energy systems are integrated accordingly using storage battery and charging and local renewable energy. These different energy system efficiencies will be scrutinized and compared in different aspects to control using algorithms on sustainability, cost, and energy system optimization. To diagnose and detect building faults Recursive Deterministic Perceptron (RDP) neural network model can be used. Detection ability can be increased using RDP model. Fault diagnosis can be done through this approach. RDP model is structured to detect faults in equipment. Energy consumption of components of buildings can be calculated using comprehensive methods by physical functions. Figure 4 shows different energy requirements and its environmental impact analysis. Different parameter are considered such as construction of building, operation, HVAC equipment, climate conditions and utility rate schedule to utilize the energy consumption data. To evaluate energy efficiency and renewable energy software tools are brought into use such as BLAST, Energy Plus.

To check energy standards of building and energy conservation measure many software can be used. Department of Energy (DOE) of the United States surveyed different tool's performance and maintain a list of all simulation tools which gets constantly updated. These tools work on the data obtained through basic principles using environmental parameters and input data of the building. Operating tools require huge work which makes it tough to perform and expensive. For usage of energy in building weather conditions plays an important role. Many forms are considered and taken into consideration over time

Figure 4. Ecological footprint of different sector

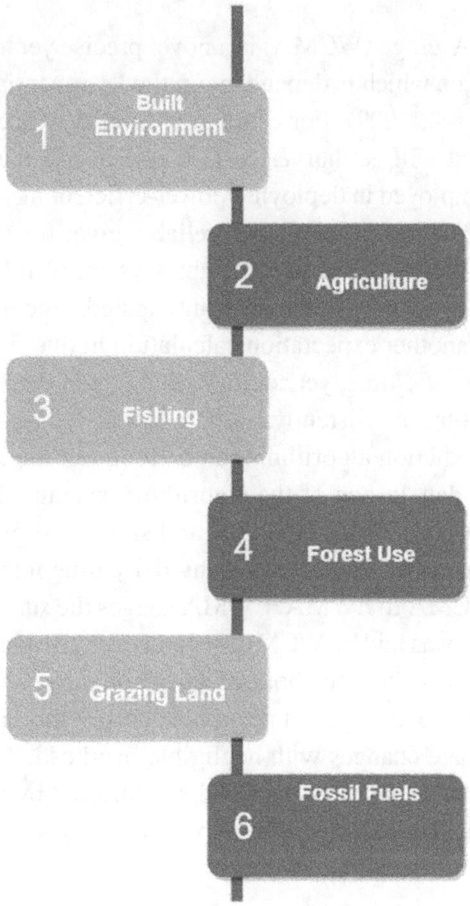

and these varying data are temperature, humidity, solar radiation and wind speed. Even to simplify energy requirement calculation in the building, according to different conditions of weather various studies has been conducted. White and Reichmuth has predicted consumption of energy buildings monthly in an attempt to use monthly average temperatures. This prediction became more effective and accurate comparison to the earlier ones of process with temperature using degree of hotness or coldness days using temperature scale.

The next one is to estimate parameters which are important in energy usage, i.e. total heat loss coefficient, gain factor and total heat capacity, which are useful in the thermal behavior analysis of building or sub-level systems. In few simple engineering models, for climate variables energy consumption can be correlated using regression and obtained an energy signature. Figure 4 represents the ecological footprint of different sectors.

A relapse strategy for taking care of estimations of both warming and cooling was proposed by Bauer and Scartezzini which can adjust the system as sun power increases. For inside and outside temperature difference a simple regression temperature function is used at each sub-level cooling load. It is easy to use and develop CDA, which shows the ability of prediction as compared to the neural network methods. It's drawback that CDA model lacks with flexibility and detail and it also requires a lot of input information. To analyze energy consumption in residential buildings CAD gets employed in the early days.

PREDICTING ENERGY HARVESTING FROM WEATHER FORECASTS

Weather Conditioned Moving Average (WCMA) is a novel precise yet low overhead, sun based vitality expectation calculation technique which is dependent on the Exponentially Weighted Moving-Average (EWMA) estimation (Piorno et al, 2009). For estimation of energy accurately WCMA algorithm can be used and that energy amount will get harvested in the form of solar energy, so that in near future wireless sensor nodes can be employed in deploying power-efficient method on solar energy harvesting.

In spite of the fact that EWMA-based calculation reliably gives the exact information for steady climate conditions, regardless of whether shady and bright days are blended, distinctive days vitality and time esteems present mistakes which are essentially anticipated. Accordingly, to keep away from this issue, researchers also present another expectation calculation in this field which not just considers the sun-oriented conditions at a specific time, yet additionally bargain the vitality consumption estimation for the changing climate conditions consistently.

A number of other solar prediction algorithms can be found in literature which are based on mean values at different time of the day. In one of the algorithms, average daily solar system performance is shown. Performance of solar system got predicted and simplified greatly using the approximation. Though it does not provide an energy guarantee during daily time intervals. One can think about the vitality forecast precision of WCMA to EWMA. EWMA gauges the sunlight-based vitality which enters the system and its establishment was laid by WCMA, regular changes additionally get described utilizing this by adjusting the progressions in both the hour of dawn and dusk, and furthermore the sun-oriented power contrast between seasons. Seasons assumes a noteworthy job in vitality forecast. Besides, this new calculation considers climate changes with negligible overhead. Then again, WCMA utilizes the information for the same hour over the early days and as indicated by the past qualities from that day it produces results with much better accuracy and help in adjusting against the real climate conditions.

RETURN ON INVESTMENT OF GREEN BUILDINGS

Before leaving the room light must be turned off. In winter all windows must be sealed. For keeping out summer heat, blinds must get close. At the macroeconomic level, energy efficient buildings will increase energy security. It will decrease the public budget, and stimulate productivity and create new jobs. As per different reports, "Energy Efficient Buildings", the return on investment can reduce the ownership costs to 50% for commercial buildings and these changes are substantial in nature and energy efficiency. These energy efficient buildings will lead to attracting more tenants and residents, thus bring down the rate of vacancies and boost the value of property. USGBC and keybridge has published a study which states that 96% of Fortune 200 companies has contributed to sustainable growth by using LEED green building. LEED green building program contributes to sustainable growth (Eicholotz et al, 2013). Solar panels and green roofs also play a great role in saving energy in both residential and commercial buildings. For proper insulation of heating and cooling, windows are used which are optimum to these climates and LED lights and systems that can be used for optimization and control for energy uses.

LEED (Leadership in Energy and Environmental Design) is the most widely used, internationally recognized third-party verification for green buildings. It was developed by the United States Green

Building Council (USGBC) (Fuerst, F, 2009). In different surveys, majority has believed that LEED has impacted positively their Return Of Investment and also contributed to save money. It is mostly used as third-party which has internationally recognized for verification of green building construction. The United States Green Building Council (USGBC) has developed this having its parent organization, the World Green Building Council. All LEED projects are registered and LEED Accredited Professionals get qualified are evaluated and examined by independent Green Building Certification Institute (GBCI).

According to green building standards costs of construction cannot be added in total or a defined percentage as it depends on certification levels and project team strategic decision. Though cost seem increasing first, while in a timescale increase in quality will define it as a reasonable investment. For high green premiums potential will be higher and benefit as well. Standard buildings which are not green or with a low certification, will neither be economic in the long term nor have ecological and societal benefits. Sustainable development life cycle will lead to a great healthy life style and its life cycle consists of savings in operation and maintenance and productive environment for residents.

In the long term, energy efficient buildings are proven a cost saving option as compared to other standard buildings. Green buildings planned well for energy saving will give benefit after a reasonable duration. These project values are higher at the beginning, but to maintain its cost efficient nature, it should maintain its quality standards. Opportunities for renting depends on certification of green buildings. Green design provides proper management of water reuse and sewer, thus cost will get reduced associated with it and provide a measurable benefit financially. Green building provides the strategy of reusing water and plumbing fixes which are water efficient and rainwater harvesting and green roofs. Water bill can get reduced as a result of water uses gets reduced. These days water prices are rising globally and its importance is increasing. There is an urgent requirement of projects working on reducing pollution and waste water be treated and get transported which incur infrastructure costs for societal benefits. Consensus through different studies has reported that in green building designed environmental health benefits are higher and work productivity got increased. It had shown various impacts related to health benefits and work productivity.

Despite the fact that this progression is transcendental for retrofits, with new development it is critical that the climate stripping for windows and entryways is appropriately introduced equitably. Guaranteeing that the climate seals are appropriately kept up throughout the years spares definitely on vitality costs. Entryway closers that are balanced and working legitimately likewise add to outstanding weatherproofing of a structure. Vitality costs has been decreased utilizing sun oriented boards. Vitality from lattice will be enhanced with sun oriented board innovation and help LEED activities and vitality costs got diminished.

As per USGBC reports, green building owners have reported a return on investment for existing buildings of 19.2% and 9.9% for new buildings. A major software company which received LEED Platinum awards stated that they have experienced a net present value return of nearly 20 to one of its initial investment. Other large firms had also reported that under 8 years they achieved the full return on their investment.

POTENTIAL BENEFITS OF ENERGY EFFICIENT GREEN BUILDINGS

There are many potential benefits of energy efficient green buildings (Ganji et al, 2015). Some are them are as follows:

Environmental

Standard building emits high pollutants and less energy efficient which if replaced by green buildings will discard negative impacts by using less energy uses, water, and other natural resources.It will have positive impact on nature by contributing through energy generation and increasing biodiversity. In Australia Green Star certification is given to green buildings has shown that it produces 62% fewer greenhouse gas emissions than average standard Australian buildings, and 51% water efficient than the other buildings which meet minimum industry requirements.

Economic

By use of efficient energy and water it can economically benefit a large group of people, thus offers financial benefits. Utility bills of water and electricity will get reduced and benefit tenants and households. Building construction will create jobs and creates property value of builders. With increase in occupancy operating costs will get reduced. Canada's green building industry generated $23.45 billion in GDP and represented nearly 300,000 full-time jobs in 2014.

Social

Economically and environmentally are not only benefits with the adoption of green buildings but also have positive impacts socially. Around health and wellbeing benefits green buildings and green offices play a major role. It has been observed that cognitive scores increased by 101 percent in green, well-ventilated offices. In the EU, energy is largely consumed in the construction industry. Thus,excelling at sustainable development in this sector will protect the environment and investments. For a green building big concern is energy efficiency. It plays a major role in green building success. Since it reduces consumptions of energy in various ways. Decrease energy of the green building using efficient design, recycled and local materials usage and recycling construction waste. For rapidly increasing population, resource utilization has come under a lot pressure. By using technology, which can process and increase efficient uses of water and energy, green building uses can reduce these pressures.

CHALLENGES IN THE CONSTRUCTION OF GREEN BUILDINGS

To manage the market for green building, it requires a lot of green building professionals. There is not much information and understanding among the people regarding this industry. Thus, lacking the professionals and awareness among people has been a stumble block for the growth of this sector. There is a misconception among people that it incurs high cost to maintain due to green premium which is above the cost of green building.

In constructing buildings and its operations there is a lot concern on water availability and its cost. For standard building water usage is high mainly in bathrooms so utilizing it properly is a big concern. A developer must find it profitable to invest in the additional cost of green features. The energy such as solar panels or other green energy addition can cost high, sobuilding resale value or renting building may not look benefitable. The construction problems which are common has activities related with the wastes of energy. Contractors and architects are working on to create a facility which can optimize uses of energy. These are difficult in dealing with cooling or heating units for doors which are frequently opened, which lets in the outside air. Since windows could also become an energy loss source, allowing the air to enter in through gaps.

Green building sector is booming while infrastructure has been stagnant. Infrastructure of green buildings must attain great efficiency (Eichholtz, Kok & Quigley, 2013). Trying new things always brings risk and uncertainty. It is always challenging to introduce people the change which involves their capital. The architectural education must be stressed on buildings which are not harmful in nature. Sustainable design must be emphasized in our education.

CONCLUSION

In this chapter, many factors related to green buildings, i.e. its need, its solutions is discussed. Many different types of sensors can be used to demand of energy production. Different wireless sensors can be used for energy usage reduction. Machine learning and different models can be used to predict energy storage and energy software tools. How investing in this sector is beneficial has also been discussed with its different parameters i.e. environmental, social, economic (Piomo, Atienza & Rosing, 2009).

REFERENCES

Baird, G., Leaman, A., & Thompson, J. (2012). A comparison of the performance of sustainable buildings with conventional buildings from the point of view of the users. *Architectural Science Review*, *55*(2), 135–144. doi:10.1080/00038628.2012.670699

Byun, J., Hong, I., Kang, B., & Park, S. (2011). A smart energy distribution and management system for renewable energy distribution and context-aware services based on user patterns and load forecasting. *IEEE Transactions on Consumer Electronics*, *57*(2), 436–444. doi:10.1109/TCE.2011.5955177

Chappell, T. W., & Corps, C. (2009). High performance green building: what's it worth. Lacey, WA: Washington State Department of Ecology.

Diamond, R., Opitz, M., & Hicks, T. et al. (2006). *Evaluating the energy performance of the first generation of LEED-certified commercial buildings (No. LBNL-59853). Lawrence Berkeley National Lab.* Berkeley, CA: LBNL.

Eichholtz, P., Kok, N., & Quigley, J. M. (2010). Doing well by doing good? Green office buildings. *The American Economic Review*, *100*(5), 2492–2509. doi:10.1257/aer.100.5.2492

Eichholtz, P., Kok, N., & Quigley, J. M. (2013). The economics of green building. *The Review of Economics and Statistics*, *95*(1), 50–63. doi:10.1162/REST_a_00291

Fowler, K. M., Rauch, E. M., Henderson, J. W., & Kora, A. R. (2010). *Re-assessing green building performance: A post occupancy evaluation of 22 GSA buildings (No. PNNL-19369). Pacific Northwest National Lab*. Richland, WA: PNNL. doi:10.2172/1029438

Fuerst, F. (2009). Building momentum: An analysis of investment trends in LEED and Energy Star-certified properties. *Journal of Retail & Leisure Property*, *8*(4), 285–297. doi:10.1057/rlp.2009.18

Fuerst, F., & McAllister, P. (2011). Green noise or green value? Measuring the effects of environmental certification on office values. *Real Estate Economics*, *39*(1), 45–69. doi:10.1111/j.1540-6229.2010.00286.x

Ganji, F., Budzisz, Ł., & Debele, F. G. et al. (2015). Greening campus WLANs: Energy-relevant usage and mobility patterns. *Computer Networks*, *78*, 164–181. doi:10.1016/j.comnet.2014.10.033

Hunt, W. D. (2008). *Literature Review of Data on the Incremental Costs to Design and Build Low-Energy Buildings (No. PNNL-17502). Pacific Northwest National Lab*. Richland, WA: PNNL. doi:10.2172/935614

Kubba, S. (2009). LEED practices, certification, and accreditation handbook. Oxford, UK: Butterworth-Heinemann.

Piorno, J. R., Bergonzini, C., Atienza, D., & Rosing, T. S. (2009, May). Prediction and management in energy harvested wireless sensor nodes. In *2009 1st International Conference on Wireless Communication, Vehicular Technology, Information Theory and Aerospace & Electronic Systems Technology* (pp. 6-10). IEEE.

Saunders, T. (2008, March). A discussion document comparing international environmental assessment methods for buildings. *BRE*.

Chapter 11
A Study of Green Building Prospects on Sustainable Management Decision Making

Subhra Mondal
https://orcid.org/0000-0003-1194-5678
Duy Tan University, Vietnam

Kalyan Kumar Sahoo
African University College of Communications (AUCC), Ghana

ABSTRACT

The world faces many environmental crises such as increased threat of climate change, the depletion of key natural resources, increasing air and water pollution, and growing levels of solid wastes. These issues are becoming the major aspects of value in real estate and a key driver in the decision-making processes. The strategic sustainability process called "the halo effect" was more worldwide which is affected by the popularity in environmental actions criteria. It showed consequence that green concept has to not only focus on technical or moral issues, but also need to base more on the economic and financial imperative.

INTRODUCTION

Nowadays, the world faces various natural crises, for instance, the extended threat of ecological change, the utilization of shared resources, growing air and water tainting and creating measurements of solid wastes (Gilbert et al., 2006). These issues are transforming into the genuine pieces of critical worth in land and a key driver in the essential authority frames. The supportability process as called "the brilliance sway" was logically by and large which is impacted by the reputation in environmental action criteria (Mansfield, 2009). Mansfield (2009) exhibited a result that green thought has focused on the particular or good issue just as the need to assemble more as for the financial and budgetary target. There has numerous

DOI: 10.4018/978-1-5225-9754-4.ch011

evidence to clear up these issues in the gathered condition and a great deal of its show skilled in current orchestrating, plan, advancement and property the administrators in the environment and the advancing utilization of nonrenewable resources. Various organizations have more participate in condition commitments by signatories to all-inclusive courses of action and traditions seeing issues, for instance, carbon outpouring, and ozone draining substance. "Property perception has been changed in the past ten years. The green structure has ended up being subject for the property business (Robinson, 2007)". Thus, they have various terms of land audit for calling "reasonable improvement, for instance, "green structure" (US term), "sensible structure" (the UK and Australia term), "supportable plan" and "monetary advancement" (Mansfield, 2009; Sayce, 2010). As a property perspective changed, the of down to earth property will stress in structure qualities and execution as impacted by the property's estimation and market regard (Lorenz, 2007). The standard three focused motivations behind supportability confirmation are money related prospering, social progress, and security. The business property publicizes that try to complete in supportability inspiration and diminishing in any realized an additional structure costs called 'green' structures (Francesco, 2008). In the latest decade, there has been a quick improvement of industrialization on the planet, especially in the making similarly as making countries like India. (Das Sharma, 2008). It has evaluated that 2030, 60 percent of the total masses will live in urban territories. From a reasonable improvement perspective, the cooperative natural association between municipal networks and the urban and rural districts are of inconceivable essentialness to the success of who and what is to come. The fast improvement realized by industrialization has led to the unconstrained headway of urban zones. The change of cultivating zone to social living arrangement and deforestation has made it difficult to keep up characteristic leveling. A quick steady in masses advancement and development in urban locales have caused across the board sullying (Das Sharma, 2008). In case urban zones are not indeed organized and managed, the nature of the air, the availability of water, waste taking care of, reusing structures and all attributes of the urban condition adding to human flourishing will be under hazard.

Green Building in India: An Emerging Business Opportunity

The Indian advancement part is creating at a rate of 9.2% as against the world typical of 5.5%. The zone is presumably going to record higher advancement in the coming years. India has advancement limits in the regions of structures, establishment improvement, and expressway adventures. The improvement of advancement industry offers power to other collecting divisions like bond, metal and steel, control, manufactured substances, etc.,

Green Building Movement in India

The Green Building advancement has expanded enormous power amid the past five years, as far back as the Green Business Center left on achieving the LEED rating for their individual inside at Hyderabad.

The 'Platinum Rating' for the Green Business Center structure has honed the accomplices of the improvement business. Today, a couple of corporate and Government affiliations are contemplating Green Buildings permanently. It has realized a spurt in the enthusiasm for green materials and apparatus.

From a modest beginning of 20,000 sq.ft of the green impression in the country in the year 2003 to an astounding 70 million sq.ft till date, green structures are all around prepared to accomplish scalar statures. Today a grouping of green structure adventures is coming up in the country - private buildings, the show centers, crisis facilities, educational associations, explore focuses, IT parks, plane terminals, government structures, and corporate working environments. The Indian materials and apparatus makers are directly looking with a test to indeed observe green features to satisfy the formative requirement for Green Buildings.

MODERN ARCHITECTURE SUPPORT TO PRESERVE ANCIENT PHILOSOPHY

India has commonly been an imperativeness powerful nation inferable from its decently starting time of money related progression, organizations, and agribusiness based economy and obliged private resources. Tolerating 80% of the structures and contraptions of 2030 yet to be amassed or bought, India has a unique opportunity to ensure imperativeness capable structures and machines from the start. With a consistently expanding number of specialists becoming acquainted with attainable practices and headways, there is a longing in the improvement of the Indian structure division in a greater essentialness viable way. The need is to see the viability issue, in the country, in a wide range to consolidate the diverse problems related to it". Regardless of the way that, there is an amazing effort in achieving a progressively important and carbon gainful economy, there is a need to foresee the potential consequence of the significant number of endeavors. "In case the issues are moved nearer beneficially and inside not all that removed future, India can be foreseen to can also cut down its imperativeness and carbon power. Sensible building in India uses dominatingly two approaches - a 'green' plan and 'alternative' structure. The two loads on the progress that is essentially development based. This refinement in viable designing can be credited to the legacy of India's opportunity period, made through dreams of the father of the nation, Mahatma Gandhi and India's first Prime Minister Jawaharlal Nehru. Gandhiji required that India's progression should happen through its Villages. He required towns to act normally needy and the lion's share rule". "Nevertheless, Pundit Nehru clutched advancement as the structure square of India's post-Independence Economic and Infrastructural scene. As such Scientifically transformed into the picture of 'soundness and headways in India.

Post-1973 Oil Crisis, western advancement subordinate plans were gotten to deal with India's everyday issues. In this philosophy, imperativeness viability was sorted out over each other concern. Also, the organization and Universities set up new research centers, for exploring non-common and sustainable power source resources and in the meantime the establishments asking about Adobe as negligible exertion decisions created. Government foundations propelled essentialness capability as a response for achieving the pragmatic plan. The remarkable features in this system were extending imperativeness capability, decreasing water usage, using a sustainable power source and reused things. The country by and by has a different course of action exercises to standard imperativeness adequacy and green structures. The state by and by has different methodology exercises to standard essentialness capability and green structures.

Energy Conservation Building Code 2007

This is the country's first structure vitality code and intends to majorly affect vitality effectiveness in structures.

Sustainable Habitat Mission

It was articulated by then PM Manmohan Singh in the year 2008; this methodology encompasses eight fundamental missions that will be looked for after as critical pieces of the arrangement of prudent headway. These fuse missions on sun situated imperativeness, improved essentialness adequacy, practical living space, protecting water, supporting the Himalayan natural framework, making a "Green India," conservative agribusiness and, finally, developing a key data arrange for ecological change.

Green Rating for Integrated Habitat Assessment (GRIHA)

This rating system was made by the Energy and Resource Institute (TERI). It is an indigenously produced rating structure wholly tuned to the climatic assortments, building deals with, existing practices of advancement and attempting to reestablish the uninvolved plan.

MANAGEMENT AND DESIGN OF GREEN BUILDINGS

Green improvement is the demonstration of structure essentialness capability in new advance through arrangement and new headways. As excitement for imperativeness capability for cost and general practicality builds up, various areas by and by requiring business structures to change following green improvement models. There is a contrasting extended enthusiasm with people who have the correct stuff to supervise and keep up these essentialness powerful systems. Essential norms, procedures, and advances which are identified with the five significant segments of green structure plan are Sustainable Site Design; Water Conservation and Quality; Energy and Environment; Indoor Environmental Quality; and Conservation of Materials and Resources. These information sponsorships of the usage of the USGBC LEED Green Building Rating System, yet based on measures and strategies rather than specific courses of action or developments, which are every now and again site express and will change from dare to broaden.

A BIRD'S EYE VIEW OF GREEN BUILDING PRACTICES

The writing survey is the chief and fundamental contraption to understanding the prerequisite for research (Creswell, 2009). The composition overview empowers the examiner to assess how much information is starting at now fathomed and explained a particular subject, and whether more research ought to be coordinated. The hidden study of these recently referenced confirmation programs revealed that a hankering to usually improve structure and advancement frames has existed among modelers and contractual workers for a significantly long time. "While Robert Watson established the forefront green improvement through the game plan of the USGBC, the longing to work incongruity with the regular living space has existed since as in front of timetable as 15BC". According to Charles Kibert, the "green structure improvement and the striking climb of the US Green Building Council and its accomplices elsewhere is an extraordinary case of beating misfortune" (Kibert, 2012, p. xv) Which in the long run began upon the course of action of the world's first genuine urban territories.

Furthermore, these undertakings changed and formed into a green structure advancement, whose primary concern was to reduce materials used, upgrade the biological impression of structures, improve a system, and decrease essentialness usage" (Willson 2008). There are different motivating parts of the grasped research consider. "In particular, the way that essentialness profitability of the structures worked amid the latest decades have added to an inefficiency inside the structure region, which at any rate adds to high ozone hurting substance releases and other natural risks and to lost imperativeness. Additional motivation originates from the way that both structure owners and building occupants finally hope to diminish their essentialness costs, particularly amidst extending imperativeness costs. Consequently, the mix of the recently referenced factors together makes inspirations for essentialness profitability in structures and fills in as a better than average convincing stage from which the coordinated examination has advanced". (Ryghaug and Sorensen, 2009).

Green structures incorporate using supportable procedures and materials in the arrangement and improvement of properties (or in existing properties rebuild). And exercises, where the fact of the matter is to cut down assignments and upkeep costs fusing essentialness capability in lighting headways and air quality, redeveloping darker field goals, or using green housetops that contemplate flood water to be reused and moreover add to the upgraded agent and understudy prosperity, comfort, and gainfulness. In any case, the benefits of green structures estimation are generally on numerous occasions more significant than the typical additional cost of construction. (Kats, 2003; Wiley, 2008; Wiley et al., 2010). USGBC is a national non-advantage enlistment affiliation that has developed the Leadership in Energy and Environmental Design (LEED) System to give standards and rating structures to green structure headways. There still have issues in measure green structure because of the liberal standard even exceptional in each country. The rule will show the beneficial green element of that building. (Robinson, 2007; May et al., 2007; Sayce 2010; Shimizu, 2010).

Moreover, the one greatest block for progressively green structure vast that is liberally more costly than conventional, plan and may not be shielded as cash sparing preferred standpoint perspective. Something different, the planner perspective is revolved around in general income. Anyway, there has been some investigation show even green structure has risen yet alternately the cost of green arrangement has dropped over the latest couple of years, which impacted from extended in planners, makers experience and made of progressions (Chang et al., 2009). "Sensible structures adventures bunch has outfitted information on cost that with for all intents and purposes no extra cost of spending extraordinary growing commonly in the relevant program (Davis Langdon, 2007; Mansfield, 2009). There still nonappearance of data and market affirmations to exhibit that the improvement cost has genuinely decreased or points of interest of its" (Robinson, 2007).

Boyd (2006) mentions that green structure in like manner display into the "Drift of Blame", used to grasp the condition in the land industry. "The occupier will effect to increase makes inland industry and cause it to move upward/slipping. Regardless, Lutzkendorf et al. (2005) fought that budgetary pros are the essential driver for the green structure industry. Something different, there appears to be the mix of components that the make influenced from theorists data in green cash related as endeavor perspective similarly as occupier enthusiasm for expanded green space" (Myers et al., 2007). An examination by Kats (2003) avows that the earlier green structure was brought into the arrangement technique, the lower the general costs of the endeavor. Subsequently, when the poor green formation occurs, the composition may not accomplish its highest limit. Even more explicitly, the structure may never recuperate the additional costs. To fight poor construction, the joint effort between specialists, legally binding laborers, architects, estimators, owners, and authoritative experts should set up. Consolidated arrangement needs to strike adequately make relevant plausible systems.

A TRIAL FOR OVERCOMING THE LACUNAE IN PREVIOUS PRACTICES

- Research in Green Building in Indian setting is outstandingly obliged.
- An extraordinary and fundamental test for our overall population is achieving viability of the world's genetic networks.
- Green building is a dynamic, rapidly creating and propelling the field, driven by a combination of rising open stresses over global natural change, cost and availability of essentialness.
- Sources impact of the developed condition on human prosperity and execution.

PERSPECTIVES FOR ENHANCING GREEN BUILDING PRACTICES

Research objectives must be clear, concise and be in the authoritative declaration. The purposes of research, consider must be in a state to consolidate that is cultivated from the examination. These goals are:

- To dismember elective essentialness makers, suppliers and constructors, see viable arrangement and advancement, related government undertakings, and "green" assessment ventures to be leverage or a block to whole deal business practices.
- To examination customer's and fashioner's tendency for green structures over darker structures.
- Two examinations the complete manual for being grasped at the national, state levels to help large gathering of reasonable portion green and imperativeness beneficial structures.
- To look at the imaginative work requirements for making fitting advances for improved green structures.

EXPERTISE INTERVENTION FOR THE GREEN BUILDING PRACTICES

The data got amid this investigation was accumulated using mixed research strategies, which is appropriate for this sort of research. This approach was completed through a survey. This instrument was created as the ideal approach to get data from large measures of specialists in associations with an orderly on elective essentialness inside a three-month range (March 01, 2017 – May 31, 2017). A couple of systems for data examination were executed after the finish of the audit. The responses to each question were evaluated only, and after that examined everything considered in the general report. Likewise, charts and tables were made to contrast with the data got for every outline question; every request has been tended to in the data examination bit of this examination.

The examination system used for this enlightening file is expressive estimations sought after by an explanation of the results. Undeniable bits of knowledge are portrayed as "connecting with an examination of data for all self-governing and ward factors in the examination". Interpretation of the results is described as "the researcher achieves surmising from the results for the examination questions, theories, and the greater noteworthiness of the results" (Creswell 2009).

TECHNICAL REPRESENTATION OF THE PRACTICES

The investigation yielded tremendous results. One thousand email addresses were recognized for study transport. Out of the 1000 messages will, in general, send the audit, 103 reviews were returned, either in full or mostly wrapped up. The response yielded a 10.03% response rate. This response rate produced a last model, gauge which makes the results reasonably generalizable to the general people, and was sufficient to research for this examination consider.

The eventual outcomes of every request are presented underneath, in the numeric solicitation, the application was requested inside the setting from the diagram. Of those 103 written responses, 55 proper investigations were submitted and like this separated. The reactions to address one seems Table 1 above. Outline individuals were given a choice to pick "Other" as a vocation, and after that fill free; Director, Principal, Architect/Principal, Principal and Director of Sustainability was filled in by audit individuals. Over segment of the respondents were creators, masters, or VPs.

The outcomes of request 1 suggest that the examination individuals' occupations inside their associations move tremendously. A broad scope of employment is recorded as included positions inside associations. Results, out of 55 hard and fast respondents, 47 fell into classes saw as unusual state positions: Architect/structure, Chief Executive Officer, Department Manager, Project Manager, and Vice President. Five individuals picked other and self-recognized as Director, Principal, and Architect. These results prescribe that out of 55 respondents to address #1, 53 hold irregular state positions. In perspective on this, it is acknowledged that the aspects grabbed all through the audit is that of official positions; a near report driven by subordinates to these positions may yield unmistakable results.

Review individuals were given the choice to pick "Other" as a discretionary activity, and after that fill free; responses to "other" included keen power inverters, water conservation, and little hydro. It should be seen that elective structure materials, photovoltaic, and co-age got the most essential rankings.

Figure 1. N=55 Question #1: "What is your job inside the organization?"

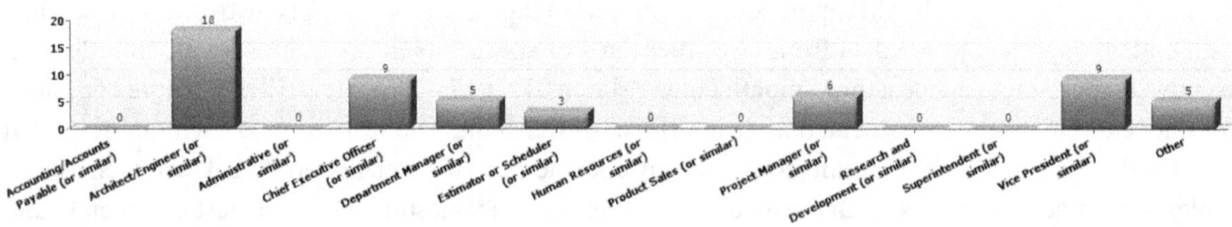

Figure 2. N=53 Question#2: "What elective vitality system(s) is/are your organization's core interest?"

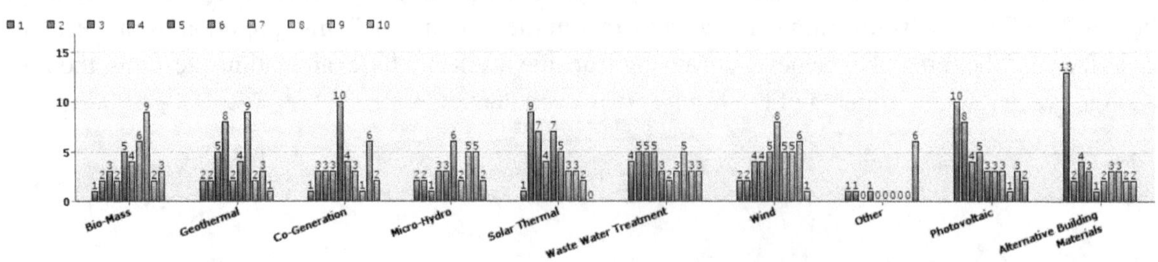

Table 1. Response listing

Serial No.	Answer	1	2	3	4	5	6	7	8	9	10	Total Responses
1	Biomass	1	2	3	2	5	4	6	9	2	3	37
2	Geothermal	2	2	5	8	2	4	9	2	3	1	38
3	Co-Generation	1	3	3	3	10	4	3	1	6	2	36
4	Micro-Hydro	2	2	1	3	3	6	2	5	5	2	31
5	SolarThermal	1	9	7	4	7	5	3	3	2	0	41
6	Waste WaterTreatment	4	5	5	5	3	2	3	5	3	3	38
7	Wind	2	2	4	4	5	8	5	5	6	1	42
8	Other	1	1	0	1	0	0	0	0	0	6	9
9	Photovoltaic	10	8	4	5	3	3	3	1	3	2	42
10	AlternativeBuildingMaterials	13	2	4	3	1	2	3	3	2	2	35
	Total	37	36	36	38	39	38	37	34	32	22	

Table 2. Question2RankOrder

RankOrderBreakout 1-mostto10-leastimportant	
Photovoltaic	1
SolarThermal	3
WasteWaterTreatment	4
Geothermal	5
Co-Generation	6
Wind	7
Micro-Hydro	8
Bio-Mass	9
Other	10

Outline individuals were drawn closer to rank 10 particular elective essentialness foci on the size of 1-10, with 1 having the most attention and ten including insignificant fixation inside their associations. The results recommend that electoral structure materials (13) and photovoltaic(10) dominate the competition with the most number 1 positions inside the outline results. Elective structure materials and photovoltaic get the most government enrichments consistently and are the most acclaimed strategies for grabbing accreditation centers inside systems, for instance, LEED and Green Globes.

The responses to address three are showed up recently referenced figure. Study individuals were given the decision to pick "Other" as a strategy for imperativeness widening, and subsequently fill free; the stunning response to "other" was "not proper." While by far most of the respondents entered "not applicable" as their approach to growing, a minority of respondents picked photovoltaic, elective structure materials, co-age and wind as electoral essentialness execution methodology inside their associations.

Figure 3. N = 33, Question# 3: "If elective vitality execution is the essential wellspring of income age inside your business, will you expand?"

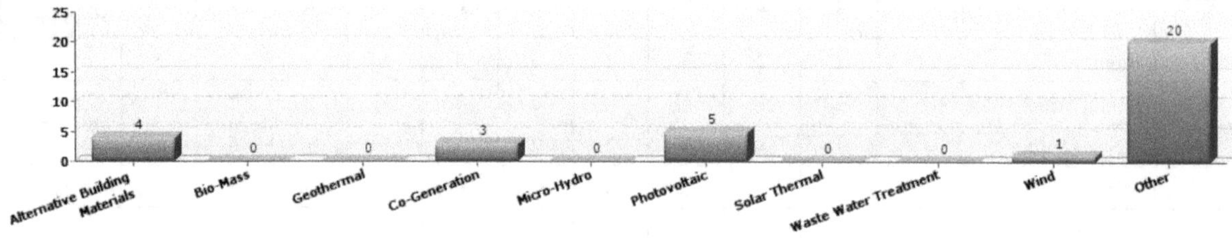

Figure 4. N=49, Question# 4: How long has your organization been centered around elective vitality items and administrations?"

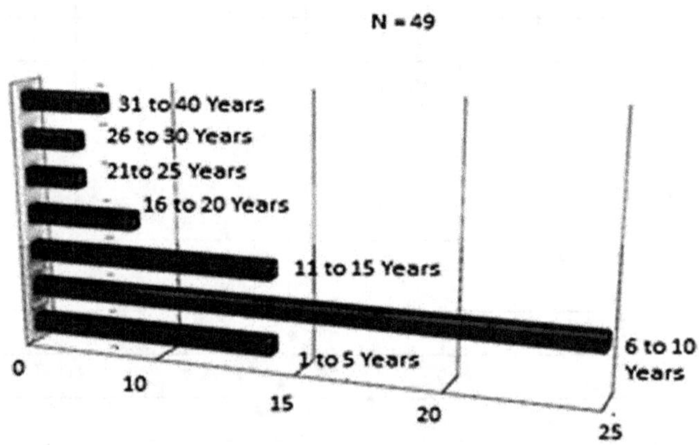

Figure 5. N= 55, Question#5: "What is your commonality with "green" evaluations programs?"

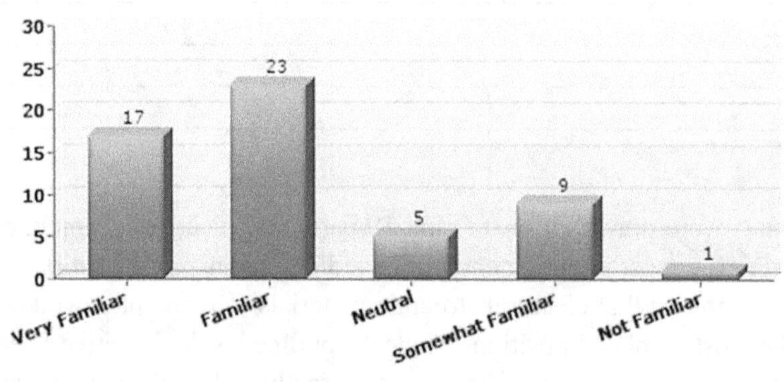

Study individuals were asked concerning whether their associations would find a perhaps diverse assortment from the thing they directly manufacture, plan, move or present. A minority of respondents replied in the positive that their association would be likely different assortment into other elective essentialness things than those correct currently used. As anybody may anticipate, electoral structure materials (4), co-age (3), photovoltaic (5) and wind (1) got the most votes as potential fields to wander into. Elective structure materials and photovoltaic are the two most consistently executed.

Figure 6. N=53. Question#6: "How likely is your organization to burn through cash on the innovative work of elective vitality items if government sponsorships were give

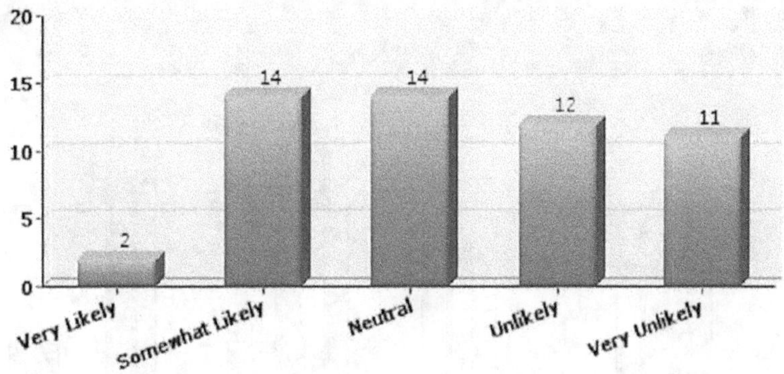

Review individuals' inclusion in the green structure industry moves comprehensively. While 20 of the 49 respondents expressed that they have been related with the green structure industry for between 6-10 years, the most consistent reactions to this request fell amidst 1-5 years (9), 6-10 years (20), and 11-15 years (9). Like this, a more extensive section (38) of the respondents have been related to green structure and green structure assessments systems someplace in the scope of 1 and 15 years.

The responses to address 5 are a lot of the respondents see themselves as Very Familiar and Familiar with green evaluations programs. A smaller piece of the respondents considers themselves Neutral, Somewhat Familiar, or Not Familiar. Survey individuals demonstrated that a lion's offer (73%) think about green structure rating projects, and view themselves as either very outstanding (31%) or current (42%). Out of the 55 individuals to respond to this request, five consider themselves fair (9%), nine somewhat conspicuous (16%), and one not ordinary (2%), for a total of 15 respondents (27%).

The responses to address 6 are showed up in Figure-. The stunning response to this request is no, respondents' associations are not inclined to consume money on R&D of elective essentialness things if government enrichments were given. Then again, two respondents exhibit it Very Likely that their association would consume money on imaginative work of elective essentialness things if government sponsorships were given. The data got from part responses being alluded to 6 suggest that even with government subsides given to R&D of elective essentialness things, an a lot of associations are simply potentially interested, and most are in no way charmed. This result may be made by the manner in which that organization apportionments don't totally deal with the costs associations cause to perform thing R&D.

The two most common purposes behind respondents' associations to develop their elective essentialness associations were Social Responsibility and Strategic Growth. Inside the Other characterization, clarifications behind structure up an electoral imperativeness point of view inside a business were Client and Code Requirement, Improved Architects, Client Demands and Client Goals. Diagram individuals' responses to address 7 reflect moving ends concerning the reasons behind making elective imperativeness things or organizations as an arrangement of activity. Social obligation made the most contribution with a 4.28 ordinary, while fundamental improvement sought after with a 4.20 typical. An association can be thought about in which social commitment and significant advancement are synonymous, and one may state dependent on one another for a green association to make, create an advantage.

There is an overwhelming response to this request is no, respondents' associations are not at risk to consume money on R&D of elective essentialness things if government blessings were given. Then

Figure 7. N=52. Question#7: "What were the explanations behind building up your elective vitality business?"

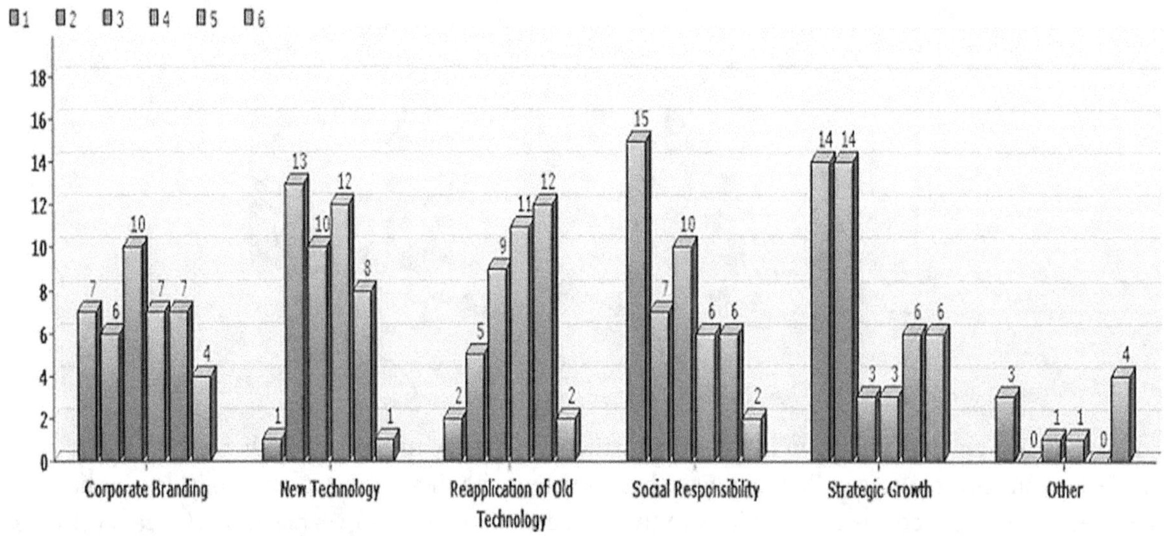

Figure 8. N=53. Question#8: "How likely is your organization to burn through cash on the innovative work of elective vitality items if government endowments were given?"

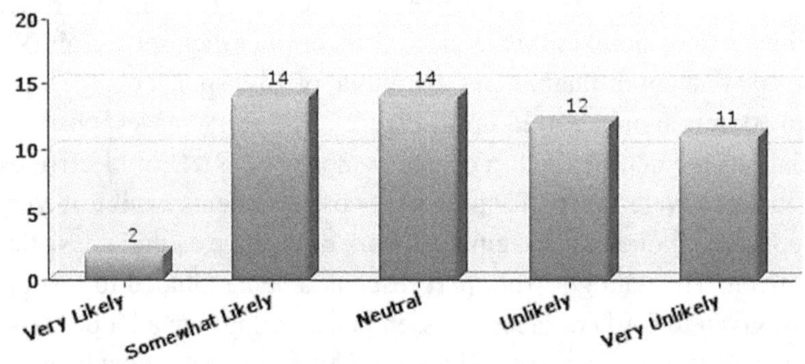

Figure 9. N=50. Question#9: "If government sponsorships were not given, how likely would your organization be to burn through cash on the innovative work of elective vitality items?"

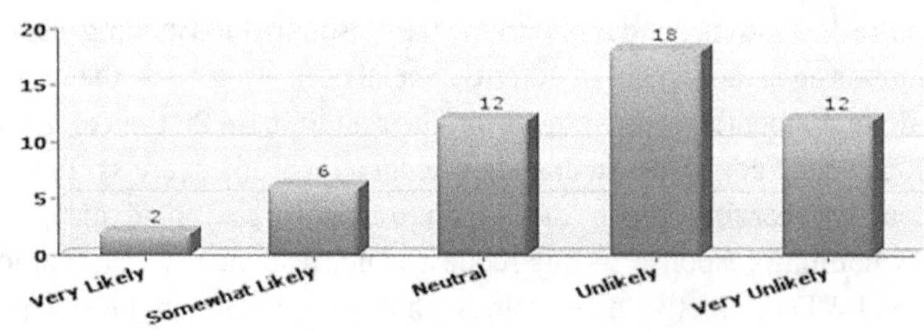

Figure 10. N=52. Question#10: "As you would see it, where do you see the main thrust of elective vitality/deals acknowledgment to be?"

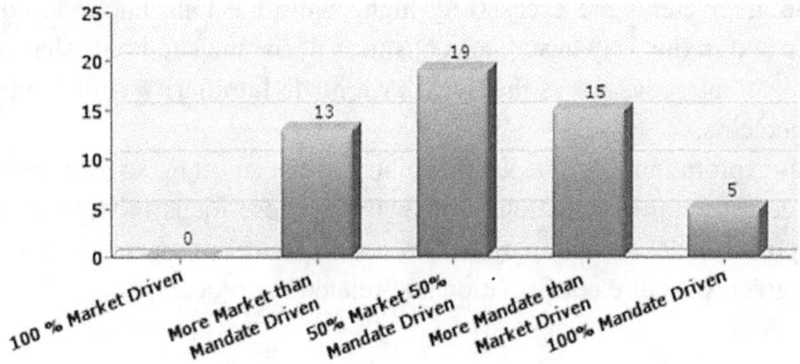

again, two respondents show it Very Likely that their association would waste money on original work of elective essentialness things if government allocations were given. The data grabbed from part responses being alluded to 12 suggest that even with government fades away given to R&D of elective essentialness things, a more significant piece of associations are simply irrelevantly charmed, and most are not at all interested. This result may be made by how that organization allotments don't deal with the costs associations achieve to perform thing R&D.

Again there is the overwhelming response to this request is no, respondents' associations are not inclined to consume money on imaginative work of elective essentialness things if government blessings were not given. Then again, two respondents show it Very Likely that their association would waste money on creative work of elective essentialness things paying little respect to whether government blessings were not given. The likelihood of study individuals to participate in R&D drops astonishingly without the game plan of government enrichments. Out of 53 respondents, 16 individuals (30%) concur that they would continue checking out R&D without government sponsorships, while 37 individuals (70%) agree that they would no uncertainty not share in R&D without the assistance of government gifts. These results point to a lukewarm hypothesis and eagerness for R&D without the aid of government allotments.

As indicated by inquiry 10, respondents' answers exhibit that the central purpose of elective essentialness is generally direction driven. None of the respondents demonstrate the fundamental catalyst of elective essentialness to be feature driven.

CONCLUSION OF CHAPTER

The comprehensive aftereffects of this examination paper recommend that green structure appraisals programs have had a general beneficial outcome on the development business. The advancement of projects has taken into consideration an open exchange to develop between structure proprietors and elective vitality suppliers. This discourse has and will keep on promoting a superior comprehension of what is expected of a green structure rating program. Taking an increasingly engaged perspective on the review results, notwithstanding, obviously the present green structure evaluations frameworks being used today require the amendment to all the more likely location the necessities of future development ventures.

The consequences of this examination propose that at present, organizations concentrating on elective vitality see green evaluations projects to be an impairment to long haul business rehearses. The expenses to execute green items are excessively high, while the long haul consequences of elective subjects are obscure and in this way unsure as a business decision. The lion's share disposition toward green structure evaluations programs is that they have made familiarity with fundamental issues and the potential arrangements.

At the end of this exploration examine, a few points have been set up as recommendations for future improvement of green structure evaluations frameworks. These focus incorporate expanded relevant research of green options, a closer examination of elective vitality appropriations and overhauled charge arrangements as to green structure confirmation and related tax reductions.

REFERENCES

Abercrombie, N., Hill, S., & Turner, B. S. (1994). Lexicon of humanism (3rd ed.). Harmondsworth, UK: Penguin.

Aye, L., Bamford, N., Charters, B., & Robinson, J. (1999, Sept.). Environmentally Sustainable Development: A Life Cycle Costing Approach for a Commercial Office Building in Melbourne, Australia. 15th Annual ARCOM Conference, Liverpool John Moores University. Association of Researchers in Construction Management, 2, 735-742.

Campagna, B., & Frey, P. (2008). The Impact of Evolving LEED Standards on Historic Preservation Projects. *Journal of Green Building.*, *3*(4), 21–31. doi:10.3992/jgb.3.4.21

Cooperman, A., Dieckmann, J., & Brodrick, J. (2012, June). Control Systems and LEED, *ASHRAE Journal*, 96-99.

Creswell, J. W. (2009). *Research Design: Qualitative, Quantitative, and Mixed Methods Approaches* (3rd ed.). Los Angeles, CA: Sage.

Ding, G. K. C. (2008). Maintainable Construction – The job of natural appraisal devices. *Diary of Environmental Management*, *56*, 451–464.

Etheridge, D. (2010). *Microsoft Excel Data Analysis* (3rd ed.). Indianapolis, IN: Wiley Publishing.

Galvan, J. L. (2009). *Composing Literature Reviews: A Guide for Students of the Social and Behavioral Sciences* (4th ed.). Los Angeles, CA: Pyrczak Publishing.

Gershenfeld, N., Samouhos, S., & Nordman, B. (2010). Canny Infrastructure for Energy Efficiency. *Science Magazine.*, *327*, 1086–1088.

Hernandez, P., & Kenny, P. (2009). From net vitality to zero vitality structures: Defining life cycle zero vitality structures (LC-ZEB). *Energy Policy*, *42*, 815–821.

Hernandez, P., & Kenny, P. (2011). Advancement of a Methodology forever cycle building vitality appraisals. *Energy Policy*, *39*, 3779–3788. doi:10.1016/j.enpol.2011.04.006

Maiona, M. (2010). *Green Leases: Operating Expense Considerations and Implications*. American Bar Association's Section of Real Property, Trust and Estate Law Spring Symposia.

Malangone, C. (2012). *Better Metrics of Green Building Needed, According to New McGraw Hill Construction SmartMarket Executive Brief on Decision Making for Green Building Investments.* McGraw Hill Construction.

Miles, M. P., & Covin, J. G. (2000). Natural Marketing: A Source of Reputational, Competitive and Financial Advantage. *Diary of Business Ethics*, *23*(3), 299–311. doi:10.1023/A:1006214509281

Miller, N., Pogue, D., & Gough, Q., & Davis, S. (2009). Green Building and Productivity. *Diary of Sustainable Real Estate*, *1*, 65–89.

Miller, N., Pogue, D., Saville, J., & Tu, C. (2010) The Operations and Management of Green Buildings in the United States. JOSRE, 2(1), 51-66.

Chapter 12
Sustainability of Green Building Practices in Residential Projects

Subhankar Das
https://orcid.org/0000-0001-7344-4583
Duy Tan University, Vietnam

ABSTRACT

A green building is a sustainable building that has minimal impacts on the environment throughout its life. For the purposes of this report, "green building" is understood to mean construction that makes efficient use of energy and resources in every aspect. This includes the production of building materials, and the design, use, and eventual demolition of a building in any sector (commercial, residential, industrial, public buildings) and at all stages, from new buildings to "retrofitting" or adapting existing ones. The construction sector, which accounts for 10% of global GDP, has direct and indirect impacts on the environment. It produces 23% of global greenhouse gas (GHG) emissions, and buildings are responsible for between 30% and 40% of all material flows. A green building is a sustainable building that has minimal impacts on the environment throughout its life.

INTRODUCTION

In the latest decade, there has been speedy, "improvement of industrialization" on the planet, especially in the making similarly as making countries like India. (Das Sharma, 2008). It assessed that 2030, 60 percent of the Aleut people would live in urban networks. From a viable propelled perspective, the "agreeable common relationship" among urban regions and the urban and commonplace areas are of exceptional critics to the flourishing of who and what is to come. The brisk advancement realized by industrialization has led to the unconstrained improvement of public domains. "The difference in agricultural land to individual home and deforestation has made it difficult to keep up fundamental equality. A speedy augmentation in masses advancement and movement in urban districts has caused far-reaching tainting" (Das Sharma, 2008). In case public networks are not genuinely orchestrated and managed, the nature of the air, the availability of water, waste dealing with, reusing structures and all qualities of the urban condition adding to "human success" will be under hazard.

DOI: 10.4018/978-1-5225-9754-4.ch012

Debilitated prosperity, respiratory illness, and sudden passing have associated with the elements of air sullying in a couple making countries. Focus pay, and states that have been starting late industrialized are experiencing new challenges identified with "increases in automated transport and industrialization, for instance augments in air and water sullying. The World Health Organization (WHO) evaluated that more than "1 billion people in Asia alone introduced to outdoors poisons that outperform the WHO rules, inciting the death of a substantial segment of a million people yearly" (United Nations Human Settlement Program, 2008: 123). Given that air poisons cause critical prosperity perils, and addition affectability in sound people, improving the air quality in urban zones will have constructive prosperity effects of all. Another load in urban networks is lacking waste organization. Deficient gathering and exchange of waste are transforming into extreme stress in urban areas, in perspective on the prosperity perils its stances to the urban masses". The lacking collection and exchange of waste are influencing on the natural network of urban regions and besides the urban condition. The majority of waste collected in making countries includes normal waste, sustenance, wood, coal, etc. "Despite the way that reusing and reuse systems have transformed into an outstanding practice in the making scene, these practices are consistently executed by the easygoing part in elusive conditions". Solid waste organization practices that have executed insufficiently can provoke an "extent of excreta and vector-related diseases" (United Nations Human Settlement Program, 2008: 126).

In urban territories we in like manner find the "heat island sway". The radiation modifies in urban locales impact the temperature dissemination. Sun arranged radiation is expanded and changed into warmth. "Blacktops, dividers, and housetops store warmth and produce longwave radiation to the sky" (United Nations Human Settlement Program, 2008:127). "The city takes any more drawn out to chill than the incorporating vegetated zones. Vegetated zones set aside more effort to cool in light of the way that the sun causes water held on the dirt and leaves to disappear, and shading of the plants keep the ground cool. The urban locales have higher temperatures than enveloping rural districts". This marvel is known as the "heat island" impact (United Nations Human Settlement Program, 2008).

Why Green Building

A green structure is a sensitive structure that influences the earth for an inconceivable length. For the purposes behind this report 'green structure' is grasped to mean advancement that makes capable use of essentials and resources in every point of view". This joins the age of structure materials, and the arrangement, utilize and conceivable pounding of a working in any part "(business, private, mechanical, open structures)" and at all stages, from new structures to 'retrofitting' or changing existing ones. The advancement zone, which speaks for 10% of overall GDP, has prompt and roundabout consequences for the earth. It produces 23% of overall ozone hurting substance (GHG) spreads, and structures are responsible for some place in the scope of 30% and 40% of each and every material stream. Grasping green structure practices would out and out diminishing these normal and resource impacts.

The life-cycle influence is supernatural in the essentials used for warmth, light and cool a structure while it is being utilized. Over its lifetime, the exemplified essentials of structure materials generally comparative influences the air. As demonstrated by the US Green Building Council, "the cost and proportion of the essentials required to build up a crucial, gainful spot of business is basically indistinguishable to that of a for the most part arranged structure, yet there is a colossal qualification in working costs".

"Green structures have diverse favorable circumstances other than the obvious natural ones. The structures are logically pleasing, and people working in them end up being progressively productive, which adds to the as a rule budgetary options". The preferred standpoint regards and rents of green structures will as a rule be higher than for conventional structures.

In numerous countries the private division is "accountable for a noteworthy piece of the advancement business, including building improvement, the officials and the supply of contraptions and imperativeness". The private section thusly has an imperative occupation in both financing and conveying green advancement and making a market. The private division can in like manner advance positive change in green structure sheets and industry bodies.

Market Size of Real Estate in India

Retail, invitingness and business land have furthermore grown, basically, giving the really essential establishment to India's growing needs. Around 2 lakh new houses to be built in 6 states under PMAY policy in next 3 years. The Indian land division has seen the high advancement, starting late with the rising looked for after for office similarly as private spaces. The land part in India is required to attract adventures worth US$ 7 billion out of 2017, which will rise further to US$ 10 billion by 2020".

POLICY FRAMEWORK FOR GREEN BUILDING IN INDIA

Regardless of the way that there is no set law concerning green structures in India, plans in existing authorization and a couple of procedural exercises see the normal pieces of the improved portion". "The twelfth Five Year Plan (FYP) hopes to hustle the appointment of codes relating to green working through the Jawaharlal Nehru National Urban Renewal Mission (JNNURM) and by interfacing cash related devolution to close-by urban specialists to the utilization of green development guidelines". The National Action Plan on Climate Change (NAPCC) and the National Mission on Sustainable Habitat address imperativeness efficiency in structures. "The organization will propel the Energy Conservation Building Code (ECBC) as a fundamental piece of urban masterminding, material for structures with a stack of 500 kilowatts (kW) or more".

Another area of sensibility is to be added to the National Building Code (NBC). The "Indian Bureau of Standards and the Building Materials and Technology Promotion Council are tending to building materials creation and measures". The Bureau of Energy Efficiency (BEE) is using benchmarks and checking to grow careless and diminishing the imperativeness eaten up by mechanical assemblies. "There are methodologies for water use in the structures of the state and common measurements. Water procuring, for example, is required in different states and urban networks. Hyderabad and Delhi have set rules for wastewater the officials".

Deliberate green structure rating systems have transformed into a notable instrument to encourage the advancement division to get supportable practices. In India, the two guideline rating systems are the Green Rating Integrated Habitat Assessment (GRIHA), made by TERI and the Ministry of New and Renewable Energy, and Leadership in Energy and Environment Design (LEED), worked by the Indian Green Building Council (IGBC). "GRIHA uses a great deal of 34 criteria to review structures with a story zone 2,500– 150,000 m2, and solidifies all the pertinent development guidelines and benchmarks, including the ECBC, and has been grasped by the organization Central Public Works Department. To date, 350 structures (1.02 million square meters) have been GRIHA selected, and 2,029 structures enrolled for LEED India accreditation".

Other rating systems for green structures in India are the Small Versatile, Affordable GRIHA, made by "TERI and Association for Development and Research of Sustainable Habitats (ADARSH) for minimal autonomous structures; the Eco-Housing rating structure created for Pune; and the Star Rating Program for Buildings of the Bureau of Energy Efficiency. The IGBC moreover has rating structures for homes, townships, exceptional money related zones, modern offices and scenes".

The major pieces of the green structure plan give emphasis on:

- Sustainable site improvement and building quality (orchestrating and land use; constraining environmental impact; the propriety of the site, plan and advancement to evaluate climatic and cataclysmic occasion factors).
- Water profitability (in materials, improvement, action and decommissioning).
- Energy profitability (in materials, improvement, action and decommissioning).
- Indoor regular quality (air, lighting, temperature).
- Reduced usage of structure materials (strength, flexibility, safeguarding/re-using).
- Consequent carbon holds reserves

To the extent approach duty, most by far of the above components can be regularly great. For example, arranging an attempting to restrain cooling requirements through its geographical zone can confine both imperatives and material use. There may, in any case, be a need to make trade offs (for instance arranging a house with lower cooling essentials may incorporate using all the all the more roofing material).

DETAIL BACKGROUND OVERVIEW

There are different prodding components of the endeavored research consider. "In particular, the way that the essential capability of the structures worked in the midst of the latest decades have added to an inefficiency inside the structure part, which in any event adds to high ozone hurting substance radiations and other common dangers and to lose imperativeness. Additional motivation originates from the way that both structure owners and building occupants in the long run endeavor diminish their imperativeness costs, particularly amidst extending essentialness costs. Hereafter, the mix of the "recently referenced segments" together make impulses for essentials efficiency in structures and fills in as a better than average enticing stage from which the drove examination has advanced". (Ryghaug and Sorensen, 2009).

Green structures incorporates using viable procedures and materials in the arrangement and improvement of properties (or in existing properties redesign) and exercises, where the fact is to cut down undertakings and upkeep costs, consolidating essentialness capability in lighting headways and air quality, redeveloping dim hued field districts, or using green housetops that consider overflow water to be reused and besides add to upgraded delegate and understudy prosperity, comfort, and proficiency. In any case, the upsides of green structures estimation are around numerous occasions greater than the ordinary additional cost of the structure it". (Kats, 2003; Wiley, 2008; Wiley et. Al, 2010)

From the extending biological concerns, affiliations overseeing supportability issues have set up business extents of "green" building level standard, for instance Building Research Establishment Environmental Assessment Method (BREEAM), Australian Building Greenhouse Rating gadget (ABGR) and the United States' Green Building Council (USGBC). USGBC is a national non-advantage investment affiliation that has developed the Leadership in Energy and Environmental Design (LEED) System to give principles and rating structures to green structure headways. There still have issues in measure green structure in light of the way that significant standard still one of a kind in each country. The standard will demonstrate green, viable components of that building. (Robinson, 2007; May et al., 2007; Sayce 2010; Shimizu, 2010)

Sustainability

Sustainability is a term with a various definitions. "Brandon (1999) out there exists definitions, in any case, all are accessible to decipher on catchphrases or articulations". In what way would manageability have the capacity to show convincing there are assorted understandings of what is supportability? There is an agreement between monetary, social, and environmental factors, anyway the key is finding balance (Figure 1). "Due to the possibility of the improvement business, green structures must be assessed and estimated in order to show that reasonability has enough unsurprising points of interest for the transitory laborers or conceivably owners".

Helping originators limit the danger to their budgetary authorities is the CSI's Green Format. The green thing database dealt with around affordable properties. "The things in the database are entered relies upon the thing's course of action, exemplified imperativeness, life-cycle properties, and undertakings identified with execution. The element of joint effort among producers and draftsmen will be uncommonly improved as makes put their things in the database" (Barista, 2008). The makers will never again need to swim through various specific sites to find the best match when distinctive endeavors have formally "attempted and deciphered the things". This prompts the prerequisite for changes on a greater scale in supportability, and not just for separated issues, similarly as, compromise earlier in the hypothetical and design stages (CII, 2006).

Statement of Research Problem

Exhibiting multi day, there is a dialog over viability in the arrangement and improvement of private structures. "Private structures are smaller scale and change from site to site and reach out to stretch out making it difficult to make cost assessments. Viable rating structures exist, yet there still is an arbitrary obligation to viability". Such assessing systems, like the Indian Green Building Council's (IGBC)

Figure 1. Three spheres of sustainability
Source: http://www.vanderbilt.edu/sustainvu/images/sustainability_spheres.png)

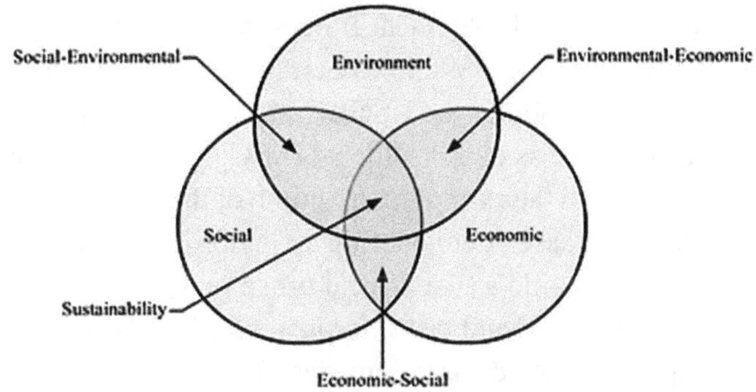

Leadership in Energy and Environmental Design for Homes (LEED-H), offer insistence levels for private structures, yet makers and architects don't utilize them. The advancement of structure green could be credited to an example in the structure; nevertheless, this could in like manner be the perspective is the destiny of the structure. In any case, for what reason are producers and other greater scale private endeavors not giving 100% of their undertakings toward green structure? Makers, organizers, and architects see spending plan, time, and costs, and supportability are routinely the least requesting to cut. The covering and multifaceted nature of the rating systems debilitates the significance behind the activities to help the earth. Regardless, there is a commitment of the designer to pass on changes to the business, and start the improvement toward green homes.

Objectives of the Chapter

The explanation behind this investigation was to give an appreciation of the essential administration process, from an engineer's point of view, on private supportability. The procedure started by requesting private designers from India to overview diverse focuses and thoughts on viability. "Private reasonability audit was requested in different measurements subject to repeat, opinion, importance, experience, and shared trait". The purpose of the examination was to recognize the feelings of dread, cost grumblings, measurements of blends, and perplexity related with private supportability in the present cabin advancement promote. This examination put together in the wake of an existing investigation with respect to the importance of rating systems and other affordable practices in the private part.

The objective of this examination was to perceive a boost for being sensible as it applies it to the private advancement industry. How the private division is completing green arranged in a helpful and legitimate way is a bit of the investigation. With the progress of advancement, being green is getting the chance to be less requesting; nonetheless, with a sticker value that facilitates the new development".

A HISTORY OF GREEN BUILDING AND POPULAR GREEN BUILDING RATING SYSTEM

LEED is definitely a singular program, yet it doesn't stay long as the primary green structure rating program opens. Distinctive ventures which perform equivalent attestation organizations are extremely unavoidable worldwide and are getting the chance to be as regularly used and standard as LEED. Green Globes, Energy Star, and High Performance Building Index contain two or three choices to LEED.

The basic review of these recently referenced affirmation programs revealed that a hankering to biologically improve the structure and advancement frames has existed among modelers and contractual workers for a significantly long time. "While Robert Watson solidified the bleeding edge green advancement through the course of action of the USGBC, the craving to work in congeniality with the ordinary living space has existed since as on schedule as 15BC". As demonstrated by Charles Colbert, the "green structure, advancement and the extraordinary climb of the US Green Building Council and its accomplices elsewhere is a fantastic case of conquering affliction" (Kibert, 2012, p. xv) Which finally began upon the improvement of the world's first critical urban networks.

With the improvement of the USGBC and tasks, for instance, LEED and Energy Star, the necessity of normally discerning arrangement and advancement was given a phase on which to make and create. The fundamental inspiration driving LEED, for example, was to give outcast affirmation of green structural shapes: ensured LEED specialists could survey the essentials usage of a structure and along these lines choose it's by and large biological impact through imperativeness illustrating. This outcast checks system incited an extended consideration regarding material life expectancy, material piece, the rate of use and common impact as accumulated and completed. Also, these tasks changed and formed into a green structure, advancement, whose basic concern was to decrease materials used, improve the natural impression of structures, upgrade system, and reduce imperativeness use" (Willson 2008).

Notwithstanding the phenomenal advances made toward common efficiency and incredible material usage, many green assessments structures don't have the consistent verification to back up their showed cases. An incredible piece of the composition review also revealed that while many green structure affirmation programs are upgraded and ending up more science-arranged in their definitions and applications, some are vivaciously reliant on optional point systems that don't generally improve the environmental surviving from an office" (Melton 2012; Stein 2004).

There are different influencing factors in the grasped research mull over. "In particular, the way that the essentials viability of the structures worked in the midst of the latest decades have added to an inefficiency inside the structure territory, which at any rate adds to high ozone exhausting substance releases and other "common perils" and to lose imperativeness. Additional motivation begins with the way that both structure owners and building occupants finally hope to reduce their essentialness costs, particularly amidst growing imperativeness costs. Subsequently, the mix of the recently referenced factors together makes driving forces for essential capability in structures and fills in as a not too bad convincing stage from which the drove investigation has advanced". (Ryghaug and Sorensen, 2009).

Green structures incorporates using sensible systems and materials in the arrangement and improvement of properties (or in existing properties redesign) and undertakings, where the fact of the matter is to cut down exercises and bolster costs fusing essentialness adequacy in lighting developments and air quality, redeveloping darker field goals, or using green housetops that consider flood water to be reused and besides add to upgraded agent and understudy prosperity, comfort, and proficiency. In any case, the benefits of green structures estimation are generally on numerous occasions greater than the ordinary additional cost of the structure it". (Kats, 2003; Wiley, 2008; Wiley et. Al, 2010).

From the growing characteristic concerns, affiliations overseeing viability issues have set up business extents of "green" building level standard, for instance Building Research Establishment Environmental Assessment Method (BREEAM), Australian Building Greenhouse Rating gadget (ABGR) and the United States' Green Building Council (USGBC). "USGBC is a national non-advantage enlistment affiliation that has developed the Leadership in Energy and Environmental Design (LEED) System to give standards and rating structures to green structural progressions. There still have issues in measure green structure in light of the way that extensive standard still interesting in each country. The standard will exhibit the green profitable element of that building". (Robinson, 2007; May et al., 2007; Sayce 2010; Shimizu, 2010)

TECHNICAL KNOW HOW OF GREEN BUILDING PRACTICES METHODS

The purpose of this examination was to "inquire about" how collecting green is completed by private producers, authoritative laborers, architects, and designers. The data accumulated in the midst of this investigation was fused using "mixed research systems" which are fitting for this sort of research. This strategy was executed through an audit overview.

A couple of systems for data examination were completed after the end of the investigation. The responses to each question were investigated only, and a while later kept an eye all in all in a general report. Likewise, charts and tables were delivered to contrast with the data got for each survey question; every request has been tended to in the data contemplate a bit of this examination.

The examination system used for this data accumulation is "obvious estimations" trailed by an interpretation of the results. Clear bits of knowledge are portrayed as "edifying investigation of data for all free and ward factors in the examination.

The examination relied upon a movement of request to gauge the evaluation of grain structure and the moves made to execute green arrangement. The request falls into out and out subjects: acknowledgment, general appraisal, a repeat of uses, hugeness, and data was accumulated from web understanding. The data assembled was explored that fell into these classes. The theory came about due to the social occasion of those parameters.

The data got from the investigation perceived the necessity for enlightenment in green structure and advancement in private assignments. The information would be useful to the home designer industry to understand the best perspectives, rehearses, and confusions in an affordable arrangement. Moreover, the examination perceived the evaluation of appropriateness, a repeat of usage of sensibility, association with green structure, acknowledgment with green thoughts, and the importance of viability inside their association.

ANALYTICAL REPRESENTATION OF HOLISTIC REPRESENTATION OF GREEN BUILDING PRACTICES

Responses from the diagram were returned through the web from designers and engineers in India. Sixteen responses were gotten of 100 appropriated by email mentioning. Of the 16 responses, seven responses were from specialists, eight responses were gotten from designers, and one respondent did not demonstrate a trade. The response rate was 16%. The response was lower than foreseen.

The normal respondent was a private maker (half), planner (43.75%), and one unspecified trade (6.25%). In this examination, an originator is portrayed as an association or person that places assets into, makes, and subdivides land to assemble and loving homes. In this investigation, a designer is portrayed as an association or person that develops or coordinates homes under an understanding or for the hypothesis. The ordinary endeavors were single family homes and mixed-use adventures. The ordinary size of private endeavors was around 1,560 sq ft. The typical yearly contracted work of the respondents was Rs 3.84 million. The extent of the volume of work was Rs 1.0 million to Rs 8.0 million consistently. Half of the respondents gave measurement information. The circumstance of the respondent was the president (62%), the VP (13%), the owner (13%), and the official (6%) of the association with one respondent (6%) not deciding a position or title (Figure 2).

Experience With Sustainable Practices

Table 1 contains responses to questions 1, 3, and 4 that related to the association of the respondents with supportable practices. The responses to these requests were used to gauge, the proportion of the sensible

Figure 2. Respondents position considered for the survey

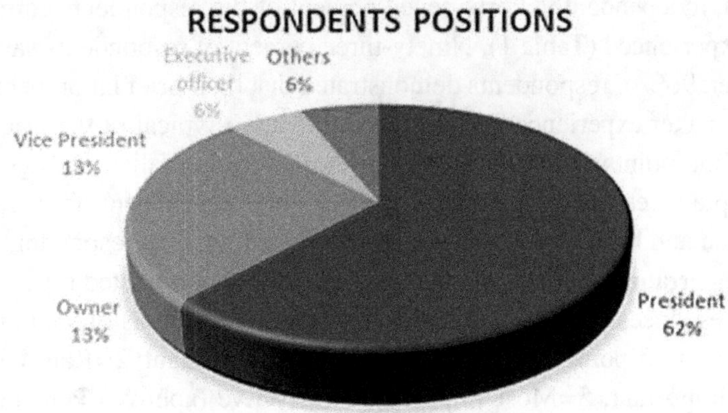

Table 1. Responses to Likert scale questions related to experience in sustainable practices for typical respondents

	No Experience (1)	Barely Experienced (2)	Somewhat Experienced (3)	Experienced (4)	Very Experienced (5)	Rating Avg.	Response Count
	0%	6%	25%	31%	**38%**		
Q1. What level of experience does your company have in green or sustainable building?	0	-1	-4	-5	**-6**	4	16
	0%	7%	20%	26%	**47%**		
Q3. Is (are) the essential designer(s) experienced with green or maintainable practices?	0	-1	-3	-4	**-7**	4.13	15
	0%	6%	12%	38%	**44%**		
Q4. Is the primary contractor experienced with green or sustainable practices?	0	-1	-2	-6	**-7**	4.19	16

use of green structure done by each respondent. Questions were introduced in a 5-point Likert rating scale plan (1=No Experience; 2=Barely Experienced; 3=Somewhat Experienced; 4=Experienced; 5=Very Experienced). Question 1 got some data about the element of association with sensible practices inside the respondent's association. Ninety-four percent of respondents had an association in supportability. Of those, 94% respondents, 38% were experienced. The rating ordinary was 4.0 or somewhat experienced in overseeing.

Questions 3 and 4 about requesting the experience from the planners and legally binding specialists inside the association, independently. Forty-seven percent of the respondents communicated that their planners were very experienced (Table 1). Ninety-three percent of respondents saw that their modelers had inclusion, however, 96% of respondents demonstrated that their brief laborers had a comprehension. The rating typical of maker experience was 4.13 and the rating typical of the brief laborer experience was 4.19. Both rating midpoints demonstrated that authoritative specialists and designers have some association with viable practices and plan. Under seven percent of respondents for the two social occasions felt that their draftsman and brief laborer have less understanding. One respondent tended to "N/A" for the request concerning arrangement involvement in light of their associated type.

Table 2 contains responses to address 2 that related to the essentials of supportable practices. The request was introduced in a 5-point Likert scale plan (1=Not Important; 2=Rarely Important; 3=Somewhat Important; 4= Important; 5=Most Important). Forty-five (approx.) Percent of respondents felt supportable practices were fundamental to their association. Of course, 6.25% felt sensible practices were not basic, and 6.25% felt reasonable practices were somewhat basic. The rating ordinary of the normal respondent was 4.13 (basic).

Practical Experiential Green Building Practices

Table 3 contains responses to questions 5, 6, 7, 8, 9, 10, 14, 15 and 18 that related to a respondent's supposed about prudent practices. Questions were exhibited in a 5-point Likert rating scale position (1=Strongly Disagree; 2=Disagree; 3=Somewhat Disagree; 4=Agree; 5=Strongly Agree). The subcategories inside the inclination social occasion of the request were appraisals on as a rule association sees (Q5), set aside extra cash related estimation of supportable practices (Q6; Q8; Q15), constructability of sensible private endeavors (Q7; Q14), and appeal of viable private errands (Q9; Q10; Q18). Question 5 asked regarding whether they believed that reasonability was viably practiced. Most of the respondents (56%) solidly agreed that their association practiced sensibility. Thirty-eight percent of respondents agreed their association successfully cleaned reasonability. The rating typical for request 5 was 4.5 (harmony).

Question 6 asked regarding whether green practices compare to extend costs. Sixty-two percent of respondents solidly agreed that efficient homes contrasted with extended costs. All respondents agreed

Table 2. Responses to Likert scale addresses identified with the significance of economical practices for regular respondents

Question	Not Important (1)	Rarely Important (2)	Somewhat Important (3)	Important (4)	Very Important (5)	Rating Avg.	Response Count
	6.25%	0%	6.25%	44.75%	**44.75%**		
Q2. How important to your company is green design or sustainable and building sustainable homes	-1	0	-1	-7	**-7**	4.13	16

Table 3. Responses to Likert scale questions related to opinion of sustainable practices for typical respondents

Question	Strongly Disagree (1)	Disagree (2)	Neutral (3)	Agree (4)	Strongly Agree (5)	Rating Avg.	Response Count
	0%	0%	6%	38%	**56%**	4.5	16
Q5. Do you concur that your organization effectively fuses green or practical structure?	0	0	-1	-6	**-9**		
	0%	0%	0%	38%	**62%**	4.63	16
Q6. Do you concur that green or feasible practicecompared to expanded expenses?	0	0	0	-6	**-10**		
	12%	31%	18%	**38%**	0%	2.81	16
Q7. Do you concur that green or maintainable plan are progressively confused to assemble?	-2	-5	-3	**-6**	0		
	12%	18%	6%	18%	**46%**	3.63	16
Q8. Do you concur that green or feasible homes ought to be sold, including some built-in costs?	-2	-3	-1	-3	**-7**		
	18%	18%	18%	**31%**	12%	3	16
Q9. Do you concur there is a developing interest for green or manageable homes?	-3	-3	-3	**-5**	-2		
	18%	0%	18%	**50%**	12%	3.38	16
Q10. Do you concur that buyer interest for feasible homes has influenced the development and additionally plan of your homes?	-3	0	-3	**-8**	-2		
	0%	6%	18%	12%	**62%**	4.31	16
Q14. Do you concur there is expanded disarray over which green benchmarks to utilize?	0	-1	-3	-2	**-10**		
	12%	25%	25%	**31%**	6%	2.94	16
Q15. Does your organization concur that rating frameworks merit the additional expenses?	-2	-4	-4	**-5**	-1		
	31%	12%	25%	18%	12%	2.69	16
Q18. Do you concur that there is a buyer inclination of green or economic homes over conventional or non-green homes?	**-5**	-2	-4	-3	-2		

or unequivocally agreed that green structure infers extended costs. Question 6 had a rating typical of 4.63 (harmony). Question 8 asked concerning whether the grain structure should be sold at a greater expense than anticipated. Forty-six percent of typical respondents determinedly agreed while 18% of respondents agreed home should move including some shrouded expenses. On the other hand, 30% of typical respondents agreed that homes should not move at a greater expense than ordinary while 12% of the average respondents immovably restrict this thought. The rating typical for request 8 was 3.63, that is, everything viewed as the respondents somewhat contrasts that homes should move at a greater expense than ordinary. Question 15 asked respondents the same they agreed that rating systems justified the extra costs. Thirty-one percent of the common respondents agreed the rating systems justified the development costs. Twelve percent of respondents accentuations vary that rating structures justified the development. Just 6%, unequivocally agreed the rating structures bolstered an extension in costs. The rating typical for request 15 was 2.94 (contradict this thought).

Question 7 asked with respect to whether the green arrangement is more confounding to work than customary structures. Thirty-eight percent of the ordinary respondents agreed that practical arrangement was progressively tangled to produce. Thirty-eight of common respondents agreed that the administration plan was progressively obfuscated, in any case, none of the respondents vehemently agreed. Of course, 12% of respondents immovably disagreed, and 31% couldn't resist negating this declaration. The rating ordinary was 2.81 (restrict this thought). Question 14 got some data about the chaos over which sensible rating structure to use. Sixty-two percent of respondents vehemently agreed that there was perplexity over the assorted rating structures open. Regardless, 6% of all respondents vary that there was confusion. The rating typical for request 14 was 4.31 (harmony).

Question 9 asked with respect to whether respondents agreed that there was a creating enthusiasm for green homes. Thirty-one percent agreed that there is a creating enthusiasm for green homes. Twelve percent of the regular respondents determinedly agreed that there is a market, and 18% unequivocally contradict this thought. The rating typical was 3.00 (reasonably agreed). Question 10 asked regarding whether respondents agreed that customer enthusiasm for green homes is changing their home structures. Twelve percent of respondents unequivocally agreed that sensibility demand is changing the structure of their homes. Half of the normal respondents agreed that there is an effect, anyway 18% vary, there was any effect of the doable purchaser demand on the structure of their homes. The rating typical was 3.38 (somewhat agreed). Question 18 asked regarding whether they agreed that there was a customer tendency for sensible living courses of action. Thirty-one percent of the average, respondents solidly vary that there was a customer tendency. 30% of respondents agreed that there was a tendency. The rating typical was 2.69 (restrict this thought).

As indicated by table 4, Question 16 assessed the repeat by using reasonable rating systems to assess the regular respondents' endeavors. Half of respondents as often as possible used a rating system for

Table 4. Responses to Likert scale questions related to opinion of implementing framework of the green economic structure

Question	Never 1	Rarely (2)	Sometimes (3)	Often (4)	Frequency 5	Rating Avg.	Response Count
	6%	12%	18%	12%	**50%**	3.88	16
Q16. How frequently is your organization effectively utilizing a rating framework for evaluating green or economical structure?	-1	-2	-3	-2	**-8**		
	0%	14%	29%	**43%**	14%		
Q17. How frequently does your organization effectively train its representatives in green methods?	0	-2	-4	**-6**	-2	3.57	14

assessing green or prudent structure. Six percent of respondents never used a rating structure, and 18% respondents a portion of the time used a reasonable home rating system. The rating typical was 3.88 (consistently). Question 17 got some data about the repeat of the respondents' association to adequately get ready agents in overseeing. Forty-three percent of respondents arranged their laborers in supportability practices normally. Twenty-nine percent of respondents a portion of the time arranged their laborers. The rating ordinary was 3.57 (sometimes).

At the point when all is said in done, both "fashioners and engineers' responses did not differ in a general sense on the parameters of experience, centrality, end, shared trait, and a repeat of supportability and green structure". Originators had a prevalent handle of sensible thoughts than producers. Also, fashioners adequately arranged their staff in administrative practices more routinely than designers, yet not basically more. "Designers and makers agreed that the cost of sensible arrangement is most basic in the midst of the advancement organized and vary that having a guaranteed essentialness originator just as an imperativeness rating structure were least basic".

The average respondent was learned about practical arrangement, with a "rating typical of 4.00 (experienced)". Regardless, engineers had higher rating midpoints for every request related to inclusion in doable practices when stood out from designers. "Producers had an ordinary of 3.75 which infers that they were somewhat experienced or experienced with viable practices".

"Every ordinary respondent, designers, and engineers agreed that building sensible home is basic to their association. In the perspective of the rating typical, engineers believed that alluring structure was the most fundamental and makers believed that an elegantly fulfilling arrangement was transcendent. In the midst of the improvement stage, architects and makers situated the elements of critics the proportional".

The ordinary respondents solidly agreed that they viably united green structure practices. The common respondent immovably varies there is a "buyer tendency for viable arrangement" and, similarly, making their homes move faster. The typical planner reasonably agreed to agree with most Likert scale questions.

CONCLUSION

Supportable structure and improvement of private endeavors will be central later on. The duty and responsibility don't merely need to start from the customer side, yet moreover from those organizing homes, building homes, and making nearby areas". The examination highlighted a review of responses from specialists and makers rely on their understanding, application, and commitment to oversee. The delineation exhibited to have essential information into the minds of the people who control how the world lives. By, the respondents do have a promise to sensitize and building green. "The two architects and designers trust practicality benefits the earth, regardless, meanwhile, don't acknowledge there is a client tendency for green homes. These irregularities in the reason incited confusion of how to produce and assess their green undertakings".

In light of the composition overview, there was an exhibited favorable position for supportable structures, yet by and mostly required clearness for the separate division, similarly as, the evidence points of interest to specialists and producers. By analyzing the responses of makers and designers on conservative and green practices in homes, it is considered a comprehension of why green improvement is not fully realized into private endeavors by different players.

REFERENCES

Aye, L., Bamford, N., & Charters, B., & Robinson, J. (1999). Environmentally Sustainable Development: A Life Cycle Costing Approach for a Commercial Office Building in Melbourne, Australia. Liverpool John Moores University. *Association of Researchers in Construction Management*, *2*, 735–742.

Bennett, F. (2006). Five Misunderstandings about Case Study Research. *Qualitative Inquiry, 12*(2), 219-245.

Boyd, T. (2006). *Can we Assess the Worth of Environmental and Social Characteristics in Investment Property?* Academic Press.

Chan, E. H. W., Qian, Q., & Lam, P. (2009). The Market for Green Building in Developed Asian Cities – the Perspective of Building Designers. *Journal of Energy Policy, 37,* 3061-3070.

Chihiro, S. (2010). *Will Green Buildings be Appropriately Valued by the Market?* Reitaku Institute of Political Economics and Social Studies, Working Report, No.40.

Davis, J., Cloutier, D., & Klein, J. (2010). Embedding Energy Efficiency in the Business of Buildings: Commercial Real Estate Contracts & Transactions. *ACEEE Summer Study on Energy Efficiency in Buildings*, *4*, 53–63.

Ellison, L., & Sayce, S. (2007). Assessing Sustainability in The Existing Commercial Property Stock Establishing Sustainability Criteria Relevant For The Commercial Property Investment Sector. *Journal of Property Management*, *25*(3), 287–304. doi:10.1108/02637470710753648

Jonathan, A. W., Justin, A. B., & Ken, H. J. (2008). Green Design and the Market for Commercial Office Space. *The Journal of Real Estate Finance and Economics*, *41*, 228–243.

Liners, C. (2007). Application of product differential semantics to quantify purchaser perceptions in housing assessment. *Building and Environment*, *42*(7), 2488–2497. doi:10.1016/j.buildenv.2006.06.012

Loeser, S., & Post, J. (2011, April 5). *A Model Energy Aligned Lease.* Retrieved from www.nyc.gov.

Lzkendorf, T. & Lorenz, D. (2005). *Sustainable Property Investment: Valuing Sustainable Buildings Through Property Performance Assessment, Building research and information*. Academic Press.

Maine, M. (2010). *Green Leases: Operating Expense Considerations & Implications*. American Bar Association's Section of Real Property, Trust and Estate Law Spring Symposia.

Mansfield, J. (2009). The Valuation of Sustainable Freehold Property: ACRE Perspective. *Journal of Corporate Real Estate*, *11*(2), 91–105. doi:10.1108/14630010910963133

Martinaitisa, V., & Kazakeviciu, E. (2007). A Two-Factor Method for Appraising Building Renovation and Energy Efficiency Improvement Projects. *Energy Policy*, *35*(1), 192–201. doi:10.1016/j.enpol.2005.11.003

Matthiessen, L., Morris, P., & Langdon, D. (2007). Cost of Green Revisited: Reexamining the Feasibility and Cost Impact of Sustainable Design in Light of increased Market Adoption. *Davis Langdon Research Journal.*

Miller, E., & Buys, L. (2008). Retrofitting Commercial Office Buildings for Sustainability: Tenants' Perspectives. *Journal of Property Investment & Finance*, *26*(6), 552–561. doi:10.1108/14635780810908398

Miller, N., Pogue, D., Saville, J., & Tu, C. (2010). The Operations & Management of Green Buildings in the United States. *JOSRE, 1,* 2.

Miller, N., Pogue, D., & Gough, Q., & Davis, S. (2009). Green Building and Productivity. *Journal of Sustainable Real Estate*, *1*, 65–89.

Mills, D. (2001). *When to Consider Distributed Systems in an Urban and Suburban Context. Water Environment Research Foundation (WERF), Casestudy: Philip Merrill environmental center building.* Chesapeake Bay Foundation.

Myers, G., Reed, R., & Robinson, J. (2007). Sustainable property–the future of the New Zealand market. Pacific Rim Property Research Journal, 14(3).

Myers, G., Reed, R. & Robinson, J. (2007, January). The relationship between sustainability and the value of office buildings. In *Proceedings of the 13th Annual Pacific Rim Real Estate Conference*. (pp. 1-22). Pacific Rim Real Estate Society.

New Building Institution. (2010, Jan. 1). Retrieved from http://www.newbuildings.org.

Solanki, A., & Nayyar, A. (2019). Green Internet of Things (G-IoT): ICT Technologies, Principles, Applications, Projects, and Challenges. In Handbook of Research on Big Data and the IoT (pp. 379-405). Hershey, PA: IGI Global.

Warren, C. (2010). Measures of Environmentally Sustainable Development and Their Effect on Property Asset Value: An Australian Perspective. *Journal of Property Management, 28*(2), 68-79.

Wiley, J., Benefield, J., & Johnson, K. (2010). Green Design and the Market for Commercial Office Space. *Journal of Real Estate Finance Econ, 41*, 228-243.

Compilation of References

10 IoT Smart Building Trends to Look Out For in 2018. (2018). [Web log post]. Retrieved from https://blog.senseware.co/2018/01/08/10-iot-smart-building-trends-look-2018

Abdullah, A. (2014). Green cloud computing: The need of the hour. *International Journal of Research in Advent Technology*, *2*(1), 316–321.

Abercrombie, N., Hill, S., & Turner, B. S. (1994). Lexicon of humanism (3rd ed.). Harmondsworth, UK: Penguin.

Abidin, N. Z., & Powmya, A. (2014). Perceptions on Motivating Factors and Future Prospects of Green Construction in Oman. *Journal of Sustainable Development*, *7*(5), 231–239. doi:10.5539/jsd.v7n5p231

Abowd, G., & Mynatt, E. D. (2000). Charting Past, Present and Future Research in Ubiquitous Computing. *ACM Transactions on Computer-Human Interaction*, *7*(1), 29–58. doi:10.1145/344949.344988

Abraham, S. (2016). Internet of Things (IoT) with Cloud Computing and Machine-to-Machine (M2M) Communication. *International Journal of Emerging Trends in Science and Technology, 3*(9), 4654-4661.

Aceto, G., Botta, A., Donato, W., & Pescape, A. (2012). Cloud Monitoring: definitions, issues and future directions. In *proceedings of the IEEE International Conference on Cloud Networking, CLOUDNET*. IEEE. 10.1109/CloudNet.2012.6483656

Adalberth, K., Almgren, A., & Petersen, E. H. (2001). Life cycle assessment of four multi-family buildings. *Intl J Low Energy Sustain Build*, *2*, 1–21.

Adaptation Fund Board. (2011). Project level results framework and baseline guideline document (No. AFB/B.14/Inf.6). Author.

Adeyeye, K., Osmani, M., & Brown, C. (2007). Energy conservation and building design: The environmental legislation push and pull factors. *Structural Survey*, *25*(5), 375–390. doi:10.1108/02630800710838428

Ahmad, T., Aibinu, A. A., & Stephan, A. (2019). Managing green building development–A review of current state of research and future directions. *Building and Environment*, *155*, 83–104. doi:10.1016/j.buildenv.2019.03.034

Ahmad, T., Thaheem, M. J., & Anwar, A. (2015). Developing a green-building design approach by selective use of systems and techniques. *Architectural Engineering and Design Management*, *12*(1), 29–50. doi:10.1080/17452007.2015.1095709

Ahn, Y. H., Pearce, A. R., Wang, Y., & Wang, G. (2013). Drivers and barriers of sustainable design and construction: The perception of green building experience. *International Journal of Sustainable Building Technology and Urban Development*, *4*(1), 35–45. doi:10.1080/2093761X.2012.759887

Aktas, B., & Ozorhon, B. (2015). Green Building Certification Process of Existing Buildings in Developing Countries: Cases from Turkey. *Journal of Management Engineering*, *31*(6), 05015002. doi:10.1061/(ASCE)ME.1943-5479.0000358

Albert. (2018). *IOT Devices Proximity Sensor.* Retrieved from https://www.infiniteinformation.com

Alcaraz, C., Najera, P., Lopez, J., & Roman, R. (2010). Wireless Sensor Networks and the Internet of Things: Do We Need a Complete Integration? In *Proceedings of the 1st International Workshop on the Security of the Internet of Things*. Tokyo, Japan: IEEE.

Al-Fuqaha, A., Mohammadi, M., & Guizani, M. (2015). Internet of Things: A Survey on Enabling Technologies, Protocols, and Applications. IEEE Communication Surveys & Tutorials, 17(4).

Altomonte, S., & Schiavon, S. (2013). Occupant satisfaction in LEED and non-LEED certified buildings. *Building and Environment*, *68*, 66–76. doi:10.1016/j.buildenv.2013.06.008

American Society for Testing and Materials. (1996). *Standard Practice for Developing Accelerated Tests to Aid Prediction of the Service Life of Building Components and Materials.* West Conshohocken, PA: Author.

Andrady, A. L. (2003). *Plastics and the Environment.* Hoboken, NJ: John Wiley & Sons.

Angelis, E. D., Ciribini, A. L. C., & Tagliabue, L. C. (2015). Smart campus demonstrator: Renovation towards a zero energy classroom building. *Procedia Engineering*, *118*, 735–743. doi:10.1016/j.proeng.2015.08.508

Annunziata, E., Testa, F., Iraldo, F., & Frey, M. (2016). Environmental responsibility in building design: An Italian regional study. *Journal of Cleaner Production*, *112*(1), 639–648. doi:10.1016/j.jclepro.2015.07.137

Anwar, F., Chaudhry, F. N., Nazeer, S., & et al, . (2016). Causes of Ozone Layer Depletion and Its Effects on Human. *Atmospheric and Climate Science*, *6*(01), 129–134. doi:10.4236/acs.2016.61011

Aparicio-Ruiz, P., Barbadilla-Martín, E., Salmerón-Lissén, J. M., & Guadix-Martín, J. (2018). Building automation system with adaptive comfort in mixed mode buildings. *Sustainable Cities and Society*, *43*, 77–85. doi:10.1016/j.scs.2018.07.028

Armbrust, M., Fox, A., Griffith, R., Joseph, A. D., Katz, R. H., Konwinski, A., & Zaharia, M. (2009). *Above the Clouds: A Berkeley View of Cloud Computing* (Rep. No. UCB/EECS-2009-28). UC Berkeley Reliable Adaptive Distributed Systems Laboratory. Retrieved from https://www2.eecs.berkeley.edu/Pubs/TechRpts/2009/EECS-2009-28.pdf

Arshad, R., Zahoor, S., & Shah, M. A. (2017). *Green IoT: An Investigation on Energy Saving Practices for 2020 and Beyond.* IEEE.

Atkinson, R., Crawford, L., & Ward, S. (2006). Fundamental uncertainties in projects and the scope of project management. *International Journal of Project Management*, *24*(8), 687–698. doi:10.1016/j.ijproman.2006.09.011

Australia's 10 Greenest Buildings. (2010, Jan. 25). Retrieved from https://sydneyarchitecturearchive.wordpress.com/2010/01/25/australias-10-greenest-buildings/

Aye, L., Bamford, N., Charters, B., & Robinson, J. (1999, Sept.). Environmentally Sustainable Development: A Life Cycle Costing Approach for a Commercial Office Building in Melbourne, Australia. 15th Annual ARCOM Conference, Liverpool John Moores University. Association of Researchers in Construction Management, 2, 735-742.

Aye, L., Bamford, N., & Charters, B., & Robinson, J. (1999). Environmentally Sustainable Development:A Life Cycle Costing Approach for a Commercial Office Building in Melbourne, Australia. Liverpool John Moores University. *Association of Researchers in Construction Management*, *2*, 735–742.

Ayres, R. U., & Leslie, W. (1996). *Industrial Ecology: Towards Closing the Materials Cycle.* Brookfield, VT: Edward Elgar Publishing.

Azar, E., & Ansari, H. A. (2017). Framework to investigate energy conservation motivation and actions of building occupants: The case of a green campus in Abu Dhabi, UAE. *Applied Energy*, *190*, 563–573. doi:10.1016/j.apenergy.2016.12.128

B.B., P.R., Saluja, P., Sharma, N., Mittal, A., & Sharma, S.V. (2012). Cloud computing for Internet of Things & sensing based applications. In *Proceedings of the International Conference on Sensing Technology*. Bangalore: ICST.

Baird, G., Leaman, A., & Thompson, J. (2012). A comparison of the performance of sustainable buildings with conventional buildings from the point of view of the users. *Architectural Science Review*, *55*(2), 135–144. doi:10.1080/00038628.2012.670699

Bajaj, R., Mishra, H. K., & Goyal, P. (2017) Design of Oxide Confined and Temperature Stable Long Wavelength Vertical Cavity Surface Emitting Laser for Optical Interconnects. *Optik, 131,* 506-514.

Bandyopadhyay, D., & Sen, J. (2011). Internet of things: Applications and challenges in technology and standardization. *Wireless Personal Communications*, *58*(1), 49–69. doi:10.100711277-011-0288-5

Barbon, G., Margolis, M., Palumbo, F., & et al,. (2016). Taking Arduino to the Internet of Things: The ASIP programming model. *Computer Communications*, *89-90*, 128–140. doi:10.1016/j.comcom.2016.03.016

Bates, O. (2012). Accounting for EnergyReliant Services within Everyday Life at Home. *Proc. Int'l Conf. Pervasive Computing*, 107–124. 10.1007/978-3-642-31205-2_8

Bennett, F. (2006). Five Misunderstandings about Case Study Research. *Qualitative Inquiry*, *12*(2), 219-245.

Berge, B., Butters, C., & Henley, F. (2009a). Fossil Oils. In B. Berge, C. Butters, & F. Henley (Eds.), *The Ecology of Building Materials* (2nd ed., pp. 139–156). Philadelphia, PA: Taylor & Francis.

Berge, B., Butters, C., & Henley, F. (2009b). Surface Materials. In B. Berge, C. Butters, & F. Henley (Eds.), *The Ecology of Building Materials* (2nd ed., pp. 299–355). Philadelphia, PA: Taylor & Francis.

Berge, B., Butters, C., & Henley, F. (2009c). Paint, varnish, stain and wax. In B. Berge, C. Butters, & F. Henley (Eds.), *The Ecology of Building Materials* (2nd ed., pp. 381–407). Philadelphia, PA: Taylor & Francis.

Berral, J. L., Goiri, I., Nguyen, T. D., Gavalda, R., Torres, J., & Bianchini, R. (2014). Building Green Cloud Services at Low Cost. In *Proceedings of the2014 IEEE 34th International Conference on Distributed Computing System.* Madrid, Spain: IEEE. 10.1109/ICDCS.2014.53

Bianchina, P. (1997). *Builder's Guide to New Materials and Techniques*. New York, NY: McGraw Hill

Borup, M., Klitkou, A., & Andersen, M. M. (2013). *Indicators of energy innovation systems and their dynamics. A review of current practice and research in the field (Radar report)*. EIS.

BossControls. (2019). *How Buildings Impact the Environment.* Retrieved Feb. 14 from https://bosscontrols.com/buildings-impact-environment/

Bostic, K. (2013, March 21). *Apple data centers now 100 percent renewable-powered, but emission rise 34 percent.* Retrieved from https://appleinsider.com/articles/13/03/21/apple-claims-data-centers-100-renewable-powered-75-for-corporate-facilities

Boucart, J. (2018). *Laser Diodes for (3D) Sensing, Swiss Photonics Meeting, Industrial 3D Vision.* Retrieved from https://www.swissphotonics.net

Bowyer, J. L. (2007). Green building programs-Are they really green? *Forest Products Journal*, *57*(9), 6.

Boyd, T. (2006). *Can we Assess the Worth of Environmental and Social Characteristics in Investment Property?* Academic Press.

Brenna, M., Falvo, M. C., & Foiadelli, F. (2012). *Sustainable Energy Microsystem (SEM): preliminary energy analysis. In 2012 IEEE PES Innovative Smart Grid Technologies* (pp. 1–6). Washington, DC: IEEE; doi:10.1109/ISGT.2012.6175735

Bünning, F., Sangi, R., & Müller, D. (2017). A Modelica library for the agent-based control of building energy systems. *Applied Energy*, *193*, 52–59. doi:10.1016/j.apenergy.2017.01.053

Byun, J., Hong, I., Kang, B., & Park, S. (2011). A smart energy distribution and management system for renewable energy distribution and context-aware services based on user patterns and load forecasting. *IEEE Transactions on Consumer Electronics*, *57*(2), 436–444. doi:10.1109/TCE.2011.5955177

Byun, J., & Park, S. (2011). Development of a self-adapting intelligent system for building energy saving and context-aware smart services. *IEEE Transactions on Consumer Electronics*, *57*(1), 90–98. doi:10.1109/TCE.2011.5735486

Campagna, B., & Frey, P. (2008). The Impact of Evolving LEED Standards on Historic Preservation Projects. *Journal of Green Building.*, *3*(4), 21–31. doi:10.3992/jgb.3.4.21

Capehart, B. L., & Brambley, M. R. (2014). *Automated Diagnostics and Analytics for Buildings*. Lilburn, GA: Fairmont Press.

Carson, R. (2009). *Silent spring*. Boston, MA: Houghton Mifflin.

CH2: Australian's Greenest Building. (2008, June 28). Retrieved from https://sydneyarchitecturearchive.wordpress.com/2008/06/28/ch2-australias-greenest-building/

Chan, E. H. W., Qian, Q., & Lam, P. (2009). The Market for Green Building in Developed Asian Cities – the Perspective of Building Designers. *Journal of Energy Policy, 37,* 3061-3070.

Chan, H. Y., Riffat, S. B., & Zhu, J. (2010). Review of passive solar heating and cooling technologies. *Renewable & Sustainable Energy Reviews*, *14*(2), 781–789. doi:10.1016/j.rser.2009.10.030

Chappell, T. W., & Corps, C. (2009). High performance green building: what's it worth. Lacey, WA: Washington State Department of Ecology.

Chasta, R., Singh, R., Gehlot, A., & et al, . (2016). A SMART Building Automation System. *International Journal of Smart Home*, *10*(9), 217–224. doi:10.14257/ijsh.2016.10.9.20

Chatterjee, A. K. (2009). Sustainable construction and green buildings on the foundation of building ecology. *Indian Concrete Journal*, *83*(5), 27–30.

Chel, A., Janssens, A., & Michel, D. (2015). Thermal performance of a nearly zero energy passive house integrated with the air-air heat exchanger and the earth–water heat exchanger. *Energy and Building*, *96*(2), 53–63. doi:10.1016/j.enbuild.2015.02.058

Chen, L., Nugent, C., & Okeyo, G. (2014). An Ontology-Based Hybrid Approach to Activity Modeling for Smart Homes. *IEEE Transactions on Human-Machine Systems*, *44*(1), 92–105. doi:10.1109/THMS.2013.2293714

Chen, X., Yang, H., & Lu, L. (2015). A comprehensive review on passive design approaches in green building rating tools. *Renewable & Sustainable Energy Reviews*, *50*, 1425–1436. doi:10.1016/j.rser.2015.06.003

Chen, Y., Paxson, V., & Katz, R. H. (2010). *What's New About Cloud Computing Security? (Rep. No. UCB/EECS-2010-5)*. Berkeley, CA: EECS Department University of California. Retrieved from https://www2.eecs.berkeley.edu/Pubs/TechRpts/2010/EECS-2010-5.html

Chihiro, S. (2010). *Will Green Buildings be Appropriately Valued by the Market?* Reitaku Institute of Political Economics and Social Studies, Working Report, No.40.

Cho, J., Kim, J., Lee, S., & Koo, J. (2016). A bi-directional systematic design approach to energy optimization for energy- efficient buildings. *Energy and Building*, *120*, 135–144. doi:10.1016/j.enbuild.2016.03.066

Chung, W., Hui, Y. V., & Lam, Y. M. (2006). Benchmarking the energy efficiency of commercial buildings. *Applied Energy*, *83*(1), 1–14. doi:10.1016/j.apenergy.2004.11.003

Churchill, E. F. (2012). From Data Divination to Data-Aware Design. *Interaction*, *19*(5), 10–13. doi:10.1145/2334184.2334188

CityMetric. (2015, April 15). *9 building materials made entirely from waste products.* Retrieved from https://www.citymetric.com/skylines/9-building-materials-made-entirely-waste-products-932

Clayman, S., Galis, A., & Mamatas, L. *(*2010*)*. Monitoring Virtual Networks with Lattice. In *Proceedings of the 2010 IEEE/IFIP Network Operations and Management Symposium Workshops*. Osaka, Japan: IEEE. 10.1109/NOMSW.2010.5486569

Clements-Croome, D. (2011). Sustainable intelligent buildings for people: A review. *Intelligent Buildings International*, *3*(2), 67–86. doi:10.1080/17508975.2013.786874

Conserve Energy Future. (2019). Top 10 sustainable construction technologies used in green construction. Retrieved from https://www.conserve-energy-future.com/top-sustainable-construction-technologies-used-green-construction.php

Cooperman, A., Dieckmann, J., & Brodrick, J. (2012, June). Control Systems and LEED, *ASHRAE Journal*, 96-99.

Coppola, L., Cerulli, T., & Salvioni, D. (2004). Sustainable development and durability of self-compacting concretes. *8th CANMET/ACI International Conference on Fly Ash, Silica Fume, Slag and Natural Pozzolans in Concrete,* 1, 29-50.

Creswell, J. W. (2009). *Research Design: Qualitative, Quantitative, and Mixed Methods Approaches* (3rd ed.). Los Angeles, CA: Sage.

Curry, E., O'Donnell, J., Corry, E, & . (2013). Linking building data in the cloud: Integrating cross-domain building data using linked data. *Advanced Engineering Informatics*, *27*(2), 206–219. doi:10.1016/j.aei.2012.10.003

Dagdougui, H., Minciardi, R., Ouammi, A., & et al, . (2012). Modeling and optimization of a hybrid system for the energy supply. *Energy Conversion and Management*, *64*, 351–363. doi:10.1016/j.enconman.2012.05.017

Dandekar, A., Kanse, R., Magar, S., Tavhare, S., & Katkar, A. (2017). IoT Based Green Building. *International Journal of Computer Science and Mobile Computing*, *6*(4), 439–443.

Danish, M. S. S., Zaheb, H., & Sabory, N. R. (2019). The Road Ahead for Municipal Solid Waste Management in the 21st Century: A Novel-standardized Simulated Paradigm. In *The 3rd International Conference on Energy and Environmental Science 2019*. Seoul, South Korea: IOP Conference Series: Earth and Environmental Science. (in press)

Danish, M. S. S. (2018). *A Managed Energy Framework for Least Developed Countries: Resilience to Energy Sustainability (PhD)*. Okinawa, Japan: University of the Ryukyus.

Danish, M. S. S., Sabory, N. R., & Ershad, A. M. (2016). The Least Developed Countries Needs for Changing the Passive Trend of Renewable Energy Exploitation to a Proactive Trend. *International Journal of Energy and Power Engineering*, *5*(6), 215–221. doi:10.11648/j.ijepe.20160506.17

Danish, M. S. S., Sabory, N. R., & Ershad, A. M. (2017). Sustainable Architecture and Urban Planning trough Exploitation of Renewable Energy. *International Journal of Sustainable and Green Energy*, *6*(3), 1–7. doi:10.11648/j.ijrse.s.2017060301.11

Danish, M. S. S., Senjyu, T., & Ibrahimi, A. M. (2019). A managed framework for energy-efficient building. *Journal of Building Engineering*, *21*, 120–128. doi:10.1016/j.jobe.2018.10.013

Danish, M. S. S., Senjyu, T., & Sabory, N. R. (2017). Afghanistan's aspirations for energy independence: Water resources and hydropower energy. *Renewable Energy*, *113*, 1276–1287. doi:10.1016/j.renene.2017.06.090

Data Center Knowledge. (n.d.). *Wind-Powered Data Centers*. Retrieved from https://www.datacenterknowledge.com/wind-powered-data-centers

David Orr. (2019). *Adam Joseph Lewis Center for Environmental Studies, Oberlin College*. Retrieved from http://www.mcdonoughpartners.com/projects/adam-joseph-lewis-center-for-environmental-studies-oberlin-college/

Davis, J., Cloutier, D., & Klein, J. (2010). Embedding Energy Efficiency in the Business of Buildings: Commercial Real Estate Contracts & Transactions. *ACEEE Summer Study on Energy Efficiency in Buildings*, *4*, 53–63.

Delsing, J. (2015). Automation Systems from IoT-Arrowhead Framework: concepts and basic architecture. *Embedded Conference Scandinavia.*

Department for Business, Energy & Industrial Strategy and Environment Agency. (2015, July 15). *CRC Energy Efficiency Scheme.* Retrieved from https://www.gov.uk/government/collections/crc-energy-efficiency-scheme#history

Devabhaktuni, V., Alam, M., & Depuru, S. S. S. R. (2013). Solar energy: Trends and enabling technologies. *Renewable & Sustainable Energy Reviews, 19*, 555–564. doi:10.1016/j.rser.2012.11.024

Diamond, R., Opitz, M., & Hicks, T. et al. (2006). *Evaluating the energy performance of the first generation of LEED-certified commercial buildings (No. LBNL-59853). Lawrence Berkeley National Lab.* Berkeley, CA: LBNL.

Ding, G. K. C. (2008). Maintainable Construction – The job of natural appraisal devices. *Diary of Environmental Management, 56*, 451–464.

Domone, P., & Illston, J. (2010). *Construction Materials: Their Nature and Behaviour* (4th ed.). London, UK: Spon Press.

Downton, P. (2013). Autoclaved aerated concrete. Retrieved from http://www.yourhome.gov.au/materials/autoclaved-aerated-concrete

DPTI, Government of South Australia. (2017). *Environmentally Sustainable Building Materials – Selection.* Retrieved from https://www.dpti.sa.gov.au/__data/assets/pdf_file/0009/293688/Environmentally_Sustainable_Building_Materials.pdf

Duverge, J. J., Rajagopalan, P., Fuller, R., & Woo, J. (2018). Energy and water benchmarks for aquatic centres in Victoria, Australia. *Energy and Building, 177*, 246–256. doi:10.1016/j.enbuild.2018.07.043

Ebeling, K. J., & Michalzik, R. (2017). VCSEL technology for imaging and sensor systems applications. *Proceedings of the 22nd Microoptics Conference (MOC2017)*, 19 – 22. 10.23919/MOC.2017.8244477

Ebeling, K. J., Michalzik, R., & Moench, H. (2018). Vertical-cavity surface-emitting laser technology applications with focus on sensors and three-dimensional imaging. *Journal of Applied Physics, 57*(8S2), 08PA02. doi:10.7567/JJAP.57.08PA02

Edwards, B. (1998). *Sustainable Architecture: European Directives and Building Design.* Oxford, UK: Architectural Press.

Ehlers, G. A., & Beaudet, J. (2006). *U.S. Patent No. 7,130,719.* Washington, DC: U.S. Patent and Trademark Office.

EIA. (2005). *Annual energy review 2004.* U.S. Department of Energy, Energy Information Administration. Retrieved from www.eia.doe.gov/emeu/aer/contents.html

Eichholtz, P., Kok, N., & Quigley, J. M. (2010). Doing well by doing good? Green office buildings. *The American Economic Review*, *100*(5), 2492–2509. doi:10.1257/aer.100.5.2492

Eichholtz, P., Kok, N., & Quigley, J. M. (2013). The economics of green building. *The Review of Economics and Statistics*, *95*(1), 50–63. doi:10.1162/REST_a_00291

Ejaz, W., Shah, G. A., Hasan, N., & Kim, H. S. (2015). Energy and throughput efficient cooperative spectrum sensing in cognitive radio sensor networks. *Emerging Telecommunications Technologies*, *26*(7), 1019–1030. doi:10.1002/ett.2803

Ellison, L., & Sayce, S. (2007). Assessing Sustainability in The Existing Commercial Property Stock Establishing Sustainability Criteria Relevant For The Commercial Property Investment Sector. *Journal of Property Management*, *25*(3), 287–304. doi:10.1108/02637470710753648

Engels, H., Pirkl, H. G., & Albers, R. (2013). Polyurethanes: Versatile Materials and Sustainable Problem Solvers for Todays Challenges. *Angewandte Chemie International Edition*, *52*(36), 9422–9441. doi:10.1002/anie.201302766 PMID:23893938

EPA. (n.d.). *Green Building.* Available at http://www.epa.gov/greenbuilding/pubs/about.htm

Erlandsson, M., & Borg, M. (2003). Generic LCA-methodology applicable for building, construction and operation services-today practice and development needs. *Building and Environment*, *38*(7), 919–938. doi:10.1016/S0360-1323(03)00031-3

Erlandsson, M., & Levin, P. (2005). Environmental Assessment of Rebuilding and Possible Performance Improvements Effect on a National Scale. *Building and Environment*, *39*, 1453–1465.

Etheridge, D. (2010). *Microsoft Excel Data Analysis* (3rd ed.). Indianapolis, IN: Wiley Publishing.

ETSI. (2011, October). *102 690: Machine-to-Machine communications (M2M). Functional architecture*. ETSI.

European Commission. (2003). EPD: The International EPD System. Retrieved from http://www.environdec.com/en/What-is-an-EPD

Evangeline, H. (2018, Jan. 4). *3D Sensing: Present and Future.* Retrieved from https://www.ledinside.com/news/2018/1/3d_sensing_present_and_future

Everett, R. (2008). The 'building colleges for the future' program. Delivering a green and intelligent building agenda. *New Review of Information Networking*, *14*(1), 3–20. doi:10.1080/13614570902953549

Fagin, R., & Wimmers, E. L. (2000). A formula for incorporating weights into scoring rules. *Theoretical Computer Science*, *239*(2), 309–338. doi:10.1016/S0304-3975(99)00224-8

Flouquet, F. (1992). Local weather correlations and bias in building parameter estimates from energy-signature models. *Energy and Building*, *19*(2), 113–123. doi:10.1016/0378-7788(92)90005-2

Forrest, W., Kaplan, J. M., & Kindler, N. (2008). *Data centers: How to cut carbon emissions and costs*. Retrieved from McKinsey & Company Website: ftp://public.dhe.ibm.com/software/uk/itsolutions/optimiseit/greenhub/carbon-managment/pov-mckinsey-report.pdf

Fowler, K. M., Rauch, E. M., Henderson, J. W., & Kora, A. R. (2010). *Re-assessing green building performance: A post occupancy evaluation of 22 GSA buildings (No. PNNL-19369). Pacific Northwest National Lab*. Richland, WA: PNNL. doi:10.2172/1029438

Frame, D., Tembo, K., & Dolan, M. J. (2011). *A community-based approach for sustainable off-grid PV systems in developing countries. In 2011 IEEE Power and Energy Society General Meeting* (pp. 1–7). Detroit, MI: IEEE. doi:10.1109/PES.2011.6039593

Frini, A. (2014). A multi-criteria multi-period approach for energy project selection in sustainable development context. In *2014 First International Conference on Green Energy ICGE 2014* (pp. 65–71). Sfax, Tunisia: IEEE. 10.1109/ICGE.2014.6835399

Fuerst, F. (2009). Building momentum: An analysis of investment trends in LEED and Energy Star-certified properties. *Journal of Retail & Leisure Property*, *8*(4), 285–297. doi:10.1057/rlp.2009.18

Fuerst, F., & McAllister, P. (2011). Green noise or green value? Measuring the effects of environmental certification on office values. *Real Estate Economics*, *39*(1), 45–69. doi:10.1111/j.1540-6229.2010.00286.x

Gabbai, A. (2015, Jan. 2). Kevin Ashton describes 'the Internet of Things': The innovator weighs in on what human life will be like a century from now. *Smithsonian Magazine,* 8-15. Retrieved from https://www.smithsonianmag.com/innovation/kevin-ashton-describes-the-internet-of-things-180953749/

Galvan, J. L. (2009). *Composing Literature Reviews: A Guide for Students of the Social and Behavioral Sciences* (4th ed.). Los Angeles, CA: Pyrczak Publishing.

Ganji, F., Budzisz, Ł., & Debele, F. G. et al. (2015). Greening campus WLANs: Energy-relevant usage and mobility patterns. *Computer Networks*, *78*, 164–181. doi:10.1016/j.comnet.2014.10.033

Garrido, M., Paulo, P., & Branco, F. (2012). Service life prediction of façade paint coating in old buildings. *Construction & Building Materials*, *29*, 394–402. doi:10.1016/j.conbuildmat.2011.10.057

Gershenfeld, N., Samouhos, S., & Nordman, B. (2010). Canny Infrastructure for Energy Efficiency. *Science Magazine.*, *327*, 1086–1088.

Gesture recognition and 3D sensing. (2019). Retrieved from https://www.jabil.com/capabilities/3d-sensing-gesture-recognition.html

GhaffarianHoseini, A. H., Dahlan, N. D., & Berardi, U. (2013). Sustainable energy performances of green buildings: A review of current theories, implementations and challenges. *Renewable & Sustainable Energy Reviews*, *25*, 1–17. doi:10.1016/j.rser.2013.01.010

Go Smart Bricks. (2017, April 12). *Rain Water Harvesting – Why, Benefits, Techniques & More*. Retrieved from http://gosmartbricks.com/rain-water-harvesting-methods-in-india/

GoSmartBricks. (2018, April 13). *Top 6 Challenges Hindering Rapid Adoption of Green Building Practices In India*. Retrieved from http://gosmartbricks.com/top-6-challenges-hindering-rapid-adoption-of-green-building-practices-in-india/

Gou, Z., Prasad, D., & Lau, S. S.-Y. (2014). Impacts of green certifications, ventilation and office types on occupant satisfaction with indoor environmental quality. *Architectural Science Review*, *57*(3), 196–206. doi:10.1080/00038628.2014.908113

Grainger. (2019). *Focus on Three Components for Effective Sustainable Building Design*. Retrieved from: https://www.grainger.com/content/supplylink-sustainable-building-design

Granjal, J., Monteiro, E., & Silva, S. J. (2015). Security for the Internet of Things: A Survey of Existing Protocols and Open Research Issues. IEEE Communication Surveys & Tutorials, 17(3).

Gray, J. (2017). Environmentally Friendly, Non-Toxic Paint. Retrieved from http://www.sustainablebuild.co.uk/nontoxicpaint.html

Green Building Brain. (2009). *Oberlin College Lewis Center*. Retrieved from https://greenbuildingbrain.org/buildings/oberlin_college_lewis_center

Gu, W., Wang, J., & Lu, S. (2017). Optimal operation for integrated energy system considering thermal inertia of district heating network and buildings. *Applied Energy, 199,* 234-246.

Gubbi, J., Buyya, R., Marusic, S., & Palaniswami, M. (2013). Internet of Things (IoT): A vision, architectural elements, and future directions. *Future Generation Computer Systems*, *29*(7), 1645–1660. doi:10.1016/j.future.2013.01.010

Gupta, S., Reynolds, M. S., & Patel, S. N. (2010). Electrisense: Single-Point Sensing Using EMI for Electrical Event Detection and Classification in the Home. *Proc. 12th ACM Int'l Conf. Ubiquitous Computing*, 139–148. 10.1145/1864349.1864375

Gustavsson, L., & Sathre, R. (2006). Variability in energy and carbon dioxide balances of wood and concrete building materials. *Building and Environment*, *41*(7), 940–951.

Haglund, E. (2015). VCSELs for high-speed, long-reach, and wavelength-multiplexed optical interconnects. Gothenburg, Sweden: Chalmers University of Technology.

Hazarika, B., & Singh, T.J. (2015). Survey Paper on Cloud Computing & Cloud Monitoring: Basics. *SSRG International Journal of Computer Science and Engineering, 2*(1), 10-15.

Hernandez, P., & Kenny, P. (2009). From net vitality to zero vitality structures: Defining life cycle zero vitality structures (LC-ZEB). *Energy Policy*, *42*, 815–821.

Hernandez, P., & Kenny, P. (2011). Advancement of a Methodology forever cycle building vitality appraisals. *Energy Policy*, *39*, 3779–3788. doi:10.1016/j.enpol.2011.04.006

Hernández-Ramos, J. L., Moreno, M. V., Bernabé, J. B., & et al, . (2015). SAFIR: Secure access framework for IoT-enabled services on smart buildings. *Journal of Computer and System Sciences*, *81*(8), 1452–1463. doi:10.1016/j.jcss.2014.12.021

He, Y., Kvan, T., Liu, M., & Li, B. (2018). How green building rating systems affect designing green. *Building and Environment*, *133*, 19–31. doi:10.1016/j.buildenv.2018.02.007

Hossain, M. F. (2018). Green science: Advanced building design technology to mitigate energy and environment. *Renewable & Sustainable Energy Reviews*, *81*, 3051–3060. doi:10.1016/j.rser.2017.08.064

Hossain, M. F. (2019). Sustainable technology for energy and environmental benign building design. *Journal of Building Engineering*, *22*, 130–139. doi:10.1016/j.jobe.2018.12.001

Howard, J. L. (2003). *The federal commitment to green building: experiences and expectations*. Washington, DC: Federal Executive, Office Of The Federal Environmental Executive.

Howe, J. C. (2010). Overview of Green Buildings. *National Wetlands Newsletter*, *33*(1), 3–14.

Hui, S. C. (2001). *HVAC Design and Operation for Green Buildings.* Shaanxi-Hong Kong Refrigeration and HVAC Seminar, Xian, China.

Hui, S. C. (2007, September). *Sustainable building technologies for hot and humid climates.* Hangzhou, China: Joint Hong Kong and Hangzhou Seminar for Sustainable Building.

Hunt, W. D. (2008). *Literature Review of Data on the Incremental Costs to Design and Build Low-Energy Buildings (No. PNNL-17502). Pacific Northwest National Lab.* Richland, WA: PNNL. doi:10.2172/935614

Huovila, P. (2007). *Buildings and climate change: status, challenges, and opportunities*. UNEP/Earthprint. Available at http://smap.ew.eea.europa.eu/media_server/files/ R/S/UNEP_Buildings_and_climate_change.pdf

Huo, X., Yu, A. T. W., & Wu, Z. (2018). An empirical study of the variables affecting site planning and design in green buildings. *Journal of Cleaner Production*, *175*, 314–323. doi:10.1016/j.jclepro.2017.12.091

Independence, E. (2007). Security act of 2007. *Public Law*, *110*(140), 19.

Indira Paryavaran Bhawan. (2011). Retrieved from http://www.moef.nic.in/ipb/home.htm

Infinity. (2016). *Infinity Benchmark*. Retrieved from http://www.infinityitpark.com/project/infinity-benchmark/

Inhabitat. (2011, Dec. 15). *Interview: Gensler's Chris Chan on the Sustainable Shanghai Tower, Asia's Tallest Skyscraper*. Retrieved from https://inhabitat.com/interview-genslers-chris-chan-on-the-sustainable-shanghai-tower-asias-tallest-skyscraper/shanghai-tower-7/

Intel IoT Smart Building. (2015). *Controlling Office Temperature with Building Robotics.* Intel Corporation. Retrieved from https://www.intel.in/content/www/in/en/internet-of-things/white-papers/iot-smart-office-building-robotics-brief.html

International Atomic Energy Agency. (2005). Energy Indicators for Sustainable Development: Guidelines and Methodologies. Vienna, Austria: IAEA.

International Energy Association. (n.d.). Retrieved from http://www.iea.org/textbase/nppdf/free/2000/Renew_main2003.pdf

International Organization for Standardization, ISO 15686-1. (2011). *Buildings and Constructed Assets. Service Life Planning. Part 1: General Principles and Framework.* Switzerland.

International Organization for Standardization. ISO 14025: 2006. (2006). Environmental labels and declarations -- Type III environmental declarations - Principles and procedures. Brussels, Belgium: CEN (European Committee for Standardisation).

International Organization for Standardization. ISO 14040: 2006. (2006). Environmental Management - life cycle assessment - principles and framework. Brussels, Belgiuim: CEN (European Committee for Standardisation).

International Organization for Standardization. ISO 14044: 2006. (2006). Environmental Management - life cycle assessment – requirements and guidelines. Brussels, Belgium: CEN (European Committee for Standardisation).

Jadhav, N.Y. (2016). *Green and Smart Building Trends.* Green Energy and Technology. Springer Science+Business Media Singapore.

Jadhav, N. Y. (2016). *Green and Smart Buildings: Advanced Technology Options.* Springer. doi:10.1007/978-981-10-1002-6

Janda, K. B. (2009). Worldwide status of energy standards for buildings: a 2009 update. *Proceedings of the ECEEE Summer Study*, 1-6.

Jeffs, E. (2017). Green Energy: Sustainable Electricity Supply with Low Environmental Impact (1st ed.). Boca Raton, FL: CRC Press.

Jensen, P. A., Maslesa, E., Berg, J. B., & Thuesen, C. (2018). 10 questions concerning sustainable building renovation. *Building and Environment*, *143*, 130–137. doi:10.1016/j.buildenv.2018.06.051

Jiang, X., Ly, M. V., & Taneja, J. (2009). Experiences with a high-fidelity wireless building energy auditing network. *Proc. 7th ACM Conf. Embedded Netw. Sensor Syst. (SenSys'09),* 13–126. 10.1145/1644038.1644050

Ji, J., Yi, H., He, W., & Pei, G. (2007). PV-Trombe wall design for buildings in composite climates. *Journal of Solar Energy Engineering*, *129*(4), 431–437. doi:10.1115/1.2770751

Jonathan, A. W., Justin, A. B., & Ken, H. J. (2008). Green Design and the Market for Commercial Office Space. *The Journal of Real Estate Finance and Economics*, *41*, 228–243.

JrB. D. (2019, Jan. 15). *Arabian Business*. Retrieved from https://www.arabianbusiness.com/construction/411344-zaha-hadid-designed-beeah-hq-to-be-ready-by-summer

Jung, G., Ganasambandam, N., & Mukerjee, T. (2012). Synchronous Parallel Processing of Big-Data Analytics Services to Optimize Performance in Federated Clouds. In *Proceedings of the 2012 IEEE Fifth International Conference on Cloud Computing*. Honolulu, HI: IEEE. 10.1109/CLOUD.2012.108

Kalogirou, S., Tripanagnostopoulos, Y., & Souliotis, M. (2005). Performance of solar systems employing collectors with colored absorber. *Energy and Building*, *37*(8), 824–835. doi:10.1016/j.enbuild.2004.10.011

Kamal, M. A. (2016). *Autoclaved Aerated Concrete (AAC): A Sustainable Building Material.* Retrieved from: https://www.masterbuilder.co.in/autoclaved-aerated-concrete-aaca-sustainable building-material-2/

Kandpal, T. C., & Garg, H. P. (2003). Financial evaluation of renewable energy technologies. Macmillan India Ltd.

Kats, G. (2003). *Green building costs and financial benefits*. Boston, MA: Massachusetts Technology Collaborative.

Katz, D., & Skopek, J. (2009). The CABA building intelligence quotient programme. *Intelligent Buildings International*, *1*(4), 277–295. doi:10.3763/inbi.2009.SI05

Kerdan, I. G., Raslan, R., & Ruyssevelt, P. (2016). An energy-based multiobjective optimisation model for energy retrofit strategies in nondomestic buildings. *Energy*, *117*, 506–522. doi:10.1016/j.energy.2016.06.041

Kibert, C. J. (2008). *Sustainable construction: green building design and delivery*. Hoboken, NJ: John Wiley and Sons.

Klein, S. J. W., & Coffey, S. (2016). Building a sustainable energy future, one community at a time. *Renewable & Sustainable Energy Reviews*, *60*, 867–880. doi:10.1016/j.rser.2016.01.129

Komal, C. (2018). *4 Reasons to Use Custom ASICs in the IoT Era.* Retrieved from https://www.dzone.com

Kotaji, S., & Loebel, O. (2010). Sustainability of Polyurethane Thermal Insulation – Performance Assessment at Building and Building Component Level. In *Central European Sustainable Building Conference*. Prague, Czech Republic. Sustainable Building Society.

Koyama, F., & Nakahama, M. (2015). Micromachined Vertical Cavity Surface Emitting Lasers athermalization, tuning, and multi wavelength Integration. IEEE Journal of Selected Topics in Quant. Electron., 21(4).

Koyama, F. (2009). VCSEL Photonics-advances and new challenges. *IEICE Electronics Express*, *6*(11), 651–672. doi:10.1587/elex.6.651

Kubba, S. (2009). LEED practices, certification, and accreditation handbook. Oxford, UK: Butterworth-Heinemann.

Kumar, S., & Buyya, R. (2012). Green Cloud Computing and Environmental Sustainability. In S. Murugesan & G. R. Gangadharan (Eds.), *Harnessing Green It: Principles and Practices* (pp. 315–339). IEEE. doi:10.1002/9781118305393.ch16

Kundur, P. (2004). Sustainable electric power systems in the 21st century: requirements, challenges and the role of new technologies. In *IEEE Power Engineering Society General Meeting, 2004* (vol. 2, pp. 2298–2299). Denver, CO: IEEE. doi:10.1109/PES.2004.1373295

Lai, C., Lu, R., & Zheng, D. (2015). Toward Secure Large-Scale Machine-to-Machine Communications in 3GPP Networks: Challenges and Solutions. *IEEE Communications Magazine Communications Standards Supplement.*

Lambert, P. (2016). Sustainability of metals and alloys in construction. In J. Khatib (Ed.), *Sustainability of Construction Materials* (2nd ed.; pp. 105–127). Sawston, UK: Woodhead Publishing. doi:10.1016/B978-0-08-100370-1.00006-8

Lambrechts, W., Gelderman, C. J., Semeijn, J., & Verhoeven, E. (2019). The role of individual sustainability competences in eco-design building projects. *Journal of Cleaner Production*, *208*, 1631–1641. doi:10.1016/j.jclepro.2018.10.084

Lawrence, T. M., Watson, R. T., Boudreau, M., & Mohammadpour, J. (2017). Data flow requirements for integrating smart Buildings and a smart grid through model predictive Control. *Procedia Engineering*, *180*, 1402–1412. doi:10.1016/j.proeng.2017.04.303

Lazarova-Molnar, S., Shaker, H. R., Mohamed, N., & Jørgensen, B. N. (2016). Fault Detection and Diagnosis for Smart Buildings: Trends and Challenges. *Proc. the 3rd MEC Int'l Conference on Big Data and Smart City (ICBDSC 2016).*

Le, K., Bianchini, R., Nguyen, T. D., Bilgir, O., & Martonosi, M. (2010). Capping the brown energy consumption of Internet services at low cost. In *Proceedings of theInternational Conference on Green Computing*. Chicago, IL: IEEE. 10.1109/GREENCOMP.2010.5598305

Leadbitter, J. (2002). PVC and sustainability. *Progress in Polymer Science*, *27*(10), 2197–2226.

Lebovits, S. C. (2010). *Geothermal Building*. Retrieved from: http://www.bu.edu/sustainability/what-were-doing/green-buildings/geothermal/

Leroy, C. (2009). Provision of LCI data in the European aluminium industry methods and examples. *The International Journal of Life Cycle Assessment*, *14*, 10–44.

Li, A., Yang, X., Kandula, S., & Zhang, M. (2010). CloudCmp: Comparing Public Cloud Providers. In Proceedings of the 2010 Internet Measurement Conference. Melbourne, Australia: Association for Computing Machinery, Inc.

Li, H., Wolf, P., & Moser, P. (2014). Impact of the quantum well gain-to-cavity etalon wavelength offset on the high temperature performance of high bit rate 980-nm VCSEL. *IEEE Journal of Selected Topics in Quantum Electronics, 50*(8), 613-612. Retrieved from ieeexplore.ieee.org/xpl/RecentIssue.jsp?punumber=2944

Li, H., Wolf, P., & Moser, P. (2015). Temperature-Stable, Energy-Efficient, and High-Bit rate Oxide-Confined 980-nm VCSELs for Optical interconnects. *IEEE Journal of Selected Topics in Quantum Electronics, 21*(6). Retrieved from ieeexplore.ieee.org/xpl/RecentIssue.jsp?punumber=2944

Li, H., Wolf, P., & Moser, P. (2017). Large bandwidth, Small Current Density, and Temperature stable 980-nm VCSELs. *IEEE Journal of Selected Topics in Quantum Electronics, 53*(6), 2400908. Retrieved from ieeexplore.ieee.org/xpl/RecentIssue.jsp?punumber=2944

Li, W., Lu, Y., & Sun, J. (2017, December). People counting based on improved gauss process regression. In *Security, Pattern Analysis, and Cybernetics (SPAC), 2017 International Conference on* (pp. 603-608). Shenzhen, China. IEEE.

Liberali, R. (2007). Renewable Energy Technologies: Long Term Research in the 6th Framework Programme 2002–2006. Project Synopses, Community Research, European Commission.

Li, D. H., Yang, L., & Lam, J. C. (2013). Zero energy buildings and sustainable development implications – A review. *Energy, 54*, 1–10. doi:10.1016/j.energy.2013.01.070

Lilis, G., Conus, G., Asadi, N., & Kayal, M. (2017). Towards the next generation of intelligent building: An assessment study of current automation and future IoT based systems with a proposal for transitional design. *Sustainable Cities and Society, 28*, 473–481. doi:10.1016/j.scs.2016.08.019

Lindner, M., Suominen, M., & Palosuo, T. (2010). ToSIA—A tool for sustainability impact assessment of forest-wood-chains. *Ecological Modelling, 221*(18), 2197–2205. doi:10.1016/j.ecolmodel.2009.08.006

Liners, C. (2007). Application of product differential semantics to quantify purchaser perceptions in housing assessment. *Building and Environment, 42*(7), 2488–2497. doi:10.1016/j.buildenv.2006.06.012

Liu, G., & Müller, D. B. (2012). Addressing sustainability in the aluminium industry: A critical review of life cycle assessments. *Journal of Cleaner Production, 35*, 108–117.

Liu, L., Kuo, S. M., & Zhou, M. (2009). Virtual sensing techniques and their applications. *IEEE International Conference on Networking, Sensing and Control*, 31-36.

Liu, M., & Mi, B. (2017). Life cycle cost analysis of energy-efficient buildings subjected to earthquakes. *Energy and Building, 154*, 581–589. doi:10.1016/j.enbuild.2017.08.056

Li, Z. (2006). A new life cycle impact assessment approach for buildings. *Building and Environment, 41*, 1414–1422.

Loeser, S., & Post, J. (2011, April 5). *A Model Energy Aligned Lease.* Retrieved from www.nyc.gov.

Love, P. E., Niedzweicki, M., Bullen, P. A., & Edwards, D. J. (2012). Achieving the Green Building Council of Australia's World Leadership Rating in an Office Building in Perth. *Journal of Construction Engineering and Management, 138*(5), 652–660. doi:10.1061/(ASCE)CO.1943-7862.0000461

Lu, C.-H., Ho, Y.-C., Chen, Y.-H., & Fu, L.-C. (2013). Hybrid User-Assisted Incremental Model Adaptation for Activity Recognition in a Dynamic Smart-Home Environment. *IEEE Transactions on Human-Machine Systems, 43*(5), 421–436. doi:10.1109/TSMC.2013.2281586

Lynch, P. (2017, Dec. 19). *Zaha Hadid Architects' Bee'ah Headquarters Tops Out in UAE.* Retrieved from https://www.archdaily.com/885744/zaha-hadid-architects-beeah-headquarters-tops-out-in-uae

Lzkendorf, T. & Lorenz, D. (2005). *Sustainable Property Investment: Valuing Sustainable Buildings Through Property Performance Assessment, Building research and information.* Academic Press.

Mahdavinejad, M. S., Rezvan, M., & Barekatain, M. (2018). Machine learning for internet of things data analysis: A survey. *Digital Communications and Networks, 4*(3), 161–175. doi:10.1016/j.dcan.2017.10.002

Maine, M. (2010). *Green Leases: Operating Expense Considerations & Implications.* American Bar Association's Section of Real Property, Trust and Estate Law Spring Symposia.

Maiona, M. (2010). *Green Leases: Operating Expense Considerations and Implications.* American Bar Association's Section of Real Property, Trust and Estate Law Spring Symposia.

Malangone, C. (2012). *Better Metrics of Green Building Needed, According to New McGraw Hill Construction SmartMarket Executive Brief on Decision Making for Green Building Investments.* McGraw Hill Construction.

Malhotra, V. M. (2004). Role of supplementary cementing materials and superplasticizers in reducing greenhouse gas emissions. In *ICFRC International Conference on Fibre Composites, High-Performance Concrete, and Smart Materials* (pp. 489-499). Indian Institute of Technology.

Manic, M., Wijayasekara, D., Amarasinghe, K., & Rodriguez-Andina, J. J. (2016). Building energy management systems: The age of intelligent and adaptive buildings. *IEEE Industrial Electronics Magazine, 10*(1), 25–39. doi:10.1109/MIE.2015.2513749

Mansfield, J. (2009). The Valuation of Sustainable Freehold Property: ACRE Perspective. *Journal of Corporate Real Estate, 11*(2), 91–105. doi:10.1108/14630010910963133

Marciniak, M., Piskorski, L., & Gebski, M. (2018). The Vertical-Cavity Surface-Emitting Laser as a Sensing Device. *Journal of Lightwave Technology, 36*(16), 3185–3192. doi:10.1109/JLT.2018.2834620

Marinakis, V., & Doukas, H. (2018). *An Advanced IoT-based System for Intelligent Energy Management in Buildings*. Athens, Greece: School of Electrical & Computer Engineering, National Technical University of Athens. doi:10.339018020610

Marteinsson, B. (2003). Durability and the factor method of ISO 15686-1. *Building Research and Information*, *31*(6), 416–426.

Martinaitisa, V., & Kazakeviciu, E. (2007). A Two-Factor Method for Appraising Building Renovation and Energy Efficiency Improvement Projects. *Energy Policy*, *35*(1), 192–201. doi:10.1016/j.enpol.2005.11.003

Mathew, S. S., Atif, Y., Sheng, Q. Z., & Maamar, Z. (2013). The Web of Things - Challenges and Enabling Technologies. In *Internet of Things and Inter-cooperative Computational Technologies for Collective Intelligence* (pp. 1–23). Berlin, Germany: Springer. doi:10.1007/978-3-642-34952-2_1

Matthiessen, L., Morris, P., & Langdon, D. (2007). Cost of Green Revisited: Reexamining the Feasibility and Cost Impact of Sustainable Design in Light of increased Market Adoption. *Davis Langdon Research Journal.*

McKenzie, M. (1978). The corrosion performance of weathering steel in highways bridges: TRRL Laboratory report 857. Wokingham, UK: Transport Research Laboratory.

McNeil, M. A., & Rue du Can, S. (2016). Energy Efficiency Indicators and Impact Metrics (No. LBNL-1005695). Lawrence Berkeley National Laboratory.

Melexis. (2017). *Time of Flight Basics.* Retrieved from https://www.melexis.com/

Meryman, H., & Silman, R. (2004). Sustainable engineering - using specifications to make it happen. *Structural Engineering International, 14*(3), 216-219.

Michaels, J. (2012). *Commercial Buildings Energy Consumption Survey: Energy Usage Summary*. Academic Press.

Miles, M. P., & Covin, J. G. (2000). Natural Marketing: A Source of Reputational, Competitive and Financial Advantage. *Diary of Business Ethics*, *23*(3), 299–311. doi:10.1023/A:1006214509281

Miller, N., Pogue, D., Saville, J., & Tu, C. (2010) The Operations and Management of Green Buildings in the United States. JOSRE, 2(1), 51-66.

Miller, N., Pogue, D., Saville, J., & Tu, C. (2010). The Operations & Management of Green Buildings in the United States. *JOSRE, 1,* 2.

Miller, R. (2014, October 14). *Super-Sizing Solar Power for Data Centers*. Retrieved from https://www.datacenterknowledge.com/archives/2014/10/14/super-sizing-solar-power-for-data-centersl

Miller, E., & Buys, L. (2008). Retrofitting Commercial Office Buildings for Sustainability: Tenants' Perspectives. *Journal of Property Investment & Finance*, *26*(6), 552–561. doi:10.1108/14635780810908398

Miller, N., Pogue, D., & Gough, Q., & Davis, S. (2009). Green Building and Productivity. *Diary of Sustainable Real Estate*, *1*, 65–89.

Miller, N., Pogue, D., & Gough, Q., & Davis, S. (2009). Green Building and Productivity. *Journal of Sustainable Real Estate*, *1*, 65–89.

Mills, D. (2001). *When to Consider Distributed Systems in an Urban and Suburban Context. Water Environment Research Foundation (WERF), Casestudy: Philip Merrill environmental center building*. Chesapeake Bay Foundation.

Minoli, D., Sohraby, K., & Kouns, J. (2017). IoT Security (IoTSec) Considerations, Requirements, and Architectures. *14th Annual IEEE Consumer Communications & Networking Conference (CCNC 2017)*.

Minoli, D., Sohraby, K., & Occhiogrosso, B. (2017). IoT considerations, requirements, and architectures for smart buildings—Energy optimization and next-generation building management systems. *IEEE Internet of Things Journal*, *4*(1), 269–283.

Moench, H., Carpaij, M., & Gerlach, P. (2016, March). VCSEL-based sensors for distance and velocity. In *Vertical-Cavity Surface-Emitting Lasers*. International Society for Optics and Photonics.

Moench, H., Frey, M., & Grabherr, M. (2017, February). VCSELs as light source for time-of-flight sensors. In Vertical-Cavity Surface-Emitting Lasers. International Society for Optics and Photonics.

Mohamed, A. (2018). *A history of cloud computing*. Retrieved from https://www.computerweekly.com/feature/A-history-of-cloud-computing

Mohanty, S., Skandhaprasaad, A. L., & Samal, S. S. (2010). *Green technology in construction. In Recent Advances in Space Technology Services and Climate Change 2010 (RSTS & CC-2010)*. IEEE.

Moser, P. (2015). *Energy-Efficient Oxide-Confined VCSELs for Optical Interconnects in Data centers and Supercomputers*. Berlin, Germany: Academic Press.

Moser, P., Lott, J.A., & Bimberg, D. (2013). Energy efficiency of directly modulated oxide-confined high bit rate 850-nm VCSELs for optical interconnects. *IEEE Journal of Selected Topics in Quantum Electronics, 19*(4). Retrieved from ieeexplore.ieee.org/xpl/RecentIssue.jsp?punumber=2944Journal

Motegi, N., Piette, M. A., Kinney, S., & Herter, K. (2003). Web-based energy information systems for energy management and demand response in commercial buildings. University of California: Lawrence Berkeley National Laboratory. doi:10.2172/823939

Murakami, S., (2003). Study on Comprehensive Assessment System for Building Environmental Efficiency Part 1-9. *Summaries of Technical Papers of Annual Meeting Architectural Institute of Japan*, 993-1006.

Myers, G., Reed, R. & Robinson, J. (2007, January). The relationship between sustainability and the value of office buildings. In *Proceedings of the 13th Annual Pacific Rim Real Estate Conference*. (pp. 1-22). Pacific Rim Real Estate Society.

Myers, G., Reed, R., & Robinson, J. (2007). Sustainable property–the future of the New Zealand market. Pacific Rim Property Research Journal, 14(3).

Naik, T. R. (2008). Sustainability of Concrete Construction. *Practice Periodical on Structural Design and Construction*, *13*(2), 98–103. doi:10.1061/(ASCE)1084-0680(2008)13:2(98)

Napier T. / Whole Building Design Guide. (2016). Construction Waste Management. Retrieved from https://www.wbdg.org/resources/construction-waste-management

Nationwide Construction. (2016, Feb. 26). *Eco-friendly Construction: 8 Advantages of Green Building*. Retrieved from https://nationwideconstruction.us/eco-friendly-construction-8-advantages-of-green-building/

Nayyar, A. (2011). Interoperability of cloud computing with web services. *International Journal of ElectroComputational World & Knowledge Interface*, *1*(1).

Nayyar, A. (2011). Private Virtual Infrastructure (PVI) Model for Cloud Computing. *International Journal of Software Engineering Research and Practices*, *1*(1), 10–14.

Nayyar, A. (2011). Private Virtual Infrastructure (PVI) Model for Cloud Computing. *International Journal Software Engineering Research and Practices*, *1*(1), 10–14.

Negi, G. P., Kapoor, H., & Kapoor, A. (2015). Green Computing: A demand of Time. *International Journal of Engineering Science*, *14*(05), 34–36.

New Building Institution. (2010, Jan. 1). Retrieved from http://www.newbuildings.org.

Nichols, M. (2016, April 22). *Green building challenges*. Retrieved from https://usgreentechnology.com/green-building-challenges

Nimje, A. R., Gaikwad, V. T., & Datir, H. N. (2013). Green Cloud Computing: A Virtualized Security Framework for Green Cloud Computing. *International Journal of Advanced Research in Computer Science and Software Engineering*, *3*(04), 642–646.

Norgate, T. E., Jahanshahi, S., & Rankin, W. J. (2007). Assessing the environmental impact of metal production processes. *Journal of Cleaner Production*, *15*(8-9), 838–848. doi:10.1016/j.jclepro.2006.06.018

Nunen, N., Hendriks, A., & Erkelens, P. (2004). Service Life as Main Aspect in Environmental Assessment. In *CIB Congress, editor. Building for the Future: The 16th CIB World Building Congress* (vol. 1, pp. 10). Toronto, Canada: Academic Press.

Ofori, G., & Kien, H. L. (2004). Translating Singapore architects' environmental awareness into decision making. *Building Research and Information*, *32*(1), 27–37. doi:10.1080/09613210210132928

Olivieri, G., Romani, A., & Neri, P. (2006). Environmental and economic analysis of aluminium recycling through life cycle assessment. *International Journal of Sustainable Development and World Ecology*, *13*(4), 269–276.

Osso, A., Walsh, T., & Gottfried, D. (1998). *Sustainable Building Technical Manual*. New York: Public Technology Inc.

Pachauri, R. K., & Reisinger, A. (2008). [Synthesis report. Contribution of Working Groups I, II and III to the fourth assessment report.]. *Climatic Change*, 2007.

Padmanabhan, G. (2018). *Methods of Rainwater Harvesting - Components, Transport and Storage*. Retrieved from https://theconstructor.org/water-resources/methods-of-rainwater-harvesting/5420/

Pan, J. (2013). Toward an energy-proportional building prospect: Evaluation and analysis of the energy consumption in a green building testbed. *Proc. IEEE Energy Tech*, 1–6. 10.1109/EnergyTech.2013.6645358

Pan, J., Jain, R., & Paul, S. (2015). An internet of things framework for smart energy in buildings: Designs, prototype, and experiments. *IEEE Internet of Things Journal*, *2*(6), 527–537. doi:10.1109/JIOT.2015.2413397

Papanek, V. (1995). *The Green Imperative Ecology and Ethics in Design and Architecture*. Singapore: Thames and Hudson.

Papantoniou, S., Kolokotsa, D., & Kalaitzakis, K. (2015). *Building optimization and control algorithms implemented in existing BEMS using a web-based energy management and control system. Energy and Buildings, 98,* 45-55.

Patel, Y. S., Mehrotra, N., & Soner, S. (2015). Green cloud computing: A review on Green IT areas for cloud computing environment. In *Proceedings of the 2015 International Conference on Futuristic Trends on Computational Analysis and Knowledge Management (ABLAZE)*. Noida: IEEE. 10.1109/ABLAZE.2015.7155006

Patterson, M. G. (1996). What is energy efficiency? Concepts, indicators and methodological issues. *Energy Policy*, *25*(5), 377–390. doi:10.1016/0301-4215(96)00017-1

Paul, S., Jain, R., Pan, J., & Bowman, M. (2008). A vision of the next generation internet: A policy-oriented perspective. *Proc. Brit. Comput. Soc. (BCS) Int. Conf. Vis. Comput. Sci.*, 1–14.

Paul, D., Chakraborty, T., Datta, S. K., & Paul, D. (2018). IoT and Machine Learning Based Prediction of Smart Building Indoor Temperature. In *4th International Conference on Computer and Information Sciences (ICCOINS)*. Kuala Lumpur, Malaysia: IEEE. 10.1109/ICCOINS.2018.8510597

Pearson, D. (1996). *The Natural House Catalog: Where to Get Everything You Need to Create an Environmentally Friendly Home*. New York, NY: Simon & Schuster.

Pearson, D. (1998). *The New Natural House Book: Creating a Healthy, Harmonious, and Ecologically Sound Home*. New York, NY: Simon & Schuster.

Perlova, E., Platonova, M., Gorshkov, A., & Rakova, X. (2015). Concept project of zero energy building. *Procedia Engineering*, *100*, 1505–1514. doi:10.1016/j.proeng.2015.01.522

Piorno, J. R., Bergonzini, C., Atienza, D., & Rosing, T. S. (2009, May). Prediction and management in energy harvested wireless sensor nodes. In *2009 1st International Conference on Wireless Communication, Vehicular Technology, Information Theory and Aerospace & Electronic Systems Technology* (pp. 6-10). IEEE.

Plageras, A., Psannis, K. E., Gupta, B., et al. (2017). Solutions for Inter-connectivity and Security in a Smart Hospital Building. In *IEEE 15th International Conference on Industrial Informatics (INDIN)*. Emden, Germany: IEEE.

Plant Connection. (2019). *Benefits of G-O2® Living Walls*. Retrieved from http://myplantconnection.com/green-wall-benefits.php

Popovic, T., Barbosa-Póvoa, A., Kraslawski, A., & Carvalho, A. (2018). Quantitative indicators for social sustainability assessment of supply chains. *Journal of Cleaner Production*, *180*, 748–768. doi:10.1016/j.jclepro.2018.01.142

Powell, J. K. (2016). *Sensors to make your home smarter and safer.* Retrieved from https://www.forbes.com/sites/jenniferhicks/2016/11/20/3d-sensors-to-make-your-home-smarter-and-safer/#2

Pramanik, P. K. D., & Choudhury, P. (2018). IoT Data Processing: The Different Archetypes and their Security & Privacy Assessments. In S. K. Shandilya, S. A. Chun, S. Shandilya, & E. Weippl (Eds.), Internet of Things (IoT) Security: Fundamentals, Techniques and Applications (pp. 37–54). Gistrup, Denmark: River Publishers.

Pramanik, P. K. D., Pal, S., & Choudhury, P. (2018). Beyond Automation: The Cognitive IoT. Artificial Intelligence Brings Sense to the Internet of Things. In A. K. Sangaiah, A. Thangavelu, & V. M. Sundaram (Eds.), Cognitive Computing for Big Data Systems Over IoT: Frameworks, Tools and Application (pp. 1–37). New York, NY: Springer. doi:10.1007/978-3-319-70688-7_1

Pramanik, P. K. D., Pal, S., Brahmachari, A., & Choudhury, P. (2018). Processing IoT Data: From Cloud to Fog. It's Time to be Down-to-Earth. In P. Karthikeyan, & M. Thangavel (Eds.), Applications of Security, Mobile, Analytic and Cloud (SMAC) Technologies for Effective Information Processing and Management (pp. 124-148). Hershey, PA: IGI Global. doi:10.4018/978-1-5225-4044-1.ch007

Pramanik, P. K. D., Pal, S., Pareek, G., Dutta, S., & Choudhury, et al. (2018). Crowd Computing: The Computing Revolution. In R. Lenart-Gansiniec (Ed.), Crowdsourcing and Knowledge Management in Contemporary Business Environments (pp. 166–198). Hershey, PA: IGI Global.

Pramanik, P. K. D., Pal, S., & Choudhury, P. (2019). Green and Sustainable High-Performance Computing with Smartphone Crowd Computing: Benefits, Enablers, and Challenges. Scalable Computing. *Practice and Experience*, *20*(2), 259–283.

Pramanik, P. K. D., Upadhyaya, B., Pal, S., & Pal, T. (2018). Internet of Things, Smart Sensors, and Pervasive Systems: Enabling the Connected and Pervasive Health Care. In N. Dey, A. Ashour, S. J. Fong, & C. Bhatt (Eds.), *Healthcare Data Analytics and Management* (pp. 1–58). Cambridge, MA: Academic Press.

Prashant, D. (2017). Greening the Cloud Computing. *International Journal of Engineering and Computer Science*, *3*(05), 5811–5814.

Qayyum, F. A., Naeem, M., Khwaja, A. S., & et al,. (2015). Appliance Scheduling Optimization in Smart Home Networks. *IEEE Access: Practical Innovations, Open Solutions*, *3*, 2176–2190. doi:10.1109/ACCESS.2015.2496117

Qin, X., Mo, Y., & Jing, L. (2016). Risk perceptions of the life-cycle of green buildings in China. *Journal of Cleaner Production*, *126*, 148–158. doi:10.1016/j.jclepro.2016.03.103

Qiu, C., Shen, H., & Chen, L. (2018). Towards green cloud computing: Demand allocation and pricing policies for cloud service brokerage. *IEEE Transactions on Big Data*, 1-1.

Quadrennial Technology Review, D. O. E. (2015). *Increasing Efficiency of Building Systems and Technologies*. Retrieved from https://www.researchgate.net/file.PostFileLoader.html?id=595770fb3d7f4b73007a48f9&assetKey=AS%3A511249461858304%401498902778200

Rademaekers, K., Yearwood, J., & Ferreira, A. (2016). Selecting Indicators to Measure Energy Poverty. Rotterdam, The Netherlands: Trinomics.

Radu, L.-D. (2017). Green Cloud Computing: A Literature Survey. *Symmetry*, *9*(12), 295. doi:10.3390ym9120295

Ragheba, A., El-Shimyb, H., & Raghebb, G. (2016). Green Architecture: A Concept of Sustainability. *Procedia: Social and Behavioral Sciences*, *216*, 778–787. doi:10.1016/j.sbspro.2015.12.075

Rak, M., Venticinque, S., M'hr, T., & Echevarria, G., & Ensal, Gorka. (2011). Cloud Application Monitoring: The mOSAIC Approach. In *Proceedings of the 2011 IEEE Third International Conference on Cloud Computing Technology and Science*. Athens: IEEE. 10.1109/CloudCom.2011.117

Regenerative design of existing buildings for net-zero energy use. (2015). *Procedia Engineering*, 72–80.

REN21. (2008). Renewables 2007 Global Status Report. Retrieved from http://www.ren21.net/pdf/RE2007_Global_Status_Report.pdf

Reportbuyer. (2018). *The VCSEL market is expected to reach USD 3.89 billion by 2023 from USD 1.78 billion in 2018, at a CAGR of 16.9% from 2018 to 2023.* Retrieved from https://www.prnewswire.com

Rivoire, S., Shah, M. A., Ranganathan, P., & Kozyrakis, C. (2007). JouleSort: A balanced energy-efficiency benchmark. In *Proceedings of the the ACM SIGMOD International Conference on Management of Data.* Beijing, China: IEEE. 10.1145/1247480.1247522

Robey, D., Ross, J. W., & Boudreau, M. C. (2002). Learning to implement enterprise systems: An exploratory study of the dialectics of change. *Journal of Management Information Systems, 19*(1), 17–46. doi:10.1080/07421222.2002.11045713

Robichaud, L. B., & Anantatmula, V. S. (2010). Greening project management practices for sustainable construction. *Journal of Management Engineering, 27*(1), 48–57. doi:10.1061/(ASCE)ME.1943-5479.0000030

Rodden, T., & Benford, S. (2003). The Evolution of Buildings and Implications for the Design of Ubiquitous Domestic Environments. *Proc. SIGCHI Conf. Human Factors in Computing Systems*, 9–16. 10.1145/642611.642615

Rode, P., Burdett, R., & Gonçalves, J. C. (2011). Buildings Investing in energy and resource efficiency. In Towards a Green Economy: Pathways to Sustainable Development and Poverty Eradication (pp. 331-373). United Nations Environment Programme.

Rosa, T. R., Betim, F. S., & Ferreira, R. Q. (2017). Development and application of a lab made apparatus using open-source "arduino" hardware for the electrochemical pretreatment of boron-doped diamond electrodes. *Electrochimica Acta, 231*, 185–189. doi:10.1016/j.electacta.2017.01.180

Rowell, R. M. (2012). Handbook of Wood Chemistry and Wood Composites (2nd ed.). New York, NY: CRC Press. doi:10.1201/b12487

Sadineni, S. B., Madala, S., & Boehm, R. F. (2011). Passive building energy savings: A review of building envelope components. *Renewable & Sustainable Energy Reviews, 15*(8), 3617–3631. doi:10.1016/j.rser.2011.07.014

Saghafi, M. D., Hosseini, T., & Teshnizi, Z. S. (2011). Recycling value of building materials in building assessment systems. *Energy and Building, 43*(11), 3181–3188. doi:10.1016/j.enbuild.2011.08.016

Saha, B. (2018). Green Computing: Current Research Trends. *International Journal on Computer Science and Engineering, 6*(03), 467–469.

Sandanasamy, D., Govindarajane, S., & Sundararajan, T. (2011). Energy-efficient materials: Route to sustainability in green buildings. In *International Conference on Green Technology and Environmental Conservation (GTEC-2011).* IEEE. 10.1109/GTEC.2011.6167702

Sartori, I., Napolitano, A., & Voss, K. (2012). Net zero energy buildings: A consistent definition framework. *Energy and Building*, *48*, 220–232. doi:10.1016/j.enbuild.2012.01.032

Saunders, T. (2008, March). A discussion document comparing international environmental assessment methods for buildings. *BRE*.

Scott, G. (2016). *How Green is Cloud Computing*. Retrieved from https://www.ewastesecurity.com/how-green-is-cloud-computing-new-study/

Selvakkumaran, S., & Limmeechokchai, B. (2011). Assessment of energy security and low carbon society scenarios in Thailand and Sri Lanka. In *2011 International Conference & Utility Exhibition on Power and Energy Systems: Issues and Prospects for Asia (ICUE)* (pp. 1–6). Pattaya, Thailand: IEEE. 10.1109/ICUEPES.2011.6497738

Shabunko, V., Lim, C. M., Brahim, S., & Mathew, S. (2014). Developing building benchmarking for Brunei Darussalam. *Energy and Building*, *85*, 79–85. doi:10.1016/j.enbuild.2014.08.047

Shaikh, P. H., Nor, N. B., Nallagownden, P., & et al, . (2014). A review on optimized control systems for building energy and comfort management of smart sustainable buildings. *Renewable & Sustainable Energy Reviews*, *34*, 409–429. doi:10.1016/j.rser.2014.03.027

Shang, W., Ding, Q., Marianantoni, A., & et al. (2014). Securing building management systems using named data networking. *IEEE Network*, *28*(3), 50–56. doi:10.1109/MNET.2014.6843232

Sharma, M. (2018). Development of a 'Green building sustainability model' for Green buildings in India. *Journal of Cleaner Production*, *190*, 538–551. doi:10.1016/j.jclepro.2018.04.154

Sheth, K.N. (2017). Water efficient technologies for green buildings. *International Journal of Engineering Innovation and Scientific Research, 1*(3), 5-10.

Siemens, A. G. (2014). *Our future depends on intelligent infrastructures.* Siemens Industry. Retrieved from https://www.siemens.com/content/dam/internet/siemens-com/us/home/company/topic-areas/intelligent-infrastructure/buildings/documents/bt-cpp-intel-infrstrctr-wp.pdf

Siemens. (2018, Nov. 6). *Unlock your city's and building's potential.* Retrieved from https://new.siemens.com/global/en/company/topic-areas/intelligent-infrastructure/buildings/digital-buildings.html

Singh, S., & Singh, N. (2015, October). Internet of Things (IoT): Security Challenges, Business Opportunities & Reference Architecture for E-commerce. In *International Conference on Green Computing and Internet of Things (ICGCIoT)* (pp. 1577-1581). IEEE.

Singhvi, V., Krause, A., & Guestrin, C. (2005). Intelligent light control using sensor networks. *Proc. 3rd Int. Conf. Embedded Netw. Sensor Syst.* 10.1145/1098918.1098942

Sinha, A., Gupta, R., & Kutnar, A. (2013). Sustainable Development and Green Buildings. *Drvna Industrija*, *64*(01), 45–53. doi:10.5552/drind.2013.1205

Skyshade Daylights. (2018). *Solar light - Pipe*. Retrieved from http://www.skyshadedaylights.com/products-solar-light-pipe-energy-solar-tube/

Smith, A. (2012). Materials from Alternative, Recycled and Secondary Sources (MARSS) 2005–2010: 6 Year Review. Retrieved from http://www.lucideon.com/newsroom/news-releases/ceram-publishes-a-6-year-review-of-the-substitute-raw-materials-used-in-uk

Solanki, A., & Nayyar, A. (2019). Green Internet of Things (G-IoT): ICT Technologies, Principles, Applications, Projects, and Challenges. In Handbook of Research on Big Data and the IoT (pp. 379-405). Hershey, PA: IGI Global.

Solanki, A., & Nayyar, A. (2019). Green Internet of Things (G-IoT): ICT Technologies, Principles, Applications, Projects, and Challenges. In Handbook of Research on Big Data and the IoT (pp. 379-405). IGI Global.

St. John, A. (1993). *The Sourcebook for Sustainable Design: a Guide to Environmentally Responsible Building Materials and Processes*. Boston, MA: The Boston Society of Architects.

State of Green. (2019). *Energy efficicency in buildings.* Retrieved from https://stateofgreen.com/en/sectors/buildings/energy-efficicency-in-buildings/

State of Green. (2019). *Green buildings as key elements of future sustainable cities.* Retrieved from https://stateofgreen.com/en/creating-smart-green-liveable-cities/green-buildings/

Strengers, Y. (2014). Smart Energy in Everyday Life: Are You Designing for Resource Man? *Interaction*, *21*(4), 24–31. doi:10.1145/2621931

Strickland, J. (2016). *How cloud computing works.* Retrieved from https://computer.howstuffworks.com/cloud-computing/cloud-computing1.htm

Sun, J., Han, B., Ekwaro-Osire, S., & Zhang, H. (2003). Design for environment: Methodologies, tools and implementation. *Journal of Integrated Design & Process Science*, *7*(1), 59–75.

Sun, Y., Wang, X., Luo, H., & Li, X. (2015). Conflict Detection Scheme Based on Formal Rule Model for Smart Building Systems. *IEEE Transactions on Human-Machine Systems*, *45*(2), 215–227. doi:10.1109/THMS.2014.2364613

SUS01/11. Stone Federation Great Britain, London, UK. (2011). *Natural Stone: The Oldest Sustainable Material.* Retrieved from http://www.stonefed.org.uk/uploads/Natural%20Stone.%20The%20oldest%20natural%20material.pdf

Sustainable Sources. (2017). Insulation: Sustainable Sources: 20 Years of Online Green Building Information. Retrieved from http://insulation.sustainablesources.com

Svensson, G., Ferro, C., & Høgevold, N. (2018). Framing the triple bottom line approach: Direct and mediation effects between economic, social and environmental elements. *Journal of Cleaner Production*, *197*, 972–991. doi:10.1016/j.jclepro.2018.06.226

Tagaza, E., & Wilson, J. L., (2004). *Green buildings: drivers and barriers - lessons learned from five Melbourne developments*. Report Prepared for Building Commission by University of Melbourne and Business Outlook and Evaluation.

Tatum, J. (2017). *VCSEL in Various Sensor Applications*. Retrieved from https://www.finisar.com

Thormark, C. (2007, September). Energy and resources, material choice and recycling potential in low energy buildings. In *CIB SB07 Sustainable Construction, Materials and Practices, Lisbon, Portugal.* (Vol. 1, pp. 759–766). IOS Press.

Tiburski, R. T., Amaral, L. A., de Matos, E., & Hessel, F. (2015, December). The Importance of a Standard Security Architecture for SOA -Based IoT Middleware. *IEEE Communications Magazine*, *53*(12), 20–26. doi:10.1109/MCOM.2015.7355580

Tolinski, M. (2011). *Plastics and Sustainability: Towards a Peaceful Coexistence between Bio-Based and Fossil Fuel-Based Plastics*. New York, NY: John Wiley & Sons.

Torcellini, P., Pless, S., Deru, M., & Crawley, D. (2006). *Zero energy buildings: A critical look at the definition*. National Renewable Energy Laboratory and Department of Energy.

Tripanagnostopoulos, Y., & Souliotis, M. (2005, May). Booster reflector contribution to performance improvement of solar collectors *Proceedings of the WREC*, 63-68.

Tripanagnostopoulos, Y., & Tselepis, S. (2003). Hybrid solar/wind PVT/WT building integrated system. *Proceedings of 2nd International European PV-Hybrid and Mini-Grid Conference,* 329-333.

Tripanagnostopoulos, Y., Nousia, T., & Souliotis, M. (2000). Low-cost improvements to building integrated air-cooled hybrid PV-Thermal systems. *Proceedings of 16th European PV Solar Energy Conference,* 1874-1877.

Tripanagnostopoulos, Y., Siabekou, Ch., & Tonui, J. K. (2005). The Fresnel lens concept for solar control of buildings. *Proceedings of International Conference PALENC-2005,* 977-982.

Tripanagnostopoulos, Y., Tselepis, S., & Souliotis, M. (2004). Practical aspects for small wind turbine applications. *Proceedings of International EWEC 2004 Conference, Track 7; Autonomous and Hybrid Systems.*

Tripanagnostopoulos, Y. (2006, July). Energy and aesthetics of building integrated RES. In *Proceedings of International Workshop on Energy Performance and Environmental Quality of Buildings. Milos, Greece*. Greece: Department of Physics, University of Patras; Retrieved from http://www.inive.org/members_area/medias/pdf/Inive%5CMilos2006%5C46_Tripanagnostopoulos_6P.pdf

Tripathi, V. (2016). A literature review of quantitative indicators to measure the quality of labor and delivery care. *International Journal of Gynaecology and Obstetrics: the Official Organ of the International Federation of Gynaecology and Obstetrics*, *132*(2), 139–145. doi:10.1016/j.ijgo.2015.07.014 PMID:26686027

Tuballa, M. L., & Abundo, M. L. (2016). A review of the development of Smart Grid technologies. *Renewable & Sustainable Energy Reviews*, *59*, 710–725. doi:10.1016/j.rser.2016.01.011

Turcotte, D., Villareal, J., & Bermingham, C. (2006). *The Benefits of Building Green.* Lowell, MA: UMass Lowell.

Tushar, W., Wijerathne, N., & Li, W. T. (2018, May). IoT for Green Building Management. *IEEE Signal Processing Magazine*.

Tushar, W., Wijerathne, N., & Li, W.-T. (2018). Internet of Things for Green Building Management: Disruptive Innovations Through Low-Cost Sensor Technology and Artificial Intelligence. *IEEE Signal Processing Magazine*, *35*(5), 100–110. doi:10.1109/MSP.2018.2842096

U. SAdministration. (2011). *The Benefits and Challenges of Green Roofs on Public and Commercial Buildings*. Tucson, AZ: The University of Arizona.

U.S. Department of Energy. (n.d.). Retrieved from http://www.eere.energy.gov/consumer/your_home/space_heating_cooling/index.cfm/mytopic=12350

U.S. Green Building Council. (2003). *Green Building Rating System for New Construction & Major Renovations.* (LEED-NCTM) Version 2.1. Available at http://www.usgbc.org/Docs/LEEDdocs/LEED RS v2-1.pdf

Uddin, M., & Abdul Rahman, A. (2011). Carbon Sustainability Framework to Reduce CO2 Emissions in Data Centers. *International Journal of Green Economics*, *5*(4), 353–369. doi:10.1504/IJGE.2011.044619

United Nations Development Programme. (n.d.). Retrieved from http://www.undp.org/energy/activities/wea/drafts-frame.html

United States Environmental Protection Agency. Indoor Air Quality (IAQ). (2017). Retrieved from: https://www.epa.gov/indoor-air-quality-iaq/introduction-indoor-air-quality

Van Hoe, B., & Bosman, E., & Missinne, J. (2012). Photonic Incremental Pressure Sensor Based on Optical Feedback in a Polymer Embedded VCSEL. IEEE Photonics Technology Letters, 24(13).

Vanderley, M. J. / UNEP Industry and Environment. (2003). *Sustainable Building and construction: On the sustainability of concrete.* Retrieved from http://www.cbcs.org.br/userfiles/bancoDeConhecimento/on%20the%20sustainability%20of%20concrete%20UNEP%20Sustainable%20concrete%20005-098.pdf

Venkata, G., Bhargav, S., Choudhary, S., & Sharma, B. (2015). Use of Renewable Energy Sources in Construction of Green Building. *International Journal for Research in Applied Science and Engineering Technology*, *3*(VII), 108–113.

Vlacheas, P., Giaffreda, R., & Stavroulaki, V. (2013). Enabling smart cities through a cognitive management framework for the internet of things. *IEEE Communications Magazine*, *51*(6), 102–111. doi:10.1109/MCOM.2013.6525602

Waas, T., Hugé, J., & Block, T. (2014). Sustainability Assessment and Indicators: Tools in a Decision-Making Strategy for Sustainable Development. *Sustainability*, *6*(9), 5512–5534. doi:10.3390u6095512

Wang, C., Schwan, K., Talwar, V., Eisenhauer, G., Hu, L., & Wolf, M. (2011). A flexible architecture integrating monitoring and analytics for managing large-scale data centers. In Proceedings of the ACM international conference on Autonomic computing. Karlsruhe, Germany: IEEE. doi:10.1145/1998582.1998605

Wang, M., Qiu, S., Dong, H., & Wang, Y. (2017). Design an IoT-based building management cloud platform for green buildings. *2017 Chinese Automation Congress (CAC)*, 5663-5667.10.1109/CAC.2017.8243793

Wang, X., Chen, Y., Lin, Q., & Zhang, W. (2011). The thinking of promotion and using green building materials. In *Proceedings of 2011 International Conference on Electronic & Mechanical Engineering and Information Technology*. IEEE. 10.1109/EMEIT.2011.6023110

Wan, K., Li, D., & Lam, J. (2011). Assessment of climate change impact on building energy use and mitigation measures in subtropical climates. *Energy*, *36*(3), 1404–1414. doi:10.1016/j.energy.2011.01.033

Warren, C. (2010). Measures of Environmentally Sustainable Development and Their Effect on Property Asset Value: An Australian Perspective. *Journal of Property Management, 28*(2), 68-79.

WBDG Sustainable Committee. (2011). *Sustainable. Whole building design guide.* Retrieved from: http://www.wbdg.org/design/sustainable.php

What is green building? (2019). Retrieved from https://www.worldgbc.org/what-green-building

Widyatmoko, I. (2016). Sustainability of Bituminous Materials. In J. Khatib (Ed.), *Sustainability of Construction Materials* (2nd ed., pp. 343–370). Sawston, UK: Woodhead Publishing.

Wikipedia. (n.d.a). Renewable energy. Retrieved from http://en.wikipedia.org/wiki/Renewable_energy#cite_note-REN21-2007-0

Wikipedia. (n.d.b). Solar chimney. Retrieved from http://en.wikipedia.org/wiki/Solar_chimney

Wiley, J., Benefield, J., & Johnson, K. (2010). Green Design and the Market for Commercial Office Space. *Journal of Real Estate Finance Econ, 41*, 228-243.

Williams. (2012). *Passive Solar Design*. Retrieved from https://sustainability.williams.edu/green-building-basics/passive-solar-design

Workplace Safety and Health Topics. (2017). Retrieved from https://www.cdc.gov/niosh/pdfs/sec_2.pdf

WRAP. (2006). Use of Recycled Glass. Retrieved from http://www.wrap.org.uk

Xu, X., & Van Dessel, S. (2008). Evaluation of a prototype active building envelope window-system. *Energy and Building*, *40*(2), 168–174. doi:10.1016/j.enbuild.2007.02.027

Yates, T., & Bourke, K. (2005). Sustainability and whole life costing for stone. *Arcamedia*, 49-52.

Younger, M., Morrow-Almeida, H. R., Vindigni, S. M., & Dannenberg, A. L. (2008). The built environment, climate change, and health: Opportunities for co-benefits. *American Journal of Preventive Medicine*, *35*(5), 517–526. doi:10.1016/j.amepre.2008.08.017 PMID:18929978

Yuan, X., Wang, X., & Zuo, J. (2013). Renewable energy in buildings in China - A review. *Renewable & Sustainable Energy Reviews*, *24*, 1–8. doi:10.1016/j.rser.2013.03.022

Zaha Hadid Architects. (2019). *Zaha Hadid Architects*. Retrieved from http://www.zaha-hadid.com/architecture/bee'ah-headquarters-sharjah-uae/

Zajac, P. (2016). Indicators and Measures of Energy Efficiency the Warehouse. In P. Zajac (Ed.), *The Energy Consumption in Refrigerated Warehouses* (pp. 23–28). Cham, Switzerland: Springer International Publishing. doi:10.1007/978-3-319-40898-9_2

Zanella, A., Bui, N., & Castellani, A. (2014). Internet of Things for Smart Cities. *IEEE Internet of Things Journal*, *1*(1), 22–32. doi:10.1109/JIOT.2014.2306328

Zhang, X. (2015). Green real estate development in China: State of art and prospect agenda – A review. *Renewable & Sustainable Energy Reviews*, *47*, 1–13. doi:10.1016/j.rser.2015.03.012

Zhang, X. L., Shen, L. Y., & Wu, Y. Z. (2011). Green strategy for gaining competitive advantage in housing development: A China study. *Journal of Cleaner Production*, *19*(1), 157–167. doi:10.1016/j.jclepro.2010.08.005

Zhao, X., Zuo, J., Wu, G., & Huang, C. (2019). A bibliometric review of green building research 2000–2016. *Architectural Science Review*, *62*(1), 74–88. doi:10.1080/00038628.2018.1485548

Related References

To continue our tradition of advancing academic research, we have compiled a list of recommended IGI Global readings. These references will provide additional information and guidance to further enrich your knowledge and assist you with your own research and future publications.

Abed, S., Khir, T., & Ben Brahim, A. (2016). Thermodynamic and Energy Study of a Regenerator in Gas Turbine Cycle and Optimization of Performances. *International Journal of Energy Optimization and Engineering*, *5*(2), 25–44. doi:10.4018/IJEOE.2016040102

Abu Bakar, W. A., Abdullah, W. N., Ali, R., & Mokhtar, W. N. (2016). Polymolybdate Supported Nano Catalyst for Desulfurization of Diesel. In T. Saleh (Ed.), *Applying Nanotechnology to the Desulfurization Process in Petroleum Engineering* (pp. 263–280). Hershey, PA: IGI Global. doi:10.4018/978-1-4666-9545-0.ch009

Addo-Tenkorang, R., Helo, P., & Kantola, J. (2016). Engineer-To-Order Product Development: A Communication Network Analysis for Supply-Chain's Sustainable Competitive Advantage. In R. Addo-Tenkorang, J. Kantola, P. Helo, & A. Shamsuzzoha (Eds.), *Supply Chain Strategies and the Engineer-to-Order Approach* (pp. 43–59). Hershey, PA: IGI Global. doi:10.4018/978-1-5225-0021-6.ch003

Adebiyi, I. D., Popoola, P. A., & Pityana, S. (2016). Mitigation of Wear Damage by Laser Surface Alloying Technique. In E. Akinlabi, R. Mahamood, & S. Akinlabi (Eds.), *Advanced Manufacturing Techniques Using Laser Material Processing* (pp. 172–196). Hershey, PA: IGI Global. doi:10.4018/978-1-5225-0329-3.ch007

Ahmad, W. (2016). Sulfur in Petroleum: Petroleum Desulfurization Techniques. In T. Saleh (Ed.), *Applying Nanotechnology to the Desulfurization Process in Petroleum Engineering* (pp. 1–52). Hershey, PA: IGI Global. doi:10.4018/978-1-4666-9545-0.ch001

Ahmed, I., Ahmad, N., Mehmood, I., Haq, I. U., Hassan, M., & Khan, M. U. (2016). Applications of Nanotechnology in Transportation Engineering. In A. Khitab & W. Anwar (Eds.), *Advanced Research on Nanotechnology for Civil Engineering Applications* (pp. 180–207). Hershey, PA: IGI Global. doi:10.4018/978-1-5225-0344-6.ch006

Aikhuele, D. (2018). A Study of Product Development Engineering and Design Reliability Concerns. *International Journal of Applied Industrial Engineering*, *5*(1), 79–89. doi:10.4018/IJAIE.2018010105

Al-Najar, B. T., & Bououdina, M. (2016). Bioinspired Nanoparticles for Efficient Drug Delivery System. In M. Bououdina (Ed.), *Emerging Research on Bioinspired Materials Engineering* (pp. 69–103). Hershey, PA: IGI Global. doi:10.4018/978-1-4666-9811-6.ch003

Al-Shebeeb, O. A., Rangaswamy, S., Gopalakrishan, B., & Devaru, D. G. (2017). Evaluation and Indexing of Process Plans Based on Electrical Demand and Energy Consumption. *International Journal of Manufacturing, Materials, and Mechanical Engineering*, *7*(3), 1–19. doi:10.4018/IJMMME.2017070101

Alexakis, H., & Makris, N. (2016). Validation of the Discrete Element Method for the Limit Stability Analysis of Masonry Arches. In V. Sarhosis, K. Bagi, J. Lemos, & G. Milani (Eds.), *Computational Modeling of Masonry Structures Using the Discrete Element Method* (pp. 292–325). Hershey, PA: IGI Global. doi:10.4018/978-1-5225-0231-9.ch012

AlMegren, H. A., Gonzalez-Cortes, S., Huang, Y., Chen, H., Qian, Y., Alkinany, M., ... Xiao, T. (2016). Preparation of Deep Hydrodesulfurzation Catalysts for Diesel Fuel using Organic Matrix Decomposition Method. In H. Al-Megren & T. Xiao (Eds.), *Petrochemical Catalyst Materials, Processes, and Emerging Technologies* (pp. 216–253). Hershey, PA: IGI Global. doi:10.4018/978-1-4666-9975-5.ch009

Alshammari, A., Kalevaru, V. N., Bagabas, A., & Martin, A. (2016). Production of Ethylene and its Commercial Importance in the Global Market. In H. Al-Megren & T. Xiao (Eds.), *Petrochemical Catalyst Materials, Processes, and Emerging Technologies* (pp. 82–115). Hershey, PA: IGI Global. doi:10.4018/978-1-4666-9975-5.ch004

Amel, M. (2016). Synthesis, Characterizations, and Biological Effects Study of Some Quinoline Family. In M. Bououdina (Ed.), *Emerging Research on Bioinspired Materials Engineering* (pp. 160–196). Hershey, PA: IGI Global. doi:10.4018/978-1-4666-9811-6.ch006

Amna, T., Haasan, M. S., Khil, M., & Hwang, I. (2016). Impact of Electrospun Biomimetic Extracellular Environment on Proliferation and Intercellular Communication of Muscle Precursor Cells: An Overview – Intercellular Communication of Muscle Precursor Cells with Extracellular Environment. In M. Bououdina (Ed.), *Emerging Research on Bioinspired Materials Engineering* (pp. 247–265). Hershey, PA: IGI Global. doi:10.4018/978-1-4666-9811-6.ch009

Amuda, M. O., Lawal, T. F., & Akinlabi, E. T. (2017). Research Progress on Rheological Behavior of AA7075 Aluminum Alloy During Hot Deformation. *International Journal of Materials Forming and Machining Processes*, *4*(1), 53–96. doi:10.4018/IJMFMP.2017010104

An, M., & Qin, Y. (2016). Challenges of Railway Safety Risk Assessment and Maintenance Decision Making. In B. Rai (Ed.), *Handbook of Research on Emerging Innovations in Rail Transportation Engineering* (pp. 173–211). Hershey, PA: IGI Global. doi:10.4018/978-1-5225-0084-1.ch009

Anil, M., Ayyildiz-Tamis, D., Tasdemir, S., Sendemir-Urkmez, A., & Gulce-Iz, S. (2016). Bioinspired Materials and Biocompatibility. In M. Bououdina (Ed.), *Emerging Research on Bioinspired Materials Engineering* (pp. 294–322). Hershey, PA: IGI Global. doi:10.4018/978-1-4666-9811-6.ch011

Armutlu, H. (2018). Intelligent Biomedical Engineering Operations by Cloud Computing Technologies. In U. Kose, G. Guraksin, & O. Deperlioglu (Eds.), *Nature-Inspired Intelligent Techniques for Solving Biomedical Engineering Problems* (pp. 297–317). Hershey, PA: IGI Global. doi:10.4018/978-1-5225-4769-3.ch015

Arokiyaraj, S., Saravanan, M., Bharanidharan, R., Islam, V. I., Bououdina, M., & Vincent, S. (2016). Green Synthesis of Metallic Nanoparticles Using Plant Compounds and Their Applications: Metallic Nanoparticles Synthesis Using Plants. In M. Bououdina (Ed.), *Emerging Research on Bioinspired Materials Engineering* (pp. 1–34). Hershey, PA: IGI Global. doi:10.4018/978-1-4666-9811-6.ch001

Atik, M., Sadek, M., & Shahrour, I. (2017). Single-Run Adaptive Pushover Procedure for Shear Wall Structures. In V. Plevris, G. Kremmyda, & Y. Fahjan (Eds.), *Performance-Based Seismic Design of Concrete Structures and Infrastructures* (pp. 59–83). Hershey, PA: IGI Global. doi:10.4018/978-1-5225-2089-4.ch003

Aydin, A., Akyol, E., Gungor, M., Kaya, A., & Tasdelen, S. (2018). Geophysical Surveys in Engineering Geology Investigations With Field Examples. In N. Ceryan (Ed.), *Handbook of Research on Trends and Digital Advances in Engineering Geology* (pp. 257–280). Hershey, PA: IGI Global. doi:10.4018/978-1-5225-2709-1.ch007

Azevedo, N. M., Lemos, J. V., & Rocha de Almeida, J. (2016). Discrete Element Particle Modelling of Stone Masonry. In V. Sarhosis, K. Bagi, J. Lemos, & G. Milani (Eds.), *Computational Modeling of Masonry Structures Using the Discrete Element Method* (pp. 146–170). Hershey, PA: IGI Global. doi:10.4018/978-1-5225-0231-9.ch007

Bamufleh, H. S., Noureldin, M. M., & El-Halwagi, M. M. (2016). Sustainable Process Integration in the Petrochemical Industries. In H. Al-Megren & T. Xiao (Eds.), *Petrochemical Catalyst Materials, Processes, and Emerging Technologies* (pp. 150–163). Hershey, PA: IGI Global. doi:10.4018/978-1-4666-9975-5.ch006

Banerjee, S., Gautam, R. K., Gautam, P. K., Jaiswal, A., & Chattopadhyaya, M. C. (2016). Recent Trends and Advancement in Nanotechnology for Water and Wastewater Treatment: Nanotechnological Approach for Water Purification. In A. Khitab & W. Anwar (Eds.), *Advanced Research on Nanotechnology for Civil Engineering Applications* (pp. 208–252). Hershey, PA: IGI Global. doi:10.4018/978-1-5225-0344-6.ch007

Bas, T. G. (2017). Nutraceutical Industry with the Collaboration of Biotechnology and Nutrigenomics Engineering: The Significance of Intellectual Property in the Entrepreneurship and Scientific Research Ecosystems. In T. Bas & J. Zhao (Eds.), *Comparative Approaches to Biotechnology Development and Use in Developed and Emerging Nations* (pp. 1–17). Hershey, PA: IGI Global. doi:10.4018/978-1-5225-1040-6.ch001

Beale, R., & André, J. (2017). *Design Solutions and Innovations in Temporary Structures*. Hershey, PA: IGI Global. doi:10.4018/978-1-5225-2199-0

Behnam, B. (2017). Simulating Post-Earthquake Fire Loading in Conventional RC Structures. In P. Samui, S. Chakraborty, & D. Kim (Eds.), *Modeling and Simulation Techniques in Structural Engineering* (pp. 425–444). Hershey, PA: IGI Global. doi:10.4018/978-1-5225-0588-4.ch015

Ben Hamida, I., Salah, S. B., Msahli, F., & Mimouni, M. F. (2018). Distribution Network Reconfiguration Using SPEA2 for Power Loss Minimization and Reliability Improvement. *International Journal of Energy Optimization and Engineering*, *7*(1), 50–65. doi:10.4018/IJEOE.2018010103

Benjamin, S. R., de Lima, F., & Rathoure, A. K. (2016). Genetically Engineered Microorganisms for Bioremediation Processes: GEMs for Bioremediaton. In A. Rathoure & V. Dhatwalia (Eds.), *Toxicity and Waste Management Using Bioremediation* (pp. 113–140). Hershey, PA: IGI Global. doi:10.4018/978-1-4666-9734-8.ch006

Bhaskar, S. V., & Kudal, H. N. (2017). Effect of TiCN and AlCrN Coating on Tribological Behaviour of Plasma-nitrided AISI 4140 Steel. *International Journal of Surface Engineering and Interdisciplinary Materials Science*, *5*(2), 1–17. doi:10.4018/IJSEIMS.2017070101

Bhowmik, S., Sahoo, P., Acharyya, S. K., Dhar, S., & Chattopadhyay, J. (2016). Effect of Microstructure Degradation on Fracture Toughness of 20MnMoNi55 Steel in DBT Region. *International Journal of Manufacturing, Materials, and Mechanical Engineering*, *6*(3), 11–27. doi:10.4018/IJMMME.2016070102

Bhutto, A. W., Abro, R., Abbas, T., Yu, G., & Chen, X. (2016). Desulphurization of Fuel Oils Using Ionic Liquids. In H. Al-Megren & T. Xiao (Eds.), *Petrochemical Catalyst Materials, Processes, and Emerging Technologies* (pp. 254–284). Hershey, PA: IGI Global. doi:10.4018/978-1-4666-9975-5.ch010

Bhuyan, D. (2018). Designing of a Twin Tube Shock Absorber: A Study in Reverse Engineering. In K. Kumar & J. Davim (Eds.), *Design and Optimization of Mechanical Engineering Products* (pp. 83–104). Hershey, PA: IGI Global. doi:10.4018/978-1-5225-3401-3.ch005

Bouloudenine, M., & Bououdina, M. (2016). Toxic Effects of Engineered Nanoparticles on Living Cells. In M. Bououdina (Ed.), *Emerging Research on Bioinspired Materials Engineering* (pp. 35–68). Hershey, PA: IGI Global. doi:10.4018/978-1-4666-9811-6.ch002

Brunetti, A., Sellaro, M., Drioli, E., & Barbieri, G. (2016). Membrane Engineering and its Role in Oil Refining and Petrochemical Industry. In H. Al-Megren & T. Xiao (Eds.), *Petrochemical Catalyst Materials, Processes, and Emerging Technologies* (pp. 116–149). Hershey, PA: IGI Global. doi:10.4018/978-1-4666-9975-5.ch005

Bügler, M., & Borrmann, A. (2016). Simulation Based Construction Project Schedule Optimization: An Overview on the State-of-the-Art. In F. Miranda & C. Abreu (Eds.), *Handbook of Research on Computational Simulation and Modeling in Engineering* (pp. 482–507). Hershey, PA: IGI Global. doi:10.4018/978-1-4666-8823-0.ch016

Calderon, F. A., Giolo, E. G., Frau, C. D., Rengel, M. G., Rodriguez, H., Tornello, M., ... Gallucci, R. (2018). Seismic Microzonation and Site Effects Detection Through Microtremors Measures: A Review. In N. Ceryan (Ed.), *Handbook of Research on Trends and Digital Advances in Engineering Geology* (pp. 326–349). Hershey, PA: IGI Global. doi:10.4018/978-1-5225-2709-1.ch009

Carmona-Murillo, J., & Valenzuela-Valdés, J. F. (2016). Motivation on Problem Based Learning. In D. Fonseca & E. Redondo (Eds.), *Handbook of Research on Applied E-Learning in Engineering and Architecture Education* (pp. 179–203). Hershey, PA: IGI Global. doi:10.4018/978-1-4666-8803-2.ch009

Ceryan, N. (2016). A Review of Soft Computing Methods Application in Rock Mechanic Engineering. In P. Samui (Ed.), *Handbook of Research on Advanced Computational Techniques for Simulation-Based Engineering* (pp. 1–70). Hershey, PA: IGI Global. doi:10.4018/978-1-4666-9479-8.ch001

Ceryan, N., & Can, N. K. (2018). Prediction of The Uniaxial Compressive Strength of Rocks Materials. In N. Ceryan (Ed.), *Handbook of Research on Trends and Digital Advances in Engineering Geology* (pp. 31–96). Hershey, PA: IGI Global. doi:10.4018/978-1-5225-2709-1.ch002

Ceryan, S. (2018). Weathering Indices Used in Evaluation of the Weathering State of Rock Material. In N. Ceryan (Ed.), *Handbook of Research on Trends and Digital Advances in Engineering Geology* (pp. 132–186). Hershey, PA: IGI Global. doi:10.4018/978-1-5225-2709-1.ch004

Chandrasekaran, S., Silva, B., Patil, A., Oo, A. M., & Campbell, M. (2016). Evaluating Engineering Students' Perceptions: The Impact of Team-Based Learning Practices in Engineering Education. *International Journal of Quality Assurance in Engineering and Technology Education, 5*(4), 42–59. doi:10.4018/IJQAETE.2016100103

Chen, H., Padilla, R. V., & Besarati, S. (2017). Supercritical Fluids and Their Applications in Power Generation. In L. Chen & Y. Iwamoto (Eds.), *Advanced Applications of Supercritical Fluids in Energy Systems* (pp. 369–402). Hershey, PA: IGI Global. doi:10.4018/978-1-5225-2047-4.ch012

Chen, L. (2017). Principles, Experiments, and Numerical Studies of Supercritical Fluid Natural Circulation System. In L. Chen & Y. Iwamoto (Eds.), *Advanced Applications of Supercritical Fluids in Energy Systems* (pp. 136–187). Hershey, PA: IGI Global. doi:10.4018/978-1-5225-2047-4.ch005

Clementi, F., Di Sciascio, G., Di Sciascio, S., & Lenci, S. (2017). Influence of the Shear-Bending Interaction on the Global Capacity of Reinforced Concrete Frames: A Brief Overview of the New Perspectives. In V. Plevris, G. Kremmyda, & Y. Fahjan (Eds.), *Performance-Based Seismic Design of Concrete Structures and Infrastructures* (pp. 84–111). Hershey, PA: IGI Global. doi:10.4018/978-1-5225-2089-4.ch004

Cortés-Polo, D., Calle-Cancho, J., Carmona-Murillo, J., & González-Sánchez, J. (2017). Future Trends in Mobile-Fixed Integration for Next Generation Networks: Classification and Analysis. *International Journal of Vehicular Telematics and Infotainment Systems*, *1*(1), 33–53. doi:10.4018/IJVTIS.2017010103

Cui, X., Zeng, S., Li, Z., Zheng, Q., Yu, X., & Han, B. (2018). Advanced Composites for Civil Engineering Infrastructures. In K. Kumar & J. Davim (Eds.), *Composites and Advanced Materials for Industrial Applications* (pp. 212–248). Hershey, PA: IGI Global. doi:10.4018/978-1-5225-5216-1.ch010

Dalgıç, S., & Kuşku, İ. (2018). Geological and Geotechnical Investigations in Tunneling. In N. Ceryan (Ed.), *Handbook of Research on Trends and Digital Advances in Engineering Geology* (pp. 482–529). Hershey, PA: IGI Global. doi:10.4018/978-1-5225-2709-1.ch014

de la Varga, D., Soto, M., Arias, C. A., van Oirschot, D., Kilian, R., Pascual, A., & Álvarez, J. A. (2017). Constructed Wetlands for Industrial Wastewater Treatment and Removal of Nutrients. In Á. Val del Río, J. Campos Gómez, & A. Mosquera Corral (Eds.), *Technologies for the Treatment and Recovery of Nutrients from Industrial Wastewater* (pp. 202–230). Hershey, PA: IGI Global. doi:10.4018/978-1-5225-1037-6.ch008

del Valle-Zermeño, R., Chimenos, J. M., & Formosa, J. (2016). Flue Gas Desulfurization: Processes and Technologies. In T. Saleh (Ed.), *Applying Nanotechnology to the Desulfurization Process in Petroleum Engineering* (pp. 337–377). Hershey, PA: IGI Global. doi:10.4018/978-1-4666-9545-0.ch011

Delgado, J. M., Henriques, A. A., & Delgado, R. M. (2016). Structural Non-Linear Models and Simulation Techniques: An Efficient Combination for Safety Evaluation of RC Structures. In F. Miranda & C. Abreu (Eds.), *Handbook of Research on Computational Simulation and Modeling in Engineering* (pp. 540–584). Hershey, PA: IGI Global. doi:10.4018/978-1-4666-8823-0.ch018

Delgado, P. S., Arêde, A., Pouca, N. V., & Costa, A. (2016). Numerical Modeling of RC Bridges for Seismic Risk Analysis. In F. Miranda & C. Abreu (Eds.), *Handbook of Research on Computational Simulation and Modeling in Engineering* (pp. 457–481). Hershey, PA: IGI Global. doi:10.4018/978-1-4666-8823-0.ch015

Deng, Y., & Liu, S. (2016). Catalysis with Room Temperature Ionic Liquids Mediated Metal Nanoparticles. In H. Al-Megren & T. Xiao (Eds.), *Petrochemical Catalyst Materials, Processes, and Emerging Technologies* (pp. 285–329). Hershey, PA: IGI Global. doi:10.4018/978-1-4666-9975-5.ch011

Deperlioglu, O. (2018). Intelligent Techniques Inspired by Nature and Used in Biomedical Engineering. In U. Kose, G. Guraksin, & O. Deperlioglu (Eds.), *Nature-Inspired Intelligent Techniques for Solving Biomedical Engineering Problems* (pp. 51–77). Hershey, PA: IGI Global. doi:10.4018/978-1-5225-4769-3.ch003

Dias, G. L., Magalhães, R. R., Ferreira, D. D., & Vitoriano, F. A. (2016). The Use of a Robotic Arm for Displacement Measurements in a Cantilever beam. *International Journal of Manufacturing, Materials, and Mechanical Engineering*, *6*(3), 45–57. doi:10.4018/IJMMME.2016070104

Dimitratos, N., Villa, A., Chan-Thaw, C. E., Hammond, C., & Prati, L. (2016). Valorisation of Glycerol to Fine Chemicals and Fuels. In H. Al-Megren & T. Xiao (Eds.), *Petrochemical Catalyst Materials, Processes, and Emerging Technologies* (pp. 352–384). Hershey, PA: IGI Global. doi:10.4018/978-1-4666-9975-5.ch013

Dixit, A. (2018). Application of Silica-Gel-Reinforced Aluminium Composite on the Piston of Internal Combustion Engine: Comparative Study of Silica-Gel-Reinforced Aluminium Composite Piston With Aluminium Alloy Piston. In K. Kumar & J. Davim (Eds.), *Composites and Advanced Materials for Industrial Applications* (pp. 63–98). Hershey, PA: IGI Global. doi:10.4018/978-1-5225-5216-1.ch004

Drei, A., Milani, G., & Sincraian, G. (2016). Application of DEM to Historic Masonries, Two Case-Studies in Portugal and Italy: Aguas Livres Aqueduct and Arch-Tympana of a Church. In V. Sarhosis, K. Bagi, J. Lemos, & G. Milani (Eds.), *Computational Modeling of Masonry Structures Using the Discrete Element Method* (pp. 326–366). Hershey, PA: IGI Global. doi:10.4018/978-1-5225-0231-9.ch013

Dutta, S., Roy, P. K., & Nandi, D. (2016). Optimal Allocation of Static Synchronous Series Compensator Controllers using Chemical Reaction Optimization for Reactive Power Dispatch. *International Journal of Energy Optimization and Engineering*, *5*(3), 43–62. doi:10.4018/IJEOE.2016070103

Dutta, S., Roy, P. K., & Nandi, D. (2016). Quasi Oppositional Teaching-Learning based Optimization for Optimal Power Flow Incorporating FACTS. *International Journal of Energy Optimization and Engineering*, *5*(2), 64–84. doi:10.4018/IJEOE.2016040104

Eloy, S., Dias, M. S., Lopes, P. F., & Vilar, E. (2016). Digital Technologies in Architecture and Engineering: Exploring an Engaged Interaction within Curricula. In D. Fonseca & E. Redondo (Eds.), *Handbook of Research on Applied E-Learning in Engineering and Architecture Education* (pp. 368–402). Hershey, PA: IGI Global. doi:10.4018/978-1-4666-8803-2.ch017

Elsayed, A. M., Dakkama, H. J., Mahmoud, S., Al-Dadah, R., & Kaialy, W. (2017). Sustainable Cooling Research Using Activated Carbon Adsorbents and Their Environmental Impact. In T. Kobayashi (Ed.), *Applied Environmental Materials Science for Sustainability* (pp. 186–221). Hershey, PA: IGI Global. doi:10.4018/978-1-5225-1971-3.ch009

Ercanoglu, M., & Sonmez, H. (2018). General Trends and New Perspectives on Landslide Mapping and Assessment Methods. In N. Ceryan (Ed.), *Handbook of Research on Trends and Digital Advances in Engineering Geology* (pp. 350–379). Hershey, PA: IGI Global. doi:10.4018/978-1-5225-2709-1.ch010

Erinosho, M. F., Akinlabi, E. T., & Pityana, S. (2016). Enhancement of Surface Integrity of Titanium Alloy with Copper by Means of Laser Metal Deposition Process. In E. Akinlabi, R. Mahamood, & S. Akinlabi (Eds.), *Advanced Manufacturing Techniques Using Laser Material Processing* (pp. 60–91). Hershey, PA: IGI Global. doi:10.4018/978-1-5225-0329-3.ch004

Farag, H., & Kishida, M. (2016). Kinetic Models for Complex Parallel–Consecutive Reactions Assessment of Reaction Network and Product Selectivity. In H. Al-Megren & T. Xiao (Eds.), *Petrochemical Catalyst Materials, Processes, and Emerging Technologies* (pp. 330–351). Hershey, PA: IGI Global. doi:10.4018/978-1-4666-9975-5.ch012

Faroz, S. A., Pujari, N. N., Rastogi, R., & Ghosh, S. (2017). Risk Analysis of Structural Engineering Systems Using Bayesian Inference. In P. Samui, S. Chakraborty, & D. Kim (Eds.), *Modeling and Simulation Techniques in Structural Engineering* (pp. 390–424). Hershey, PA: IGI Global. doi:10.4018/978-1-5225-0588-4.ch014

Fernando, P. R., Hamigah, T., Disne, S., Wickramasingha, G. G., & Sutharshan, A. (2018). The Evaluation of Engineering Properties of Low Cost Concrete Blocks by Partial Doping of Sand with Sawdust: Low Cost Sawdust Concrete Block. *International Journal of Strategic Engineering, 1*(2), 26–42. doi:10.4018/IJoSE.2018070103

Fragiadakis, M., Stefanou, I., & Psycharis, I. N. (2016). Vulnerability Assessment of Damaged Classical Multidrum Columns. In V. Sarhosis, K. Bagi, J. Lemos, & G. Milani (Eds.), *Computational Modeling of Masonry Structures Using the Discrete Element Method* (pp. 235–253). Hershey, PA: IGI Global. doi:10.4018/978-1-5225-0231-9.ch010

Gaines, T. W., Williams, K. R., & Wagener, K. B. (2016). ADMET: Functionalized Polyolefins. In H. Al-Megren & T. Xiao (Eds.), *Petrochemical Catalyst Materials, Processes, and Emerging Technologies* (pp. 1–21). Hershey, PA: IGI Global. doi:10.4018/978-1-4666-9975-5.ch001

Garg, H. (2016). Bi-Criteria Optimization for Finding the Optimal Replacement Interval for Maintaining the Performance of the Process Industries. In P. Vasant, G. Weber, & V. Dieu (Eds.), *Handbook of Research on Modern Optimization Algorithms and Applications in Engineering and Economics* (pp. 643–675). Hershey, PA: IGI Global. doi:10.4018/978-1-4666-9644-0.ch025

Gaspar, P. D., Dinho da Silva, P., Gonçalves, J. P., & Carneiro, R. (2016). Computational Modelling and Simulation to Assist the Improvement of Thermal Performance and Energy Efficiency in Industrial Engineering Systems: Application to Cold Stores. In F. Miranda & C. Abreu (Eds.), *Handbook of Research on Computational Simulation and Modeling in Engineering* (pp. 1–68). Hershey, PA: IGI Global. doi:10.4018/978-1-4666-8823-0.ch001

Ge, H., Tang, M., & Wen, X. (2016). Ni/ZnO Nano Sorbent for Reactive Adsorption Desulfurization of Refinery Oil Streams. In T. Saleh (Ed.), *Applying Nanotechnology to the Desulfurization Process in Petroleum Engineering* (pp. 216–239). Hershey, PA: IGI Global. doi:10.4018/978-1-4666-9545-0.ch007

Ghosh, S., Mitra, S., Ghosh, S., & Chakraborty, S. (2017). Seismic Reliability Analysis in the Framework of Metamodelling Based Monte Carlo Simulation. In P. Samui, S. Chakraborty, & D. Kim (Eds.), *Modeling and Simulation Techniques in Structural Engineering* (pp. 192–208). Hershey, PA: IGI Global. doi:10.4018/978-1-5225-0588-4.ch006

Gil, M., & Otero, B. (2017). Learning Engineering Skills through Creativity and Collaboration: A Game-Based Proposal. In R. Alexandre Peixoto de Queirós & M. Pinto (Eds.), *Gamification-Based E-Learning Strategies for Computer Programming Education* (pp. 14–29). Hershey, PA: IGI Global. doi:10.4018/978-1-5225-1034-5.ch002

Gill, J., Ayre, M., & Mills, J. (2017). Revisioning the Engineering Profession: How to Make It Happen! In M. Gray & K. Thomas (Eds.), *Strategies for Increasing Diversity in Engineering Majors and Careers* (pp. 156–175). Hershey, PA: IGI Global. doi:10.4018/978-1-5225-2212-6.ch008

Gopal, S., & Al-Hazmi, M. H. (2016). Advances in Catalytic Technologies for Selective Oxidation of Lower Alkanes. In H. Al-Megren & T. Xiao (Eds.), *Petrochemical Catalyst Materials, Processes, and Emerging Technologies* (pp. 22–52). Hershey, PA: IGI Global. doi:10.4018/978-1-4666-9975-5.ch002

Goyal, N., Ram, M., Bhardwaj, A., & Kumar, A. (2016). Thermal Power Plant Modelling with Fault Coverage Stochastically. *International Journal of Manufacturing, Materials, and Mechanical Engineering*, *6*(3), 28–44. doi:10.4018/IJMMME.2016070103

Goyal, N., Ram, M., & Kumar, P. (2017). Welding Process under Fault Coverage Approach for Reliability and MTTF. In M. Ram & J. Davim (Eds.), *Mathematical Concepts and Applications in Mechanical Engineering and Mechatronics* (pp. 222–245). Hershey, PA: IGI Global. doi:10.4018/978-1-5225-1639-2.ch011

Gray, M., & Lundy, C. (2017). Engineering Study Abroad: High Impact Strategy for Increasing Access. In M. Gray & K. Thomas (Eds.), *Strategies for Increasing Diversity in Engineering Majors and Careers* (pp. 42–59). Hershey, PA: IGI Global. doi:10.4018/978-1-5225-2212-6.ch003

Guha, D., Roy, P. K., & Banerjee, S. (2016). Application of Modified Biogeography Based Optimization in AGC of an Interconnected Multi-Unit Multi-Source AC-DC Linked Power System. *International Journal of Energy Optimization and Engineering*, *5*(3), 1–18. doi:10.4018/IJEOE.2016070101

Guha, D., Roy, P. K., & Banerjee, S. (2016). Grey Wolf Optimization to Solve Load Frequency Control of an Interconnected Power System: GWO Used to Solve LFC Problem. *International Journal of Energy Optimization and Engineering*, *5*(4), 62–83. doi:10.4018/IJEOE.2016100104

Gupta, A. K., Dey, A., & Mukhopadhyay, A. K. (2016). Micromechanical and Finite Element Modeling for Composites. In S. Datta & J. Davim (Eds.), *Computational Approaches to Materials Design: Theoretical and Practical Aspects* (pp. 101–162). Hershey, PA: IGI Global. doi:10.4018/978-1-5225-0290-6.ch005

Guraksin, G. E. (2018). Internet of Things and Nature-Inspired Intelligent Techniques for the Future of Biomedical Engineering. In U. Kose, G. Guraksin, & O. Deperlioglu (Eds.), *Nature-Inspired Intelligent Techniques for Solving Biomedical Engineering Problems* (pp. 263–282). Hershey, PA: IGI Global. doi:10.4018/978-1-5225-4769-3.ch013

Hansman, C. A. (2016). Developing Mentoring Programs in Engineering and Technology Education. *International Journal of Quality Assurance in Engineering and Technology Education*, *5*(2), 1–15. doi:10.4018/IJQAETE.2016040101

Hasan, U., Chegenizadeh, A., & Nikraz, H. (2016). Nanotechnology Future and Present in Construction Industry: Applications in Geotechnical Engineering. In A. Khitab & W. Anwar (Eds.), *Advanced Research on Nanotechnology for Civil Engineering Applications* (pp. 141–179). Hershey, PA: IGI Global. doi:10.4018/978-1-5225-0344-6.ch005

Hejazi, T., & Akbari, L. (2017). A Multiresponse Optimization Model for Statistical Design of Processes with Discrete Variables. In M. Ram & J. Davim (Eds.), *Mathematical Concepts and Applications in Mechanical Engineering and Mechatronics* (pp. 17–37). Hershey, PA: IGI Global. doi:10.4018/978-1-5225-1639-2.ch002

Hejazi, T., & Hejazi, A. (2017). Monte Carlo Simulation for Reliability-Based Design of Automotive Complex Subsystems. In M. Ram & J. Davim (Eds.), *Mathematical Concepts and Applications in Mechanical Engineering and Mechatronics* (pp. 177–200). Hershey, PA: IGI Global. doi:10.4018/978-1-5225-1639-2.ch009

Hejazi, T., & Poursabbagh, H. (2017). Reliability Analysis of Engineering Systems: An Accelerated Life Testing for Boiler Tubes. In M. Ram & J. Davim (Eds.), *Mathematical Concepts and Applications in Mechanical Engineering and Mechatronics* (pp. 154–176). Hershey, PA: IGI Global. doi:10.4018/978-1-5225-1639-2.ch008

Henao, J., & Sotelo, O. (2018). Surface Engineering at High Temperature: Thermal Cycling and Corrosion Resistance. In A. Pakseresht (Ed.), *Production, Properties, and Applications of High Temperature Coatings* (pp. 131–159). Hershey, PA: IGI Global. doi:10.4018/978-1-5225-4194-3.ch006

Huirache-Acuña, R., Alonso-Nuñez, G., Rivera-Muñoz, E. M., Gutierrez, O., & Pawelec, B. (2016). Trimetallic Sulfide Catalysts for Hydrodesulfurization. In T. Saleh (Ed.), *Applying Nanotechnology to the Desulfurization Process in Petroleum Engineering* (pp. 240–262). Hershey, PA: IGI Global. doi:10.4018/978-1-4666-9545-0.ch008

Ilori, O. O., Adetan, D. A., & Umoru, L. E. (2017). Effect of Cutting Parameters on the Surface Residual Stress of Face-Milled Pearlitic Ductile Iron. *International Journal of Materials Forming and Machining Processes*, *4*(1), 38–52. doi:10.4018/IJMFMP.2017010103

Imam, M. H., Tasadduq, I. A., Ahmad, A., Aldosari, F., & Khan, H. (2017). Automated Generation of Course Improvement Plans Using Expert System. *International Journal of Quality Assurance in Engineering and Technology Education*, *6*(1), 1–12. doi:10.4018/IJQAETE.2017010101

Injeti, S. K., & Kumar, T. V. (2018). A WDO Framework for Optimal Deployment of DGs and DSCs in a Radial Distribution System Under Daily Load Pattern to Improve Techno-Economic Benefits. *International Journal of Energy Optimization and Engineering*, *7*(2), 1–38. doi:10.4018/IJEOE.2018040101

Ishii, N., Anami, K., & Knisely, C. W. (2018). *Dynamic Stability of Hydraulic Gates and Engineering for Flood Prevention*. Hershey, PA: IGI Global. doi:10.4018/978-1-5225-3079-4

J., J., Chowdhury, S., Goyal, P., Samui, P., & Dalkiliç, Y. (2016). Determination of Bearing Capacity of Shallow Foundation Using Soft Computing. In P. Saxena, D. Singh, & M. Pant (Eds.), *Problem Solving and Uncertainty Modeling through Optimization and Soft Computing Applications* (pp. 292-328). Hershey, PA: IGI Global. doi:10.4018/978-1-4666-9885-7.ch014

Jagan, J., Gundlapalli, P., & Samui, P. (2016). Utilization of Classification Techniques for the Determination of Liquefaction Susceptibility of Soils. In S. Bhattacharyya, P. Banerjee, D. Majumdar, & P. Dutta (Eds.), *Handbook of Research on Advanced Hybrid Intelligent Techniques and Applications* (pp. 124–160). Hershey, PA: IGI Global. doi:10.4018/978-1-4666-9474-3.ch005

Jayapalan, S. (2018). A Review of Chemical Treatments on Natural Fibers-Based Hybrid Composites for Engineering Applications. In K. Kumar & J. Davim (Eds.), *Composites and Advanced Materials for Industrial Applications* (pp. 16–37). Hershey, PA: IGI Global. doi:10.4018/978-1-5225-5216-1.ch002

Jeet, K., & Dhir, R. (2016). Software Module Clustering Using Bio-Inspired Algorithms. In P. Vasant, G. Weber, & V. Dieu (Eds.), *Handbook of Research on Modern Optimization Algorithms and Applications in Engineering and Economics* (pp. 445–470). Hershey, PA: IGI Global. doi:10.4018/978-1-4666-9644-0.ch017

Joshi, S. D., & Talange, D. B. (2016). Fault Tolerant Control for a Fractional Order AUV System. *International Journal of Energy Optimization and Engineering*, *5*(2), 1–24. doi:10.4018/IJEOE.2016040101

Julião, D., Ribeiro, S., de Castro, B., Cunha-Silva, L., & Balula, S. S. (2016). Polyoxometalates-Based Nanocatalysts for Production of Sulfur-Free Diesel. In T. Saleh (Ed.), *Applying Nanotechnology to the Desulfurization Process in Petroleum Engineering* (pp. 426–458). Hershey, PA: IGI Global. doi:10.4018/978-1-4666-9545-0.ch014

Kamthan, P. (2016). On the Nature of Collaborations in Agile Software Engineering Course Projects. *International Journal of Quality Assurance in Engineering and Technology Education*, *5*(2), 42–59. doi:10.4018/IJQAETE.2016040104

Karaman, O., Celik, C., & Urkmez, A. S. (2016). Self-Assembled Biomimetic Scaffolds for Bone Tissue Engineering. In M. Bououdina (Ed.), *Emerging Research on Bioinspired Materials Engineering* (pp. 104–132). Hershey, PA: IGI Global. doi:10.4018/978-1-4666-9811-6.ch004

Karkalos, N. E., Markopoulos, A. P., & Dossis, M. F. (2017). Optimal Model Parameters of Inverse Kinematics Solution of a 3R Robotic Manipulator Using ANN Models. *International Journal of Manufacturing, Materials, and Mechanical Engineering*, 7(3), 20–40. doi:10.4018/IJMMME.2017070102

Kesimal, A., Karaman, K., Cihangir, F., & Ercikdi, B. (2018). Excavatability Assessment of Rock Masses for Geotechnical Studies. In N. Ceryan (Ed.), *Handbook of Research on Trends and Digital Advances in Engineering Geology* (pp. 231–256). Hershey, PA: IGI Global. doi:10.4018/978-1-5225-2709-1.ch006

Khanh, D. V., Vasant, P. M., Elamvazuthi, I., & Dieu, V. N. (2016). Multi-Objective Optimization of Two-Stage Thermo-Electric Cooler Using Differential Evolution: MO Optimization of TEC Using DE. In F. Miranda & C. Abreu (Eds.), *Handbook of Research on Computational Simulation and Modeling in Engineering* (pp. 139–170). Hershey, PA: IGI Global. doi:10.4018/978-1-4666-8823-0.ch004

Kim, D., Hassan, M. K., Chang, S., & Bigdeli, Y. (2016). Nonlinear Vibration Control of 3D Irregular Structures Subjected to Seismic Loads. In P. Samui (Ed.), *Handbook of Research on Advanced Computational Techniques for Simulation-Based Engineering* (pp. 103–119). Hershey, PA: IGI Global. doi:10.4018/978-1-4666-9479-8.ch003

Knoflacher, H. (2017). The Role of Engineers and Their Tools in the Transport Sector after Paradigm Change: From Assumptions and Extrapolations to Science. In H. Knoflacher & E. Ocalir-Akunal (Eds.), *Engineering Tools and Solutions for Sustainable Transportation Planning* (pp. 1–29). Hershey, PA: IGI Global. doi:10.4018/978-1-5225-2116-7.ch001

Kose, U. (2018). Towards an Intelligent Biomedical Engineering With Nature-Inspired Artificial Intelligence Techniques. In U. Kose, G. Guraksin, & O. Deperlioglu (Eds.), *Nature-Inspired Intelligent Techniques for Solving Biomedical Engineering Problems* (pp. 1–26). Hershey, PA: IGI Global. doi:10.4018/978-1-5225-4769-3.ch001

Kostić, S. (2018). A Review on Enhanced Stability Analyses of Soil Slopes Using Statistical Design. In N. Ceryan (Ed.), *Handbook of Research on Trends and Digital Advances in Engineering Geology* (pp. 446–481). Hershey, PA: IGI Global. doi:10.4018/978-1-5225-2709-1.ch013

Kumar, A., Patil, P. P., & Prajapati, Y. K. (2018). *Advanced Numerical Simulations in Mechanical Engineering*. Hershey, PA: IGI Global. doi:10.4018/978-1-5225-3722-9

Kumar, G. R., Rajyalakshmi, G., & Manupati, V. K. (2017). Surface Micro Patterning of Aluminium Reinforced Composite through Laser Peening. *International Journal of Manufacturing, Materials, and Mechanical Engineering*, 7(4), 15–27. doi:10.4018/IJMMME.2017100102

Kumari, N., & Kumar, K. (2018). Fabrication of Orthotic Calipers With Epoxy-Based Green Composite. In K. Kumar & J. Davim (Eds.), *Composites and Advanced Materials for Industrial Applications* (pp. 157–176). Hershey, PA: IGI Global. doi:10.4018/978-1-5225-5216-1.ch008

Kuppusamy, R. R. (2018). Development of Aerospace Composite Structures Through Vacuum-Enhanced Resin Transfer Moulding Technology (VERTMTy): Vacuum-Enhanced Resin Transfer Moulding. In K. Kumar & J. Davim (Eds.), *Composites and Advanced Materials for Industrial Applications* (pp. 99–111). Hershey, PA: IGI Global. doi:10.4018/978-1-5225-5216-1.ch005

Lemos, J. V. (2016). The Basis for Masonry Analysis with UDEC and 3DEC. In V. Sarhosis, K. Bagi, J. Lemos, & G. Milani (Eds.), *Computational Modeling of Masonry Structures Using the Discrete Element Method* (pp. 61–89). Hershey, PA: IGI Global. doi:10.4018/978-1-5225-0231-9.ch003

Loy, J., Howell, S., & Cooper, R. (2017). Engineering Teams: Supporting Diversity in Engineering Education. In M. Gray & K. Thomas (Eds.), *Strategies for Increasing Diversity in Engineering Majors and Careers* (pp. 106–129). Hershey, PA: IGI Global. doi:10.4018/978-1-5225-2212-6.ch006

Macher, G., Armengaud, E., Kreiner, C., Brenner, E., Schmittner, C., Ma, Z., ... Krammer, M. (2018). Integration of Security in the Development Lifecycle of Dependable Automotive CPS. In N. Druml, A. Genser, A. Krieg, M. Menghin, & A. Hoeller (Eds.), *Solutions for Cyber-Physical Systems Ubiquity* (pp. 383–423). Hershey, PA: IGI Global. doi:10.4018/978-1-5225-2845-6.ch015

Maghsoodlou, S., & Poreskandar, S. (2016). Controlling Electrospinning Jet Using Microscopic Model for Ideal Tissue Engineering Scaffolds. *International Journal of Chemoinformatics and Chemical Engineering*, *5*(2), 1–16. doi:10.4018/IJCCE.2016070101

Mahendramani, G., & Lakshmana Swamy, N. (2018). Effect of Weld Groove Area on Distortion of Butt Welded Joints in Submerged Arc Welding. *International Journal of Manufacturing, Materials, and Mechanical Engineering*, *8*(2), 33–44. doi:10.4018/IJMMME.2018040103

Maiti, S. (2016). Engineered Gellan Polysaccharides in the Design of Controlled Drug Delivery Systems. In M. Bououdina (Ed.), *Emerging Research on Bioinspired Materials Engineering* (pp. 266–293). Hershey, PA: IGI Global. doi:10.4018/978-1-4666-9811-6.ch010

Majumdar, J. D., Weisheit, A., & Manna, I. (2016). Laser Surface Processing for Tailoring of Properties by Optimization of Microstructure. In E. Akinlabi, R. Mahamood, & S. Akinlabi (Eds.), *Advanced Manufacturing Techniques Using Laser Material Processing* (pp. 121–171). Hershey, PA: IGI Global. doi:10.4018/978-1-5225-0329-3.ch006

Maldonado-Macías, A. A., García-Alcaraz, J. L., Hernández-Arellano, J. L., & Cortes-Robles, G. (2016). An Ergonomic Compatibility Perspective on the Selection of Advanced Manufacturing Technology: A Case Study for CNC Vertical Machining Centers. In G. Alor-Hernández, C. Sánchez-Ramírez, & J. García-Alcaraz (Eds.), *Handbook of Research on Managerial Strategies for Achieving Optimal Performance in Industrial Processes* (pp. 137–165). Hershey, PA: IGI Global. doi:10.4018/978-1-5225-0130-5.ch008

Mamaghani, I. H. (2016). Application of Discrete Finite Element Method for Analysis of Unreinforced Masonry Structures. In V. Sarhosis, K. Bagi, J. Lemos, & G. Milani (Eds.), *Computational Modeling of Masonry Structures Using the Discrete Element Method* (pp. 440–458). Hershey, PA: IGI Global. doi:10.4018/978-1-5225-0231-9.ch017

Mansor, M. R., Sapuan, S. M., Salim, M. A., Akop, M. Z., Musthafah, M. T., & Shaharuzaman, M. A. (2016). Concurrent Design of Green Composites. In D. Verma, S. Jain, X. Zhang, & P. Gope (Eds.), *Green Approaches to Biocomposite Materials Science and Engineering* (pp. 48–75). Hershey, PA: IGI Global. doi:10.4018/978-1-5225-0424-5.ch003

Mansouri, I., & Esmaeili, E. (2016). Nanotechnology Applications in the Construction Industry. In A. Khitab & W. Anwar (Eds.), *Advanced Research on Nanotechnology for Civil Engineering Applications* (pp. 111–140). Hershey, PA: IGI Global. doi:10.4018/978-1-5225-0344-6.ch004

Manzoor, A. (2016). MOOCs for Enhancing Engineering Education. In D. Fonseca & E. Redondo (Eds.), *Handbook of Research on Applied E-Learning in Engineering and Architecture Education* (pp. 204–223). Hershey, PA: IGI Global. doi:10.4018/978-1-4666-8803-2.ch010

Martin, A., Kalevaru, V. N., & Radnik, J. (2016). Palladium in Heterogeneous Oxidation Catalysis. In H. Al-Megren & T. Xiao (Eds.), *Petrochemical Catalyst Materials, Processes, and Emerging Technologies* (pp. 53–81). Hershey, PA: IGI Global. doi:10.4018/978-1-4666-9975-5.ch003

Melnyczuk, J. M., & Palchoudhury, S. (2016). Introduction to Bio-Inspired Hydrogel and Their Application: Hydrogels. In M. Bououdina (Ed.), *Emerging Research on Bioinspired Materials Engineering* (pp. 133–159). Hershey, PA: IGI Global. doi:10.4018/978-1-4666-9811-6.ch005

Mitra-Kirtley, S., Mullins, O. C., & Pomerantz, A. E. (2016). Sulfur and Nitrogen Chemical Speciation in Crude Oils and Related Carbonaceous Materials. In T. Saleh (Ed.), *Applying Nanotechnology to the Desulfurization Process in Petroleum Engineering* (pp. 53–83). Hershey, PA: IGI Global. doi:10.4018/978-1-4666-9545-0.ch002

Moalosi, R., Uziak, J., & Oladiran, M. T. (2016). Using Blended Learning Approach to Deliver Courses in An Engineering Programme. *International Journal of Quality Assurance in Engineering and Technology Education*, *5*(1), 23–39. doi:10.4018/IJQAETE.2016010103

Mohammadzadeh, S., & Kim, Y. (2017). Nonlinear System Identification of Smart Buildings. In P. Samui, S. Chakraborty, & D. Kim (Eds.), *Modeling and Simulation Techniques in Structural Engineering* (pp. 328–347). Hershey, PA: IGI Global. doi:10.4018/978-1-5225-0588-4.ch011

Mohanty, I., & Bhattacherjee, D. (2016). Artificial Neural Network and Its Application in Steel Industry. In S. Datta & J. Davim (Eds.), *Computational Approaches to Materials Design: Theoretical and Practical Aspects* (pp. 267–300). Hershey, PA: IGI Global. doi:10.4018/978-1-5225-0290-6.ch010

Mohebkhah, A., & Sarhosis, V. (2016). Discrete Element Modeling of Masonry-Infilled Frames. In V. Sarhosis, K. Bagi, J. Lemos, & G. Milani (Eds.), *Computational Modeling of Masonry Structures Using the Discrete Element Method* (pp. 200–234). Hershey, PA: IGI Global. doi:10.4018/978-1-5225-0231-9.ch009

Molina, G. J., Aktaruzzaman, F., Soloiu, V., & Rahman, M. (2017). Design and Testing of a Jet-Impingement Instrument to Study Surface-Modification Effects by Nanofluids. *International Journal of Surface Engineering and Interdisciplinary Materials Science*, *5*(2), 43–61. doi:10.4018/IJSEIMS.2017070104

Montalvan-Sorrosa, D., de los Cobos-Vasconcelos, D., & Gonzalez-Sanchez, A. (2016). Nanotechnology Applied to the Biodesulfurization of Fossil Fuels and Spent Caustic Streams. In T. Saleh (Ed.), *Applying Nanotechnology to the Desulfurization Process in Petroleum Engineering* (pp. 378–389). Hershey, PA: IGI Global. doi:10.4018/978-1-4666-9545-0.ch012

Montillet, J., Yu, K., Bonenberg, L. K., & Roberts, G. W. (2016). Optimization Algorithms in Local and Global Positioning. In P. Vasant, G. Weber, & V. Dieu (Eds.), *Handbook of Research on Modern Optimization Algorithms and Applications in Engineering and Economics* (pp. 1–53). Hershey, PA: IGI Global. doi:10.4018/978-1-4666-9644-0.ch001

Moreira, F., & Ferreira, M. J. (2016). Teaching and Learning Requirements Engineering Based on Mobile Devices and Cloud: A Case Study. In D. Fonseca & E. Redondo (Eds.), *Handbook of Research on Applied E-Learning in Engineering and Architecture Education* (pp. 237–262). Hershey, PA: IGI Global. doi:10.4018/978-1-4666-8803-2.ch012

Mukherjee, A., Saeed, R. A., Dutta, S., & Naskar, M. K. (2017). Fault Tracking Framework for Software-Defined Networking (SDN). In C. Singhal & S. De (Eds.), *Resource Allocation in Next-Generation Broadband Wireless Access Networks* (pp. 247–272). Hershey, PA: IGI Global. doi:10.4018/978-1-5225-2023-8.ch011

Mukhopadhyay, A., Barman, T. K., & Sahoo, P. (2018). Electroless Nickel Coatings for High Temperature Applications. In K. Kumar & J. Davim (Eds.), *Composites and Advanced Materials for Industrial Applications* (pp. 297–331). Hershey, PA: IGI Global. doi:10.4018/978-1-5225-5216-1.ch013

Náprstek, J., & Fischer, C. (2017). Dynamic Stability and Post-Critical Processes of Slender Auto-Parametric Systems. In V. Plevris, G. Kremmyda, & Y. Fahjan (Eds.), *Performance-Based Seismic Design of Concrete Structures and Infrastructures* (pp. 128–171). Hershey, PA: IGI Global. doi:10.4018/978-1-5225-2089-4.ch006

Nautiyal, L., Shivach, P., & Ram, M. (2018). Optimal Designs by Means of Genetic Algorithms. In M. Ram & J. Davim (Eds.), *Soft Computing Techniques and Applications in Mechanical Engineering* (pp. 151–161). Hershey, PA: IGI Global. doi:10.4018/978-1-5225-3035-0.ch007

Nazir, R. (2017). Advanced Nanomaterials for Water Engineering and Treatment: Nano-Metal Oxides and Their Nanocomposites. In T. Saleh (Ed.), *Advanced Nanomaterials for Water Engineering, Treatment, and Hydraulics* (pp. 84–126). Hershey, PA: IGI Global. doi:10.4018/978-1-5225-2136-5.ch005

Nogueira, A. F., Ribeiro, J. C., Fernández de Vega, F., & Zenha-Rela, M. A. (2018). Evolutionary Approaches to Test Data Generation for Object-Oriented Software: Overview of Techniques and Tools. In M. Khosrow-Pour, D.B.A. (Ed.), Incorporating Nature-Inspired Paradigms in Computational Applications (pp. 162-194). Hershey, PA: IGI Global. doi:10.4018/978-1-5225-5020-4.ch006

Nunes, J. F., Moreira, P. M., & Tavares, J. M. (2016). Human Motion Analysis and Simulation Tools: A Survey. In F. Miranda & C. Abreu (Eds.), *Handbook of Research on Computational Simulation and Modeling in Engineering* (pp. 359–388). Hershey, PA: IGI Global. doi:10.4018/978-1-4666-8823-0.ch012

Ogunlaja, A. S., & Tshentu, Z. R. (2016). Molecularly Imprinted Polymer Nanofibers for Adsorptive Desulfurization. In T. Saleh (Ed.), *Applying Nanotechnology to the Desulfurization Process in Petroleum Engineering* (pp. 281–336). Hershey, PA: IGI Global. doi:10.4018/978-1-4666-9545-0.ch010

Ong, P., & Kohshelan, S. (2016). Performances of Adaptive Cuckoo Search Algorithm in Engineering Optimization. In P. Vasant, G. Weber, & V. Dieu (Eds.), *Handbook of Research on Modern Optimization Algorithms and Applications in Engineering and Economics* (pp. 676–699). Hershey, PA: IGI Global. doi:10.4018/978-1-4666-9644-0.ch026

Osho, M. B. (2018). Industrial Enzyme Technology: Potential Applications. In S. Bharati & P. Chaurasia (Eds.), *Research Advancements in Pharmaceutical, Nutritional, and Industrial Enzymology* (pp. 375–394). Hershey, PA: IGI Global. doi:10.4018/978-1-5225-5237-6.ch017

Padmaja, P., & Marutheswar, G. (2017). Certain Investigation on Secured Data Transmission in Wireless Sensor Networks. *International Journal of Mobile Computing and Multimedia Communications, 8*(1), 48–61. doi:10.4018/IJMCMC.2017010104

Paixão, S. M., Silva, T. P., Arez, B. F., & Alves, L. (2016). Advances in the Reduction of the Costs Inherent to Fossil Fuels' Biodesulfurization towards Its Potential Industrial Application. In T. Saleh (Ed.), *Applying Nanotechnology to the Desulfurization Process in Petroleum Engineering* (pp. 390–425). Hershey, PA: IGI Global. doi:10.4018/978-1-4666-9545-0.ch013

Palmer, S., & Hall, W. (2017). An Evaluation of Group Work in First-Year Engineering Design Education. In R. Tucker (Ed.), *Collaboration and Student Engagement in Design Education* (pp. 145–168). Hershey, PA: IGI Global. doi:10.4018/978-1-5225-0726-0.ch007

Panneer, R. (2017). Effect of Composition of Fibers on Properties of Hybrid Composites. *International Journal of Manufacturing, Materials, and Mechanical Engineering, 7*(4), 28–43. doi:10.4018/IJMMME.2017100103

Parker, J. (2016). Hubble's Expanding Universe: A Model for Quality in Technology Infused engineering and Technology Education. *International Journal of Quality Assurance in Engineering and Technology Education*, *5*(2), 16–29. doi:10.4018/IJQAETE.2016040102

Paul, S., & Roy, P. (2018). Optimal Design of Power System Stabilizer Using a Novel Evolutionary Algorithm. *International Journal of Energy Optimization and Engineering*, *7*(3), 24–46. doi:10.4018/IJEOE.2018070102

Pavaloiu, A. (2018). Artificial Intelligence Ethics in Biomedical-Engineering-Oriented Problems. In U. Kose, G. Guraksin, & O. Deperlioglu (Eds.), *Nature-Inspired Intelligent Techniques for Solving Biomedical Engineering Problems* (pp. 219–231). Hershey, PA: IGI Global. doi:10.4018/978-1-5225-4769-3.ch010

Peña, F. (2016). A Semi-Discrete Approach for the Numerical Simulation of Freestanding Blocks. In V. Sarhosis, K. Bagi, J. Lemos, & G. Milani (Eds.), *Computational Modeling of Masonry Structures Using the Discrete Element Method* (pp. 416–439). Hershey, PA: IGI Global. doi:10.4018/978-1-5225-0231-9.ch016

Penchovsky, R., & Traykovska, M. (2016). Synthetic Approaches to Biology: Engineering Gene Control Circuits, Synthesizing, and Editing Genomes. In M. Bououdina (Ed.), *Emerging Research on Bioinspired Materials Engineering* (pp. 323–351). Hershey, PA: IGI Global. doi:10.4018/978-1-4666-9811-6.ch012

Pieroni, A., & Iazeolla, G. (2016). Engineering QoS and Energy Saving in the Delivery of ICT Services. In P. Vasant & N. Voropai (Eds.), *Sustaining Power Resources through Energy Optimization and Engineering* (pp. 208–226). Hershey, PA: IGI Global. doi:10.4018/978-1-4666-9755-3.ch009

Pioro, I., Mahdi, M., & Popov, R. (2017). Application of Supercritical Pressures in Power Engineering. In L. Chen & Y. Iwamoto (Eds.), *Advanced Applications of Supercritical Fluids in Energy Systems* (pp. 404–457). Hershey, PA: IGI Global. doi:10.4018/978-1-5225-2047-4.ch013

Plaksina, T., & Gildin, E. (2017). Rigorous Integrated Evolutionary Workflow for Optimal Exploitation of Unconventional Gas Assets. *International Journal of Energy Optimization and Engineering*, *6*(1), 101–122. doi:10.4018/IJEOE.2017010106

Puppala, A. J., Bheemasetti, T. V., Zou, H., Yu, X., Pedarla, A., & Cai, G. (2016). Spatial Variability Analysis of Soil Properties using Geostatistics. In P. Samui (Ed.), *Handbook of Research on Advanced Computational Techniques for Simulation-Based Engineering* (pp. 195–226). Hershey, PA: IGI Global. doi:10.4018/978-1-4666-9479-8.ch008

Ramdani, N., & Azibi, M. (2018). Polymer Composite Materials for Microelectronics Packaging Applications: Composites for Microelectronics Packaging. In K. Kumar & J. Davim (Eds.), *Composites and Advanced Materials for Industrial Applications* (pp. 177–211). Hershey, PA: IGI Global. doi:10.4018/978-1-5225-5216-1.ch009

Ramesh, M., Garg, R., & Subrahmanyam, G. V. (2017). Investigation of Influence of Quenching and Annealing on the Plane Fracture Toughness and Brittle to Ductile Transition Temperature of the Zinc Coated Structural Steel Materials. *International Journal of Surface Engineering and Interdisciplinary Materials Science*, *5*(2), 33–42. doi:10.4018/IJSEIMS.2017070103

Razavi, A. M., & Ahmad, R. (2016). Agile Software Development Challenges in Implementation and Adoption: Focusing on Large and Distributed Settings – Past Experiences, Emergent Topics. In I. Ghani, D. Jawawi, S. Dorairaj, & A. Sidky (Eds.), *Emerging Innovations in Agile Software Development* (pp. 175–207). Hershey, PA: IGI Global. doi:10.4018/978-1-4666-9858-1.ch010

Reccia, E., Cecchi, A., & Milani, G. (2016). FEM/DEM Approach for the Analysis of Masonry Arch Bridges. In V. Sarhosis, K. Bagi, J. Lemos, & G. Milani (Eds.), *Computational Modeling of Masonry Structures Using the Discrete Element Method* (pp. 367–392). Hershey, PA: IGI Global. doi:10.4018/978-1-5225-0231-9.ch014

Ro, H. K., & McIntosh, K. (2016). Constructing Conducive Environment for Women of Color in Engineering Undergraduate Education. In U. Thomas & J. Drake (Eds.), *Critical Research on Sexism and Racism in STEM Fields* (pp. 23–48). Hershey, PA: IGI Global. doi:10.4018/978-1-5225-0174-9.ch002

Rodulfo-Baechler, S. M. (2016). Dual Role of Perovskite Hollow Fiber Membrane in the Methane Oxidation Reactions. In H. Al-Megren & T. Xiao (Eds.), *Petrochemical Catalyst Materials, Processes, and Emerging Technologies* (pp. 385–430). Hershey, PA: IGI Global. doi:10.4018/978-1-4666-9975-5.ch014

Rudolf, S., Biryuk, V. V., & Volov, V. (2018). Vortex Effect, Vortex Power: Technology of Vortex Power Engineering. In V. Kharchenko & P. Vasant (Eds.), *Handbook of Research on Renewable Energy and Electric Resources for Sustainable Rural Development* (pp. 500–533). Hershey, PA: IGI Global. doi:10.4018/978-1-5225-3867-7.ch021

Sah, A., Bhadula, S. J., Dumka, A., & Rawat, S. (2018). A Software Engineering Perspective for Development of Enterprise Applications. In A. Elçi (Ed.), *Handbook of Research on Contemporary Perspectives on Web-Based Systems* (pp. 1–23). Hershey, PA: IGI Global. doi:10.4018/978-1-5225-5384-7.ch001

Sahoo, P., & Roy, S. (2017). Tribological Behavior of Electroless Ni-P, Ni-P-W and Ni-P-Cu Coatings: A Comparison. *International Journal of Surface Engineering and Interdisciplinary Materials Science*, *5*(1), 1–15. doi:10.4018/IJSEIMS.2017010101

Sahoo, S. (2018). Laminated Composite Hypar Shells as Roofing Units: Static and Dynamic Behavior. In K. Kumar & J. Davim (Eds.), *Composites and Advanced Materials for Industrial Applications* (pp. 249–269). Hershey, PA: IGI Global. doi:10.4018/978-1-5225-5216-1.ch011

Sahu, H., & Hungyo, M. (2018). Introduction to SDN and NFV. In A. Dumka (Ed.), *Innovations in Software-Defined Networking and Network Functions Virtualization* (pp. 1–25). Hershey, PA: IGI Global. doi:10.4018/978-1-5225-3640-6.ch001

Saikia, P., Bharadwaj, S. K., & Miah, A. T. (2016). Peroxovanadates and Its Bio-Mimicking Relation with Vanadium Haloperoxidases. In M. Bououdina (Ed.), *Emerging Research on Bioinspired Materials Engineering* (pp. 197–219). Hershey, PA: IGI Global. doi:10.4018/978-1-4666-9811-6.ch007

Saladino, R., Botta, G., & Crucianelli, M. (2016). Advances in Nanotechnology Transition Metal Catalysts in Oxidative Desulfurization (ODS) Processes: Nanotechnology Applied to ODS Processing. In T. Saleh (Ed.), *Applying Nanotechnology to the Desulfurization Process in Petroleum Engineering* (pp. 180–215). Hershey, PA: IGI Global. doi:10.4018/978-1-4666-9545-0.ch006

Saleh, T. A., Danmaliki, G. I., & Shuaib, T. D. (2016). Nanocomposites and Hybrid Materials for Adsorptive Desulfurization. In T. Saleh (Ed.), *Applying Nanotechnology to the Desulfurization Process in Petroleum Engineering* (pp. 129–153). Hershey, PA: IGI Global. doi:10.4018/978-1-4666-9545-0.ch004

Saleh, T. A., Shuaib, T. D., Danmaliki, G. I., & Al-Daous, M. A. (2016). Carbon-Based Nanomaterials for Desulfurization: Classification, Preparation, and Evaluation. In T. Saleh (Ed.), *Applying Nanotechnology to the Desulfurization Process in Petroleum Engineering* (pp. 154–179). Hershey, PA: IGI Global. doi:10.4018/978-1-4666-9545-0.ch005

Salem, A. M., & Shmelova, T. (2018). Intelligent Expert Decision Support Systems: Methodologies, Applications, and Challenges. In T. Shmelova, Y. Sikirda, N. Rizun, A. Salem, & Y. Kovalyov (Eds.), *Socio-Technical Decision Support in Air Navigation Systems: Emerging Research and Opportunities* (pp. 215–242). Hershey, PA: IGI Global. doi:10.4018/978-1-5225-3108-1.ch007

Samal, M. (2017). FE Analysis and Experimental Investigation of Cracked and Un-Cracked Thin-Walled Tubular Components to Evaluate Mechanical and Fracture Properties. In P. Samui, S. Chakraborty, & D. Kim (Eds.), *Modeling and Simulation Techniques in Structural Engineering* (pp. 266–293). Hershey, PA: IGI Global. doi:10.4018/978-1-5225-0588-4.ch009

Samal, M., & Balakrishnan, K. (2017). Experiments on a Ring Tension Setup and FE Analysis to Evaluate Transverse Mechanical Properties of Tubular Components. In P. Samui, S. Chakraborty, & D. Kim (Eds.), *Modeling and Simulation Techniques in Structural Engineering* (pp. 91–115). Hershey, PA: IGI Global. doi:10.4018/978-1-5225-0588-4.ch004

Santhanakumar, M., Adalarasan, R., & Rajmohan, M. (2016). An Investigation in Abrasive Waterjet Cutting of Al6061/SiC/Al2O3 Composite Using Principal Component Based Response Surface Methodology. *International Journal of Manufacturing, Materials, and Mechanical Engineering*, *6*(4), 30–47. doi:10.4018/IJMMME.2016100103

Sareen, N., & Bhattacharya, S. (2016). Cleaner Energy Fuels: Hydrodesulfurization and Beyond. In T. Saleh (Ed.), *Applying Nanotechnology to the Desulfurization Process in Petroleum Engineering* (pp. 84–128). Hershey, PA: IGI Global. doi:10.4018/978-1-4666-9545-0.ch003

Sarhosis, V. (2016). Micro-Modeling Options for Masonry. In V. Sarhosis, K. Bagi, J. Lemos, & G. Milani (Eds.), *Computational Modeling of Masonry Structures Using the Discrete Element Method* (pp. 28–60). Hershey, PA: IGI Global. doi:10.4018/978-1-5225-0231-9.ch002

Sarhosis, V., Oliveira, D. V., & Lourenco, P. B. (2016). On the Mechanical Behavior of Masonry. In V. Sarhosis, K. Bagi, J. Lemos, & G. Milani (Eds.), *Computational Modeling of Masonry Structures Using the Discrete Element Method* (pp. 1–27). Hershey, PA: IGI Global. doi:10.4018/978-1-5225-0231-9.ch001

Satyam, N. (2016). Liquefaction Modelling of Granular Soils using Discrete Element Method. In P. Samui (Ed.), *Handbook of Research on Advanced Computational Techniques for Simulation-Based Engineering* (pp. 381–441). Hershey, PA: IGI Global. doi:10.4018/978-1-4666-9479-8.ch015

Sawant, S. (2018). Deep Learning and Biomedical Engineering. In U. Kose, G. Guraksin, & O. Deperlioglu (Eds.), *Nature-Inspired Intelligent Techniques for Solving Biomedical Engineering Problems* (pp. 283–296). Hershey, PA: IGI Global. doi:10.4018/978-1-5225-4769-3.ch014

Sezgin, H., & Berkalp, O. B. (2018). Textile-Reinforced Composites for the Automotive Industry. In K. Kumar & J. Davim (Eds.), *Composites and Advanced Materials for Industrial Applications* (pp. 129–156). Hershey, PA: IGI Global. doi:10.4018/978-1-5225-5216-1.ch007

Shah, M. Z., Gazder, U., Bhatti, M. S., & Hussain, M. (2018). Comparative Performance Evaluation of Effects of Modifier in Asphaltic Concrete Mix. *International Journal of Strategic Engineering*, *1*(2), 13–25. doi:10.4018/IJoSE.2018070102

Shah, V. S., Shah, H. R., & Samui, P. (2016). Application of Meta-Models (MPMR and ELM) for Determining OMC, MDD and Soaked CBR Value of Soil. In S. Bhattacharyya, P. Banerjee, D. Majumdar, & P. Dutta (Eds.), *Handbook of Research on Advanced Hybrid Intelligent Techniques and Applications* (pp. 454–482). Hershey, PA: IGI Global. doi:10.4018/978-1-4666-9474-3.ch015

Sharma, N., & Kumar, K. (2018). Fabrication of Porous NiTi Alloy Using Organic Binders. In K. Kumar & J. Davim (Eds.), *Composites and Advanced Materials for Industrial Applications* (pp. 38–62). Hershey, PA: IGI Global. doi:10.4018/978-1-5225-5216-1.ch003

Sharma, T. K. (2016). Application of Shuffled Frog Leaping Algorithm in Software Project Scheduling. In P. Saxena, D. Singh, & M. Pant (Eds.), *Problem Solving and Uncertainty Modeling through Optimization and Soft Computing Applications* (pp. 225–238). Hershey, PA: IGI Global. doi:10.4018/978-1-4666-9885-7.ch011

Shivach, P., Nautiyal, L., & Ram, M. (2018). Applying Multi-Objective Optimization Algorithms to Mechanical Engineering. In M. Ram & J. Davim (Eds.), *Soft Computing Techniques and Applications in Mechanical Engineering* (pp. 287–301). Hershey, PA: IGI Global. doi:10.4018/978-1-5225-3035-0.ch014

Shmelova, T. (2018). Stochastic Methods for Estimation and Problem Solving in Engineering: Stochastic Methods of Decision Making in Aviation. In S. Kadry (Ed.), *Stochastic Methods for Estimation and Problem Solving in Engineering* (pp. 139–160). Hershey, PA: IGI Global. doi:10.4018/978-1-5225-5045-7.ch006

Shukla, R., Anapagaddi, R., Singh, A. K., Allen, J. K., Panchal, J. H., & Mistree, F. (2016). Integrated Computational Materials Engineering for Determining the Set Points of Unit Operations for Production of a Steel Product Mix. In S. Datta & J. Davim (Eds.), *Computational Approaches to Materials Design: Theoretical and Practical Aspects* (pp. 163–191). Hershey, PA: IGI Global. doi:10.4018/978-1-5225-0290-6.ch006

Siero González, L. R., & Romo Vázquez, A. (2017). Didactic Sequences Teaching Mathematics for Engineers With Focus on Differential Equations. In M. Ramírez-Montoya (Ed.), *Handbook of Research on Driving STEM Learning With Educational Technologies* (pp. 129–151). Hershey, PA: IGI Global. doi:10.4018/978-1-5225-2026-9.ch007

Singh, R., & Dutta, S. (2018). Visible Light Active Nanocomposites for Photocatalytic Applications. In K. Kumar & J. Davim (Eds.), *Composites and Advanced Materials for Industrial Applications* (pp. 270–296). Hershey, PA: IGI Global. doi:10.4018/978-1-5225-5216-1.ch012

Singh, R., & Lou, H. H. (2016). Safety and Efficiency Enhancement in LNG Terminals. In H. Al-Megren & T. Xiao (Eds.), *Petrochemical Catalyst Materials, Processes, and Emerging Technologies* (pp. 164–176). Hershey, PA: IGI Global. doi:10.4018/978-1-4666-9975-5.ch007

Sözbilir, H., Özkaymak, Ç., Uzel, B., & Sümer, Ö. (2018). Criteria for Surface Rupture Microzonation of Active Faults for Earthquake Hazards in Urban Areas. In N. Ceryan (Ed.), *Handbook of Research on Trends and Digital Advances in Engineering Geology* (pp. 187–230). Hershey, PA: IGI Global. doi:10.4018/978-1-5225-2709-1.ch005

Stanciu, I. (2018). Stochastic Methods in Microsystems Engineering. In S. Kadry (Ed.), *Stochastic Methods for Estimation and Problem Solving in Engineering* (pp. 161–176). Hershey, PA: IGI Global. doi:10.4018/978-1-5225-5045-7.ch007

Strebkov, D., Nekrasov, A., Trubnikov, V., & Nekrasov, A. (2018). Single-Wire Resonant Electric Power Systems for Renewable-Based Electric Grid. In V. Kharchenko & P. Vasant (Eds.), *Handbook of Research on Renewable Energy and Electric Resources for Sustainable Rural Development* (pp. 449–474). Hershey, PA: IGI Global. doi:10.4018/978-1-5225-3867-7.ch019

Subburaman, D., Jagan, J., Dalkiliç, Y., & Samui, P. (2016). Reliability Analysis of Slope Using MPMR, GRNN and GPR. In F. Miranda & C. Abreu (Eds.), *Handbook of Research on Computational Simulation and Modeling in Engineering* (pp. 208–224). Hershey, PA: IGI Global. doi:10.4018/978-1-4666-8823-0.ch007

Sun, J., Wan, S., Lin, J., & Wang, Y. (2016). Advances in Catalytic Conversion of Syngas to Ethanol and Higher Alcohols. In H. Al-Megren & T. Xiao (Eds.), *Petrochemical Catalyst Materials, Processes, and Emerging Technologies* (pp. 177–215). Hershey, PA: IGI Global. doi:10.4018/978-1-4666-9975-5.ch008

Tüdeş, Ş., Kumlu, K. B., & Ceryan, S. (2018). Integration Between Urban Planning and Natural Hazards For Resilient City. In N. Ceryan (Ed.), *Handbook of Research on Trends and Digital Advances in Engineering Geology* (pp. 591–630). Hershey, PA: IGI Global. doi:10.4018/978-1-5225-2709-1.ch017

Tyukhov, I., Rezk, H., & Vasant, P. (2016). Modern Optimization Algorithms and Applications in Solar Photovoltaic Engineering. In P. Vasant & N. Voropai (Eds.), *Sustaining Power Resources through Energy Optimization and Engineering* (pp. 390–445). Hershey, PA: IGI Global. doi:10.4018/978-1-4666-9755-3.ch016

Ulamis, K. (2018). Soil Liquefaction Assessment by Anisotropic Cyclic Triaxial Test. In N. Ceryan (Ed.), *Handbook of Research on Trends and Digital Advances in Engineering Geology* (pp. 631–664). Hershey, PA: IGI Global. doi:10.4018/978-1-5225-2709-1.ch018

Umar, M. A., Tenuche, S. S., Yusuf, S. A., Abdulsalami, A. O., & Kufena, A. M. (2016). Usability Engineering in Agile Software Development Processes. In I. Ghani, D. Jawawi, S. Dorairaj, & A. Sidky (Eds.), *Emerging Innovations in Agile Software Development* (pp. 208–221). Hershey, PA: IGI Global. doi:10.4018/978-1-4666-9858-1.ch011

Üzüm, O., & Çakır, Ö. A. (2016). A Bio-Inspired Phenomena in Cementitious Materials: Self-Healing. In M. Bououdina (Ed.), *Emerging Research on Bioinspired Materials Engineering* (pp. 220–246). Hershey, PA: IGI Global. doi:10.4018/978-1-4666-9811-6.ch008

Valente, M., & Milani, G. (2017). Seismic Assessment and Retrofitting of an Under-Designed RC Frame Through a Displacement-Based Approach. In V. Plevris, G. Kremmyda, & Y. Fahjan (Eds.), *Performance-Based Seismic Design of Concrete Structures and Infrastructures* (pp. 36–58). Hershey, PA: IGI Global. doi:10.4018/978-1-5225-2089-4.ch002

Vasant, P. (2018). A General Medical Diagnosis System Formed by Artificial Neural Networks and Swarm Intelligence Techniques. In U. Kose, G. Guraksin, & O. Deperlioglu (Eds.), *Nature-Inspired Intelligent Techniques for Solving Biomedical Engineering Problems* (pp. 130–145). Hershey, PA: IGI Global. doi:10.4018/978-1-5225-4769-3.ch006

Vergara, D., Lorenzo, M., & Rubio, M. (2016). On the Use of Virtual Environments in Engineering Education. *International Journal of Quality Assurance in Engineering and Technology Education*, *5*(2), 30–41. doi:10.4018/IJQAETE.2016040103

Verrollot, J., Tolonen, A., Harkonen, J., & Haapasalo, H. J. (2018). Challenges and Enablers for Rapid Product Development. *International Journal of Applied Industrial Engineering*, *5*(1), 25–49. doi:10.4018/IJAIE.2018010102

Wagner, C., & Ryan, C. (2016). Physical and Digital Integration Strategies of Electronic Device Supply Chains and Their Applicability to ETO Supply Chains. In R. Addo-Tenkorang, J. Kantola, P. Helo, & A. Shamsuzzoha (Eds.), *Supply Chain Strategies and the Engineer-to-Order Approach* (pp. 224–245). Hershey, PA: IGI Global. doi:10.4018/978-1-5225-0021-6.ch011

Wang, Z., Wu, P., Lan, L., & Ji, S. (2016). Preparation, Characterization and Desulfurization of the Supported Nickel Phosphide Catalysts. In H. Al-Megren & T. Xiao (Eds.), *Petrochemical Catalyst Materials, Processes, and Emerging Technologies* (pp. 431–458). Hershey, PA: IGI Global. doi:10.4018/978-1-4666-9975-5.ch015

Yardimci, A. G., & Karpuz, C. (2018). Fuzzy Rock Mass Rating: Soft-Computing-Aided Preliminary Stability Analysis of Weak Rock Slopes. In N. Ceryan (Ed.), *Handbook of Research on Trends and Digital Advances in Engineering Geology* (pp. 97–131). Hershey, PA: IGI Global. doi:10.4018/978-1-5225-2709-1.ch003

Zhang, L., Ding, S., Sun, S., Han, B., Yu, X., & Ou, J. (2016). Nano-Scale Behavior and Nano-Modification of Cement and Concrete Materials. In A. Khitab & W. Anwar (Eds.), *Advanced Research on Nanotechnology for Civil Engineering Applications* (pp. 28–79). Hershey, PA: IGI Global. doi:10.4018/978-1-5225-0344-6.ch002

Zindani, D., & Kumar, K. (2018). Industrial Applications of Polymer Composite Materials. In K. Kumar & J. Davim (Eds.), *Composites and Advanced Materials for Industrial Applications* (pp. 1–15). Hershey, PA: IGI Global. doi:10.4018/978-1-5225-5216-1.ch001

Zindani, D., Maity, S. R., & Bhowmik, S. (2018). A Decision-Making Approach for Material Selection of Polymeric Composite Bumper Beam. In K. Kumar & J. Davim (Eds.), *Composites and Advanced Materials for Industrial Applications* (pp. 112–128). Hershey, PA: IGI Global. doi:10.4018/978-1-5225-5216-1.ch006

About the Contributors

Arun Solanki is working in the Department of Computer Science, Gautam Buddha University, Greater Noida, India, where he has been since 2009. He has worked as member examination, admission, sports council and other university teams from time to time. He received M. Tech. from YMCA University, Faridabad, Haryana, India. He received his Ph.D. in Computer Science and Engineering from the Gautam Buddha University in 2014. More than 50 M.Tech. dissertations are submitted under his guidance. His research interests span Expert System, Machine Learning and Search Engines. He has published many research articles in SCI/Scopus indexed International journal. He has participated in many international conferences. He has been technical and advisory committee members of many conferences. He has organized several FDP, Conferences, Workshops, and Seminars. He has chaired many sessions at International Conferences. Arun Solanki is working as Associate Editor in International Journal of Web-Based Learning and Teaching Technologies (IJWLTT) IGI Global publisher. He has been working as Guest Editor for special issues in Recent Patents on Computer Science, Bentham Science Publishers. He is the editor of many Books with a reputed publisher like IGI Global, CRC and AAP. He is working as the reviewer in Springer, IGI Global, Elsevier, and other reputed publisher journals.

Anand Nayyar (Academician, Researcher, Author, Writer, Inventor, Innovator, Scientist, Consultant and Orator) received MCA from Punjabi University, Patiala, in 2008 with Gold Medal and Distinction. He received his M. Phil in Computer Science in 2009 in 1st Division from V.M.U, Tamil, Nadu. He received his M. Tech in Information Technology in 2011 with Distinction cum University Topper. He received his MBA in Information Systems from Sikkim Manipal University with 1st Division. He has done his Ph. D. in Computer Science from Desh Bhagat University, Mandi Gobindgarh in 2017 in Wireless Sensor Networks, Swarm Intelligence and Network Simulation. (Topper in Course Work). He has published 260+ Research Papers in various National and International Conferences and National and International Referred Journals (Scopus, Thomson Reuters, SCI, SCIE) and even Reputed Computer/IT Magazines. He has published 18 Books in different subjects of Computer Science with National and International publishers. Awarded with 20+ National Awards for Teaching, Research, Innovation and Scientist Activities. Member of 50+ Research Organizations as Senior/Associate and Life Member including IEEE, ACM, CSI, etc.

* * *

Jafar Alzubi received a B.Sc (Hons) in Electrical Engineering, majoring in Electronics and Communications from the University of Engineering and Technology, Lahore, Pakistan, in 2001. In 2005, he received M.Sc. (Hons) in Electrical and Computer Engineering from New York Institute of Technology, New York, NY. Between 2005-2008, he became a full-time lecturer in the School of Engineering at Al-Balqa Applied University – Jordan. In 2008, He joined the Wireless Communications Research Laboratory at Swansea University - UK, completed his PhD in Advanced Telecommunications Engineering in June 2012. He is now an associate professor at Computer Engineering department, Al-Balqa Applied University - Jordan. His research interests include and focus on Elliptic curves cryptography and cryptosystems, classifications and detection of web scams, using Algebraic-Geometric theory in channel coding for wireless networks.

Aysem Berrin Cakmakli is an instructor in Department of Architecture at METU. She has been teaching for twenty years in the field of Architecture and Building Science. She received her B.Arch., M.Sc. and Ph.D. degrees from the Middle East Technical University, Faculty of Architecture in Building Science. Her master thesis is about sustainable architecture and sustainable building materials and their effects on human and environment health. The title of her Ph.D. thesis is "Life Cycle Assessment of Building Materials in Hotel Refurbishment Projects: A Case Study in Ankara". Her main fields of interest are ecological and sustainable architecture and building materials, energy efficiency and life cycle assessment of buildings, building performance evaluations and lighting in architecture.

Mir Sayed Shah Danish earns double master degrees, in Energy & Electrical Engineering (MSc. 2015), and Business Administration (MBA, 2016), and PhD. in Sustainable Energy (2018) from India and Japan respectively. Since 2004, he has been actively involved in the industry of Energy & Electrical Engineering in practical works, academic researches, and teaching. Besides other professional duties, he has delivered various projects at different scales as team lead and project manager. Currently, he is doing as Assistant Professor (Specially Appointed Teaching Associate) in the field of Energy Engineering with University of the Ryukyus, Japan. He has remained an active member with several academic societies (namely: IEEE Senior Member, IET, IAES, IRED, PMI).

Subhankar Das has a Ph.D. in Marketing, UGC NET (Qualifying exam for Assistant Professor) qualified faculty having 6 SCOPUS indexed publications in management to date. Also, two SCOPUS indexed acceptance (Inderscience & Espacios publications) and five research papers in review (All are in SCOPUS with Sage, Inderscience & Espacios publications) to date. He has qualified IELTS (International English Language Testing System) with overall band of 6.5 in 2017. He has an authored Book in review for Digital marketing with CRC Press Routledge, India. He is being chosen as a Guest editor for a special issue with Int. Journal of Environmental Engineering (Inderscience). In the last year, He has won 3 International Awards in his field of study & research from reputed organisations. He also got certified by IIM Lucknow, University of British Columbia & IIM Bangalore, Manipal & Google for Case based pedagogy and digital branding. He is also a Member of Center of Excellence for Tata Institute of Social Science TISS-SVE for 'Management & Entrepreneurship and Professional Skills' Vertical, Mumbai. Apart from above research work, he also has an international case, three journals, nine conference proceeding publications in national repute. He has attended 12 FDP & 17 conferences and association with 9 associations with various International & national governing bodies.

Pijush Kanti Dutta Pramanik, having more than 10 years of teaching experience, is a dedicated academician with a passion for imparting knowledge. After graduating in Physics from Burdwan University, West Bengal, he continued acquiring a range of professional qualifications in the core and allied fields of Computer Science and Information Technology, viz. MIT, MCA, MBA (IT), MTech (CSE), and MPhil (CS). At present, he is registered as a Ph.D. candidate in the department of CSE at National Institute of Technology, Durgapur, India. He is actively engaged in research in the domains of the Internet of Things, Grid Computing, Fog Computing, Crowd Computing, Smartphone Computing, and Recommendation Systems. He has authored and co-authored a number of research articles and book chapters which are published (or to be published) by reputed publishers such as IEEE, Springer, CRC Press, Elsevier, InderScience, River Publishers, and IGI Global. For details, please refer to his Google Scholar profile at https://scholar.google.co.in/citations?hl=en&user=KqefJxsAAAAJ.

Vaibhav Goel is a student of Bharati Vidyapeeth's College of Engineering, New Delhi. He is currently pursuing B.Tech in Computer Science & Engineering branch. He is very passionate about everything he does.

Dhanalekshmi Gopinathan has completed her Ph.D Degree in the area of Computer Science from Jaypee Institute of Information Technology, Noida. She received her B.Tech and M.Tech in Computer Science and Engineering from National Institute of Information Technology, Calicut, formerly known as Regional Engineering College. Her research interest covers Database Management Systems, Artificial Intelligence, Automata Theory and Compiler Design. She has 15 years of teaching experience. She has worked as Scientist 'B' in CDAC- Trivandrum, Kerala. She has seven research publications in reputed peer-reviewed International Journals and International Conferences.

Priyanka Goyal is working at school of ICT, Gautam Buddha University, Greater Noida. She completed her PhD degree in on-chip optical interconnects. Priyanka has completed her M.Tech. in VLSI design and B.Tech. in Electronics and Communication Engineering from Kurukshetra University. She has published many research papers in international and national journals and conferences like SCI indexed, scopus and IEEE conference.

Gurjit Kaur is an Associate Professor in the Department of Electronics & Communication Engineering at the Delhi Technological University (DTU), Delhi, India. She has been a topper throughout her academic education. As a testimonial to the same, she has been awarded by Chief Minister S. Prakash Singh Badal for being the topper in Punjab state. After that, she was awarded a 'Gold Medal' by former President of India Dr. A. P. J. Abdul Kalam for being the overall topper of the Punjab Technical University, Jalandhar in B.Tech program. She also received an Honour by Guru Harkrishan Education Society for being topper among all the colleges and all the disciplines of PTU, Jalandhar. She then proceeded to PEC University of Technology, Chandigarh, to complete her M. Tech in 2003 and also earned her Ph.D. degree from Panjab University, Chandigarh in 2010 with distinction. She has spent over 16 years of her academic career towards research and teaching in the field of Electronics and Communication in well-reputed institutes like PEC University of Technology, Punjab University, Jaypee Institute of Information and Technology, Gautam Buddha University, and Delhi Technological University, Delhi. During this academic tenure, she has guided four Ph.D. students and more than 50 M.Tech students. Her research interests mainly include Optical CDMA, Wireless Communication system, high-speed interconnect and IOT. She has also authored three books at the international and national level. Her two books, i.e., "Handbook of Research on Big Data and the IoT" and "Examining Cloud Computing Technologies through the Internet of Things," were published by IGI Global, International Publisher of Progressive Information Science and Technology Research, in 2018 and 2017 respectively. She also authored a national level book entitled "Optical Communication," which was published by Galgotia Publications in 2005. Apart from that, Dr. Kaur contributed various book chapters in books by well-renowned publishers, i.e., IGI Global, Springer, CRC, etc. She has presented her research work as short courses/tutorials in many national and international conferences. In recognition of her outstanding contribution as an active researcher, Dr. Kaur also received Bharat Vikas Award by Institute of Self Reliance in National Seminar on Diversity of Cultural and Social Environment at Bhubaneswar, Odisha, India & Prof. Indira Parikh 50 Women in Education Leader Award in World Education Congress, Mumbai, India in 2017. Her career accomplishments over the past 16 years include over 90 peer-reviewed publications, including 55 journal publications, 35 conference publications, and over ten invited talks/seminars. Due to her pioneer research, her name is also listed in Who's Who in Science and Engineering Directory, USA, and her biography was also published by International Biographic Center, Great Britain, as 2000 Outstanding Scientists. She also worked as a convener for two international conferences i.e. International ICIAICT 2012 which was organized by CSI, Noida Chapter, and International conference EPPICTM 2012, which was held in collaboration with MTMI, USA, University of Maryland Eastern Shore, USA and Frostburg State University, USA at School of Information and Communication Technology, Gautam Buddha University, Greater Noida, India. She has delivered many expert talks on the emerging trends in optical for all domain applications at various universities and she served as a reviewer of multiple journals like IEEE transactions on communication, etc.

Rajalakshmi Krishnamurthi obtained her Ph. D in Computer Science and Engineering from Jaypee Institute of Information Technology Noida, India. She did her M. E (CSE), and B. E. (EEE) from Bharathiar University, Coimbatore, India. She is currently an Assistant Professor (Senior Grade) with the Department of Computer Science and Engineering, Jaypee Institute of Information Technology, Noida. Her current research interests include Mobile computing, Wireless networks, Cloud Computing, Pervasive Computing.

Suman Kumar is working as M.Tech Scholar in Delhi Technical University, Delhi. His research interest includes the VLSI circuits design and IoT.

Subhra Mondral is a Ph.D. in management with specialization in retail & customer satisfaction, having 10+ years of teaching experience. She is having 3 Scopus indexed publication in international journals. Her domain subjects include Brand Management, Retail Marketing, Marketing Management, Strategic Management, service marketing & General management. Her areas of research interests are Customer satisfaction & Brand Loyalty in Retail Industry. She has published several research papers in refereed peer-reviewed Journals of National repute. She has also presented several research papers at International and National seminars. She is having ten years of progressive teaching experience across with proven abilities in classroom teaching, management of the departmental activity.

Bulbul Mukherjee is an Assistant Professor in the department of Computer Science & Engineering at Bengal Institute of Technology, India. She has more than 10 years of teaching experience in the field of Computer Science & Engineering. She holds M. Tech degree from Maulana Abul Kalam Azad University of Technology, West Bengal, India. Her primary research area includes Software Engineering, Iot, Ubiquitous Computing, Learning Technology.

Saurabh Pal is an Assistant Professor in the Department of Computer Science & Engineering at Bengal Institute of Technology, India. He has more than 10 years of teaching experience in the field of Computer Science and Engineering. He holds a B.Sc from the University of Calcutta, M.Sc-IT from Allahabad Agricultural Institute, and M.Tech-IT from Jadavpur University, India. He is pursuing Ph.D. in Computer Science & Eng from National Institute of Technology, Durgapur, India. He has published his research in many International Journals, Books and Conferences. His primary research area includes education technology, mobile computing, context-aware computing, semantic web, and natural language processing.

Tanmoy Pal has an MTech and BTech in Computer Science & Engineering, and 10 years of work experience in Academics as Assistant Professor. He is working as an Assistant Professor in the department of CSE in Bengal Institute of Technology, Kolkata. Research areas of interest: IoT, medical image processing, machine learning.

Rudra Rameshwar is Management Faculty working in Operations Management, Energy and Sustainability Area at LMTSoM, Thapar Institute of Engg. & Tech. (Deemed to be University) Patiala, Dera Bassi Campus, Mohali, India. He is serving as the Coordinator, Summer Internships MBA & Placements and Coordinator, Environment and Sustainability Club. He has been active as Resource Person in various FDP, National Seminars,Workshops, Conferences, Short Term Courses organized by National Universities/Colleges in following areas: Sustainable Development, Renewable Energy Systems, Sustainable Operations Management, Research Methodology, Energy Management, Strategic Management, Business Entrepreneurship, Business Statistics, Research Design and SPSS Data Analysis. He received his Ph.D. degree in 'Strategic-Technology Management' and M. Tech. degree in 'Alternate Hydro Energy Systems' at I. I. T. Roorkee. His research interests cover the Business Research Methods, Statistics for Management, Quantitative Techniques, Technology Management and Operations, Research Methodology, Business Entrepreneurship, Sustainable Business Strategy, Qualitative & Quantitative Research, etc. He has published 47 research papers ranging from Science & Technology to Management and won awards.

Jusleen Rekhi is a student of Bharati Vidyapeeth's College of Engineering, New Delhi. She is currently pursuing B.Tech in Computer Science & Engineering branch.

Kalyan Sahoo is Vice-President of Academic affairs at African University College of Communications (AUCC), Accra, Ghana.

Tomonobu Senjyu was born in Saga, Japan, in 1963. He received the B.S. and M.S. degrees from the University of the Ryukyus, Okinawa, Japan, in 1986 and 1988, respectively, and the Ph.D. degree from Nagoya University, Aichi, Japan, in 1994, all in electrical engineering. Since 1988, he has been with the Department of Electrical and Electronics Engineering, Faculty of Engineering, University of the Ryukyus, where he is currently working as a Professor. His research interests include stability of ac machines, advanced control of electrical machines, renewable energy, and power electronics. Prof. Senjyu is a member of the Institution of Engineering and Technology of U.K., and the Institute of Electrical Engineers of Japan. He is also a Chartered Engineer in the U.K.

Deepak Kumar Sharma received the B.Tech. in computer science from G. G. S. Indraprastha University and M. E. in Computer Technology and Applications from University of Delhi, India. He has been working as Assistant Professor in the Division of Information Technology, Netaji Subhas Institute of Technology, University of Delhi, India, since February 2011. From July 2004 to January 2011, he worked as a Senior Lecturer at Maharaja Agrasen Institute of Technology (MAIT), Delhi, India. His current research interests include opportunistic networks, wireless ad hoc networks, and sensor networks.

Abhinav Singh is currently pursuing B.tech in IT from NSIT. He has written various research chapters on varied topics like Cloud Computing, Machine Learning and the Semantic Web. His area of interest lies in Cloud Computing. He is passionate about Football and his favourite player is Cristiano Ronaldo.

Prachi Singh is doing B.Tech. + M.Tech. in Electronics and Communication Engineering and M.Tech. in wireless communication and networking from Gautam Buddha University. She is doing her research work on VCSEL and IOT.

Simar Preet Singh, a research scholar at Thapar Institute of Engineering and Technology, Patiala, was an Assistant Professor at DAV University, Jalandhar and also a Microsoft Professional. Apart from this, he is having certifications like Microsoft Certified System Engineer (MCSE), Microsoft Certified Technology Specialist (MCTS) and Core Java. He also underwent training programme for VB.Net and Cisco Certified Network Associates (CCNA). He worked with Infosys Limited for nearly two years. He has presented many research papers in various National and International Conferences in India and abroad. His areas of interests include Cloud Computing, Fog Computing, IoT, Bigdata and Machine Learning.

Utsha Sinha is pursuing bachelors of engineering in IT from NSUT.

Pradeep Tomar has been working as Assistant Professor in the School of Information and Communication Technology, Gautam Buddha University, Greater Noida, U.P., India, since 2009. Dr. Tomar earned his Ph.D. from MDU, Rohtak, Haryana, India. Before joining Gautam Buddha University, he worked as a Software Engineer in a multi-national company, Noida, and lecturer in M. D. University, Rohtak, Haryana, and Kurukshetra University, Kurukshetra, Haryana. Dr. Tomar has excellent teaching, research and software development experience as well as vast administrative experience at university level on various posts like research coordinator, examination coordinator, admission coordinator, programme coordinator, time table coordinators, Proctor and hostel warden. Dr. Tomar is also a member of Computer Society of India (CSI), Indian Society for Technical Education (ISTE), Indian Science Congress Association (ISCA), International Association of Computer Science and Information Technology (IACSIT) and International Association of Engineers (IAENG). Dr. Tomar qualified the National Eligibility Test (NET) for Lecturership in Computer Applications in 2003, Microsoft Certified Professional (MCP) in 2008, SUN Certified Java Programmer (SCJP) for the Java platform, standard edition 5.0 in 2008 and qualified the IBM Certified Database Associate - DB2 9 Fundamentals in 2010. Dr. Tomar was awarded Bharat Jyoti Award by India International Friendship Society in the field of Technology in 2012 and Bharat Vikas Award by Institute of Self Reliance in National Seminar on Diversity of Cultural and Social Environment at Bhubaneswar, Odisha, in 2017. He was awarded the Best Computer Faculty award by Govt. of Pondicherry and ASDF society. His biography is published in Who's Who Reference Asia, Volume II. Dr. Tomar has been awarded distinguished Research Award from Institute for Global Business Research for his work in "A Web-Based Stock Selection Decision Support System for Investment Portfolio Management in 2018. Tomar, and he delivered expert talks at FDP, workshops, national and international conferences. Dr. Tomar has organized three conferences: one national conference with COMMUNE group and two international conferences, in which one international ICIAICT 2012 was organized by CSI, Noida Chapter and second international conference 2012 EPPICTM was organized in collaboration with MTMI, USA, University of Maryland Eastern Shore, USA and Frostburg State University, USA at School of Information and Communication Technology, Gautam Buddha University, Greater Noida, India. Apart from teaching, he is running a programming club for ICT students and he is guiding various research scholars in the areas of software engineering, reusability of code, soft computing technique, Big Data, and IoT. His major current research interest is in Component-Based Software Engineering. He is working as Co-investigator in a sponsored research project in High throughput design, synthesis and validation of TALENs for targeted Genome Engineering, funded by Department of Biotechnology, Ministry of Science and Technology Government of India New. Two books "Teaching of Mathematics" and "Communication and Information Technology" at national levels, two books "Examining Cloud Computing Technologies through the Internet of Things (IoT)" and Dr. Tomar has authored handbook of Research on Big Data and the IoT at international level. He has also contributed more than 100 papers/articles in national/international journals and conferences. He served as a member of the editorial board and reviewer for various Journals and national/international conferences.

Index

H

I

K

L

M

N

P

Q

R

S

V

W

Z

www.ingramcontent.com/pod-product-compliance
Lightning Source LLC
LaVergne TN
LVHW060422300926
849162LV00006B/1501